Lichen Secondary Metabolites

Branislav Ranković
Editor

Lichen Secondary Metabolites

Bioactive Properties and Pharmaceutical Potential

Second Edition

Springer

Editor
Branislav Ranković
Faculty of Science
University of Kragujevac
Kragujevac, Serbia

ISBN 978-3-030-16813-1 ISBN 978-3-030-16814-8 (eBook)
https://doi.org/10.1007/978-3-030-16814-8

This Springer imprint is published by the registered company Springer Nature Switzerland AG.
The registered company address is: Gewerbestrasse 11, 6330 Cham, Switzerland

Contents

Lichens as a Potential Source of Bioactive Secondary Metabolites

1

Branislav Ranković and Marijana Kosanić

Abstract

Lichens are complex symbiotic associations between fungi and algae which are important constituents of many ecosystems. The production of various unique extracellular secondary metabolites known as lichen substances is the result of this symbiosis. These compounds exist within the thalli and typically form crystals on the surface of the fungal hyphae. Thus far, more than 800 secondary metabolites of lichens have been discovered, most of them being exclusively present in lichens. In recent date, lichens have been taken up for many researches concerning the phytochemical and pharmaceutical applications. Lichens and their secondary metabolites have many pharmaceutical roles, primarily including antimicrobial, antioxidant, antiviral, anticancer, antigenotoxic, anti-inflammatory, analgesic, and antipyretic activities. Hence, the present study was undertaken to explain the lichens as the important potential sources of bioactive secondary metabolites.

1.1 The Lichens: Lichenized Fungi

Lichens are association between fungi (*mycobiont*) and photoautotrophic, algal partners (*photobionts*). Since the mycobiont is unique in the symbiotic association and usually dominates the association, lichens are traditionally classified as a life-form of fungi. About 18,500 different lichen species have been described all over the world. The fungal partners are mostly (98%) *Ascomycota* (Honegger 1991; Gilbert 2000), and the others belong to the *Basidiomycota* and anamorphic fungi. Approximately 21% of all fungi are able to act as a mycobiont (Honegger 1991); thus lichens form the largest

B. Ranković (✉) · M. Kosanić
Department of Biology, Faculty of Science, University of Kragujevac, Kragujevac, Serbia
e-mail: rankovic@kg.ac.rs

B. Ranković (ed.), *Lichen Secondary Metabolites*,
https://doi.org/10.1007/978-3-030-16814-8_1

mutualistic group among fungi. Only 40 genera are involved as photosynthetic partners in lichen formation: 25 algae and 15 cyanobacteria (Kirk et al. 2008).

Most common photobionts are the genera *Trebouxia*, *Trentepohlia*, and *Nostoc*. The genera *Trebouxia* and *Trentepohlia* are of eukaryotic structure and belong to the green algae, while the genus *Nostoc* belongs to the prokaryotic cyanobacteria. Eukaryotic photobionts are sometimes called **phycobionts** (90% of lichens), while cyanobacterial photobionts are termed **cyanobionts** (10% of lichens).

In lichen associations both partners have benefit. The mycobiont has two principal roles in the lichen symbiosis: to protect the photobiont from exposure to intense sunlight and desiccation and to absorb mineral nutrients from the underlying surface or from minute traces of atmospheric contaminants. The photobiont also has two roles: to synthesize organic nutrients from carbon dioxide and, in the case of cyanobacteria, to produce ammonium (and then organic nitrogen compounds) from N_2 gas, by nitrogen fixation. In some ecosystems such as desert soils, tundra heaths, and Douglas-fir forests of the Pacific Northwest of the USA, lichens can provide the major input of nitrogen which supports other forms of life (Hale 1983; Nash 1996).

Thus, through the lichen partnership, the photobionts are protected and able to grow in conditions in which they could not grow alone; they also benefit from the highly efficient uptake of mineral nutrients by the lichen fungi. The fungi, in turn, obtain sugars and in some cases organic nitrogen from the photosynthetic partner, enabling them to grow in environments deficient in organic nutrients (Hale 1983; Nash 1996).

A lichen thallus usually consists of layers such as an upper and lower cortex, algal layer, and medulla (Fig. 1.1). The layers differ in thickness and are better developed in some species than in others. Fungal hyphae make up most of a thallus; the photobionts are cells only a small percentage (about 7%) of the total volume (Ahmadijan 1993). There are three main types of thalli: crustose, foliose, and fruticose (Fig. 1.2). A crustose thallus lacks a lower cortex and is generally considered to be the most primitive type. Thalli of the *Lepraria* species do not have layers but consist only of powdery granules. Many crustose lichens stick tightly to the substratum and appear to be painted on it. Squamules are a specialized type of

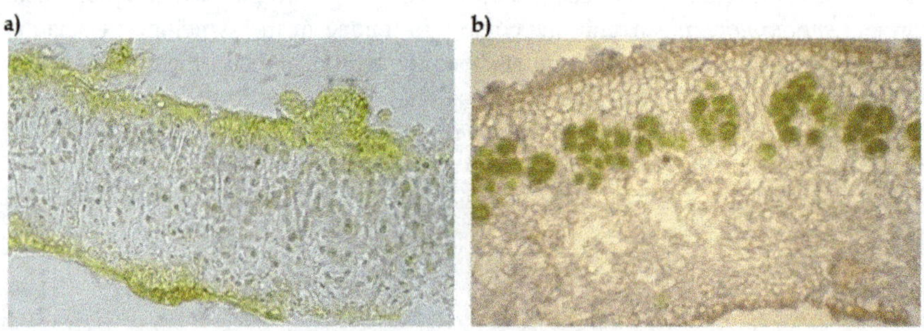

Fig. 1.1 A vertical section of lichen thallus showing anatomical features. (**a**) Homoiomerous thallus; (**b**) heteromerous thallus

a) b) c)

Fig. 1.2 Main morphological groups of lichen thallus. (**a**) Crustose; (**b**) foliose; (**c**) fruticose

crustose thallus and are attached at only one and to the substratum. A foliose thallus has an upper and lower cortex, an algal layer, and medulla and is usually loosely attached to the substrate by hair-like structures called rhizines. Some foliose lichens have thalli that are attached to the substrate by only one central point. Fruticose thalli are upright or hanging, round or flat, and often highly branched. The layers of a fruticose thallus may surround a central thick cord, as in *Usnea*, or a hollow space as in some *Cladonia* species (Ahmadijan 1993; Nash 1996).

Some lichens, such as *Collema*, which have *Nostoc* as a photobiont, do not form a well-organized thallus (Fig. 1.1). In these cases, fungal hyphae grow inside the thick gelatinous sheaths of the photobiont, which makes up much of the thallus (Paracer and Ahmadijan 2000).

In nature, lichens grow very slowly. Their radial growth is measured in millimeters per year. Hale (1983) made the following generalization about growth rates of lichens: most foliose species grow 0.5–4 mm per year, fruticose species 1.5–5 mm per year, and crustose species 0.5–2 mm per year.

Lichens grow practically everywhere—on and within rocks, on soil and tree bark, and on almost any inanimate object. They grow in deserts and in tropical rainforests, where they occur on living leaves of plants and ferns. They have been found on the shells of tortoises in the Galapagos Islands and on large weevils in New Guinea. In the dry valleys of Antartica, endolithic lichens, such as *Buellia* and *Lecidea*, grow inside sandstone crevices. *Dermatocarpon fluviatile* and *Hydrothyria venosa* grow in freshwater streams, and species of *Verrucaria* are common in the intertidal zones of rocky, ocean shores. *Verrucaria serpuloides* is permanently submerged marine lichen that grows on stones and rocks 4–10 m below mean low tide off the coast of the Antarctic Peninsula. *Verrucaria tavaresiae* is another unusual marine lichen with brown algal photobiont. Douglas Larson has estimated that about 8% of the earth's terrestrial surface is dominated by lichens. Lichens abound in areas with high annual humidity, such as the fog belt zones of Chile and Baja California. Extensive lichen populations also grow in the cool, northern forests of the world, where hundreds of miles of forest floor are covered with thick carpets of reindeer lichens (*Cladonia*). Trees along the coasts of the northwestern United States may be blanketed with beard lichens such as *Alectoria* and *Usnea* (Hale 1983; Moe 1997; Cox 2003).

Lichens with organized thalli do not grow well in areas that are continuously wet, such as tropical rainforests. Only poorly organized species of *Lepraria* and leaf-inhabiting lichens are found in these regions. *Lecanora conizaeoides* and *Lecanora dispersa* colonize trees and gravestones in industrial cities and towns, but most lichens cannot tolerate the polluted atmosphere and persistent dryness of urban areas. The sensitivity of lichens to atmospheric pollutants such as sulfur dioxide, ozone, and fluorides has made them valuable indicators of pollution in cities and industrial regions (Richardson 1992).

Interactions between the symbiotic partners allow lichens that live in unusual environments (Bačkor and Fahselt 2008). Lichens are able to survive in extreme ecological conditions; they can adapt to extreme temperatures, drought, inundation, salinity, high concentrations of air pollutants, and nutrient-poor, highly nitrified environments (Nash 2008). Despite this extreme range of ecological adaptations, most lichens are sensitive to changes of their preferred ecological conditions and can hardly grow in nonnative habitats.

1.2 Lichen Metabolites

Specific conditions in which lichens live are the reason of production of many metabolites that they provide good protection against various negative physical and biological influences. Metabolites synthesized by lichens were divided into two groups: primary and secondary (Lawrey 1986).

The primary (intracellular) metabolites include proteins, amino acids, carotenoids, polysaccharides, and vitamins. They are generally soluble in water and can be easily isolated from the lichens by boiling water. Some of the primary metabolites are produced by fungi and some by algae. Most of these metabolites are the nonspecific and also may occur in free-living fungi, algae, and higher plants. Presence of free amino acids is similar to their presence in the plants. In general, the amount of nitrogen compounds is between 1.6 and 11.4% dry weight of the lichen thallus (Hale 1983). Carotenoids are metabolic products of both symbionts and are in the range 1.5–24 mg/g dry weight of the talus. Among the carotenoids in the lichens talus identified were β-carotene epoxide, α-cryptoxanthin, lutein, astaxanthin. Polysaccharides and related compounds are present in the lichen in an amount of 3–5% dry weight of the talus (Culberson 1970). Among vitamins, lichen contains ascorbic acid, biotin, α-tocopherol, nicotinic acid, pantothenic acid, riboflavin, thiamine, and folic acid. Vitamins were identified as metabolic products which biosynthesize alga, until the mushrooms are poor sources of these compounds (Hale 1983).

The majority of organic compounds found in lichens are secondary metabolites. Lichens may contain substantial amounts of secondary metabolites, usually between 0.1 and 10% of the dry weight but sometimes up to 30% (Galun and Shomer-Ilan 1988; Stocker-Wörgötter 2008; Solhaug et al. 2009). More than 800 secondary metabolites known are from lichens, most are unique to these organisms, and only a small minority occurs in the other fungi or higher plants. All of the secondary substances in lichens are of fungal origin. These substances are the crystals deposited on the surface of the hyphae, which are poorly soluble in water and usually can be isolated from the lichens by organic solvents (Bačkorová et al. 2012).

Once formed, the lichen secondary metabolites appear to be extremely stable. Very old herbarium specimens of lichens show no significant reduction in concentrations of lichen substances (Rundel 1978). In addition, the production of secondary compounds is genetically controlled (Culberson and Culberson 2001) and in some instances is correlated with morphology and geography in individuals at the species and genus levels (Egan 1986; Zhou et al. 2006).

Histologically, lichen secondary metabolites are deposited in either the cortex or, more commonly, the medulla. The most usual cortical compounds are usnic acid and atranorin, but anthraquinones, pulvinic acid derivates, and xanthones may also occur here. All of these, with the exception of atranorin, are typically pigmented. The great majority of lichen substances are deposited in the medulla. Lichen substances are often expressed differentially in the layers of lichen thallus, and typical cortical substances can be distinguished from compounds usually found only in the medulla. This could be linked with their biological function: the cortical compounds are regarded as a kind of light filter, which is apparently not a function of compounds below the algal layers (Marques 2013).

Lichens synthesize significant amounts of substances only in permissive physiological stages. As a consequence, the production in axenic cultures can differ substantially from that in nature. Mycobionts grown without their photobionts synthesize specific secondary lichen compounds under certain conditions (Culberson and Armaleo 1992; Mattsson 1994; Stocker-Wörgötter and Elix 2002; Fazio et al. 2007; Hager et al. 2008) but can also produce substances that are different from the metabolites found in symbiosis (Yoshimura et al. 1994; Brunauer et al. 2007). For example, natural lichen *Lecanora dispersa* contains 2,7-dichlorolichexanthone as the major secondary compound, but cultured spore isolates, growing without the alga, produced pannarin and related depsidones instead (Leuckert et al. 1990). Pannarin could not be confirmed of the natural source lichen, but the biosynthetic potential for this depsidone was proven for the species by a herbarium survey. Another similar result is the production of atranorin in *Usnea hirta*, when cultured in a modified LB medium (Kinoshita 1993), since this compound is not present throughout the entire genus *Usnea*.

Each lichen mycobiont prefers specially adapted culture conditions (such as nutrient medium, added sugars or polyols, pH, temperature, light, stress) to produce the specific secondary metabolites (Hager et al. 2008). Similarly, lichen "tissue" cultures, in many cases, can produce secondary substances (Yamamoto et al. 1985, 1993), but the chemistry is usually different from the chemosyndrome of the corresponding natural lichen thalli (Yamamoto et al. 1993). Lichenized *Basidiomycotina* do not contain lichen substances (Lumbsch 1998).

The distribution patterns of secondary metabolites are usually taxon-specific and, therefore, have been widely used in lichen taxonomy and systematics (Nordin et al. 2007; Fehrer et al. 2008; Nelsen and Gargas 2008). However, it has been shown that the similarities in secondary chemistry may not necessarily indicate close phylogenetic relationships (Nelsen and Gargas 2008). For example, *Xanthoparmelia* and *Neofuscelia*, two closely related genera of the *Parmeliaceae*, differ primarily in the pigments of their upper cortex: while *Xanthoparmelia* contains crystallized usnic acid, the closely related genus *Neofuscelia* is characterized by melanin-like cortical pigments. Moreover, other members of the *Parmeliaceae* may exclusively contain atranorin in the cortex.

Secondary metabolites are not absolutely essential for the survival and growth of lichens (Bentley 1999), and the functions of these compounds in the lichen symbioses are still poorly understood (Hager et al. 2008). However, it is important that they may help to protect the thalli against herbivores, pathogens, competitors, and external abiotic factors, such as high UV irradiation. Also, lichens have impact interactions with their environment.

Production of secondary metabolites in lichens is complex and variously influenced by environmental factors, including light, UV exposure, elevation, temperature fluctuations, and seasonality.

Although numerous studies have attempted to correlate the variation of metabolite concentrations in lichens under different light levels, the results have often been contradictory. Armaleo et al. (2008) reported a positive correlation between the amount of light available to *Parmotrema hypotropum* annually and the concentration of atranorin in the thallus; however, there was a decrease in norstictic acid concentration under the same conditions. In contrast, Stephenson and Rundel (1979) did not find any correlations between atranorin levels and light intensity for *Letharia vulpina*. Bjerke and Dahl (2002) reported a positive correlation between thallus concentration of usnic acid and light intensity in *Ramalina siliquosa* under culture conditions, but there was no correlation between light intensity and salazinic acid concentration in the same species. Reasons for conflicting patterns in the production of lichen substances in response to varying light conditions in lichens remain unclear. Bjerke et al. (2004, 2005) have suggested that these conflicting results may be due, at least in part, to the integrative effects of multiple variables across differing habitat conditions.

Understanding how lichen metabolite concentrations vary in response to UV exposure may provide important insights into how lichens thrive in these harsh environments. One of the primary functions of lichen secondary compounds is to protect the algal layer from intense light levels, especially in the ultraviolet range (Waring 2008). For example, lichen-derived phenolic compounds including depsides, depsidones, usnic acid, and pulvinic acid derivates are effective UV-absorbing compounds. Extensive field experiments have assessed the effects of UV on the concentration of lichen phenolics with pattern of increasing concentrations under high UV-A conditions (BeGora and Fahselt 2001). On the other hand, UV-B radiation appears to have a negative effect on lichen metabolite concentrations (BeGora and Fahselt 2001). Increased levels of lichen metabolites under high UV-A conditions may be due to greater production of photosynthates which in turn support increased production of lichen compounds (BeGora and Fahselt 2001), while lower production of lichen compounds in the presence of UV-B may be due to the production of extracellular degradative enzymes or electronic transitions within molecules caused by exposure to UV-B (Fessenden and Fessenden 1986).

Various studies have shown that differences in geographic location and changing seasons can influence the concentration of secondary metabolites in some lichen species. Bjerke et al. (2004) showed variation in the usnic acid content of

Flavocetraria nivalis thalli collected from 25 sites in northwestern Spitsbergen, Norway. In addition to the influence of geographic location, the phenolic content of lichens has also been reported to vary by season. For example, BeGora and Fahselt (2001) reported the lowest levels of usnic acid for *Cladonia mitis* during the spring and summer with the highest concentrations occurring in the late winter. The lower levels of usnic acid reported for summer months may be due to drought and heat stress which have a depressing effect on metabolic activity (Bjerke et al. 2005). Another reason could be a function of higher levels of UV-A or the ratio of UV-B to UV-A during the summer. However, Bjerke et al. (2005) found increased levels of usnic acid in *Flavocetraria nivalis* in the late spring and early summer and generally lower levels during the autumn and winter which they believe may be due to higher levels of precipitation during the spring and summer months.

Studies have also demonstrated that the production and/or content of secondary metabolites in lichens varied with changes in elevation. For example, it has been reported for *Umbilicaria americana* that lichen secondary metabolite production decreases with increasing elevation (Swanson et al. 1996). In contrast, Rubio et al. (2002) found increasing levels of rhizocarpic acid in *Acarospora schleicheri* with increasing elevation. The direct effects of elevation on the production of phenolic compounds by lichens may in fact be complicated by the dynamic interaction of various environmental factors.

Production of lichen compound may fluctuate with the with temperature changes. For example, the concentration of salazinic acid in *Ramalina siliquosa* was positively correlated with mean annual temperature (Hamada 1982). Under seminatural conditions, the amount of gyrophoric acid and methyl gyrophorate was reported to increase by an average of 6.7–22.3 times, respectively, in *Peltigera didactyla* with an increase in temperature of 3 °C (Bjerke et al. 2003). Hamada (1991) observed a steady increase in the concentration of usnic acid and 4-*O*-dimethylbarbatic acid in mycobiont cultures from *Ramalina siliquosa* when the temperature of the culture was increased to 12 °C for usnic acid and 15 °C for 4-*O*-dimethylbarbatic acid, with higher temperatures resulting in a decrease in the content of both components. In contrast, the content of 4-*O*-dimethylbarbatic acid in intact lichen increased with temperature up to 19 °C (Hamada 1984), suggesting that the optimal temperature for the production of lichen substances is generally lower in the isolated mycobiont when compared with intact lichen. Increased concentrations in lichen phenolics may be due to increased photosynthetic activity related to increased temperature. Culberson et al. (1983) found the concentrations some phenols in *Cladonia cristatella*, in a controlled phytotron experiment, were higher at lower temperature. Tundra lichens are typically metabolically active at near-zero of subzero temperature. Similarly, Bjerke et al. (2004) found that in *Flavocetraria nivalis*, the concentration of usnic acid was highest at the sites with the lowest temperatures. Based on the abovementioned, it can be concluded that the variation in temperature shows important influence on the levels of secondary metabolites in lichens.

1.3 Secondary Metabolite Production in Lichens: The Important Pathways

Lichen secondary metabolites are derived from three chemical pathways: **acetate-polymalonate pathway, shikimic acid pathway, and mevalonic acid pathway** (Fig. 1.3).

Acetate-polymalonate pathway includes the most common lichen compounds such as:

- Secondary aliphatic acids, esters, and related derivates
- Mononuclear phenolic compounds
- Depsides, tridepsides, and benzyl esters
- Depsidones and diphenyl esters
- Depsones
- Dibenzofurans, usnic acids, and derivates
- Anthraquinones and biogenetically related xanthones
- Chromones
- Naphthoquinones
- Xanthones

Mevalonic acid pathway includes:

- Di-, sester- and triterpenes
- Steroids

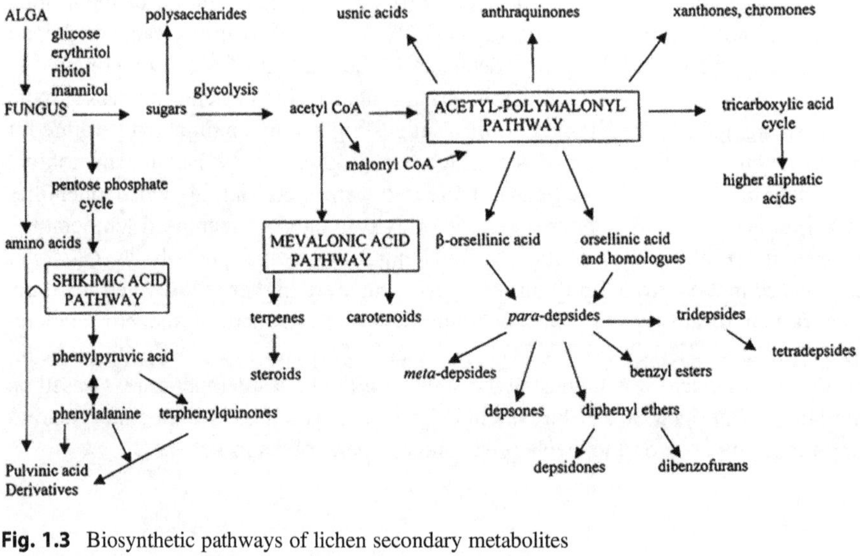

Fig. 1.3 Biosynthetic pathways of lichen secondary metabolites

Shikimic acid pathway includes:

- Terphenylquinones
- Pulvinic acid derivates

Apart from compounds derived from these common pathways, which are found throughout all major lichen groups, there are also some unusual compound classes among these organisms, for example, arthogalin, a cyclic depsipeptide (Huneck and Himmelreich 1995), and other amino acid-derived compounds such as the cytotoxic scabrosin esters isolated from *Xanthoparmelia scabrosa* (Ernst-Russell et al. 1999). Uncommon features are also detected in residues of common substance classes, in the form of other intramolecular arrangements, or in the binding with other compounds such as sugars. Recently, Řezanka and Guschina (1999) described many unusual compounds as brominated depsidones, brominated acetylenic fatty acids (Řezanka and Dembitsky 1999), and monotetrahydrofuranic acetogenin derivates (Řezanka et al. 2004). Series of g-lactonic aliphatic acid glycosides were also identified (Řezanka and Guschina 2000, 2001a, b), some of them forming a macrolactone ring (gobienins: Řezanka and Guschina 2001c). Glycosides are not frequently encountered in lichens, yet recent reports revealed the presence of xanthone glucosides (umbilicaxanthosides: Řezanka et al. 2003) and the mycosporine collemin A from *Collema cristatum* (Torres et al. 2004).

1.3.1 Lichen Secondary Metabolites of the Acetate-Polymalonate Pathway

Acetate-polymalonate pathway utilizes acetyl-CoA and malonyl-CoA which are derivates of coenzyme A. Of the lichen secondary metabolites, which are formed in this pathway, aromatic products are especially well represented, the most characteristic being formed by the bonding of two or three orcinol or β-orcinol-type phenolic units through ester, ether, and carbon-carbon linkages. The large majority of depsides, depsidones, dibenzofurans, usnic acid, and depsones all appear to be produced by such mechanisms, and all are peculiar to lichens. Other aromatic compounds such as the chromones, xanthones, and anthraquinones are probably formed by internal cyclization of a single, folded polyketide chain and are often identical or analogous to products of non-lichen-forming fungi or higher plants. In addition to the compounds of known chemical structure, many of the unknown structures are given common names and assigned to compound class, because they are frequently encountered and easily recognized by microchemical methods.

1.3.1.1 Aromatic Products Arising from Intermolecular Esterification or Oxidative Coupling

The commonest phenolic acid units derived by the acetate-polymalonate pathway and combined to form the characteristic lichen substances are of two types: the orcinol-type units and β-orcinol-type units. While compounds formed from these two types of units are similar in many ways, differences in their structure and especially in their distribution among the lichens suggest that the usual tendency to consider the orcinol and β-orcinol compounds separately probably has a biosynthetic justification.

The Orcinol Series The most common fate of acetate-polymalonate-derived phenolic acids in lichens is intermolecular esterification of two or three similar or identical units. For example, the carboxylic acid of one unit is joined to the hydroxyl *para* to the carboxylic acid of the second unit. Such esterifications lead to the *para*-depsides. A second esterification reaction leads to tridepsides. If an ester linkage joins the first unit to a position *meta* to the carboxylic acid of the second ring, a *meta*-depside results.

Known compounds of *para*-depsides are anziaic acid, confluentic acid, diploschistesic acid, erythrin, evernic acid, glomelliferic acid, hiascic acid, imbricaric acid, tumidulin, microphyllinic acid, obtusatic acid, olivetoric acid, perlatolic acid, planaic acid, sphaerophorin, divaricatic acid, and lecanoric acid. Known *meta*-depsides are boninic acid, cryptochlorophaeic acid, sekikaic acid, homosekikaic acid, merochlorophaeic acid, novochlorophaeic acid, paludosic acid, ramalinolic acid, scrobiculin, thamnolic acid. Tridepsides include tenuiorin, umbilicaric acid, and gyrophoric acid.

Most known orcinol-type depsidones have an *a*- or a β-keto group in the side chain of the first ring. It is well known that this functional group has a strong effect upon the ester linkage between the two rings, since enol lactones form readily. Oxidative cyclization of depsides to depsidones usually joins the 2-hydroxyl of ring A and the 5-position of ring B.

The best known depsidones are physodic acid, lobaric acid, norlobaridone, grayanic acid, alectoronic acid, diploicin, 4-O-methylphysodic acid, α-collatolic acid, lividic acid, and variolaric acid.

Depsone (picrolichenic acid) has a methylated 2-hydroxyl, and carbon-carbon coupling links the 1-position of the A ring and the 5-position of the B ring.

β-Orcinol Series The orcinol-type compounds discussed in the previous section form a closely related series of substances in which changes in the length and oxidation state of the 6-alkyl substituents are major sources of variation among phenolic units. The compounds synthesized by various combinations of these units show secondary modifications attributable to O-methylation, chlorination, decarboxylation, and lactonization. The β-orcinol compounds may undergo all of the same secondary reactions, but the commonest variation is in the oxidation state of the C_1 substituents at the 3- and 6-positions of the phenolic acid units.

Of the 16 units theoretically derivable by changing the oxidation states of the C_1 substituents, only 8 are represented among the known compounds. Six *para*-and three *meta*-depsides are derived from only three β-orcinol units, all of which have the original fully reduced methyl group at the 6-position and either a methyl, an aldehydic, or more rarely a carboxylic acid group at the 3-position. All the *para*-depsides have the same B ring substituted with methyls at both the 3- and the 6-positions. The C_1 substituents of the B rings of *meta*-depsides and depsidones are usually more highly oxidized than those of *para*-depsides.

The best known *para*-depsides are atranorin, baeomycesic acid, barbatic acid, chloroatranorin, diffractaic acid, squamatic acid, and 4-O-demethylbarbatic acid, while *meta*-depsides include decarboxythamnolic acid, haemathamnolic acid, hypothamnolic acid, and thamnolic acid.

Except in gangaleoidin, the A ring of all β-orcinol depsidones has either a 3-aldehydic substituent like the A ring of the *para*-depside atranorin or a 3-methyl substituent like the B ring of atranorin. The B rings of depsidones are much more variable, however, and some show oxidation of the 6-methyl substituent, a condition never observed among the phenolic acid units of the known *para*- and *meta*-depsides. Known compounds of depsidones are fumarprotocetraric acid, galbinic acid, gangaleoidin, hypoprotocetraric acid, myriocarpic acid, norstictic acid, pannarin, physodalic acid, protocetraric acid, psoromic acid, salazinic acid, stictic acid, vicanicin, and virensic acid.

One benzyl ester, barbatolic acid, is known in three usneaceous genera producing a number of the usual β-orcinol esters previously described. The A ring of barbatolic acid is like the A ring of atranorin, but the B ring is substituted at the 6-position with the hydroxymethyl group involved in the ester linkage.

Dibenzofurans The phenolic units involved in the production of the true lichen dibenzofurans are derived by the orsellinic acid-type cyclization. The dibenzofurans appear to form by carbon-carbon coupling and cyclodehydration of two such acetate-polymalonate-derived phenolic acid units. Two types of lichen dibenzofurans can be distinguished by the orientation of the aromatic rings. The first type is presently known only in *Cladoniaceae*; the second type is known in the *Lecideaceae*, *Lecanoraceae*, *Roccellaceae*, and the genus *Lepraria*. In all cases, the 5- and 4-positions of one phenolic unit, which becomes ring C in the dibenzofuran structure, are involved in the furan ring. The lichen dibenzofurans are dendroidin, didymic acid, pannaric acid, porphyrilic acid, schizopeltic acid, and strepsilin.

Usnic Acid The best known lichen products are the usnic acid, yellow pigments produced in the upper cortex of many species from phylogenetically widely separated families. These pigments are known only in lichens. The usnic acids appear, therefore, to be among the compounds which are most characteristic of lichens, are apparently not synthesized in cultures of the isolated fungal components, and are formed by reactions uniting two or three phenolic units.

It was first isolated by the German scientist W. Knop in 1844 and first synthesized between 1933 and 1937 by Curd and Robertson. Usnic acid was identified in many genera of lichens including *Usnea*, *Cladonia*, *Lecanora*, *Ramalina*, *Evernia*, *Parmelia*, and *Alectoria*.

Usnic acid possesses a wide range of interesting biological properties. It is a potent antibiotic effective against Gram-positive bacteria, including *Staphylococcus*, *Strep tococcus*, and *Pneumococcus*, other bacteria such as *Mycobacterium tuberculosis*, and some pathogenic fungi. It also exhibits antiviral, antiprotozoal, antimitotic, anti-inflammatory, and analgesic activity. Other characteristics, like ultraviolet absorption, preserving properties, antigrowth, antiherbivore, and anti-insect properties, have also been demonstrated.

Cyclization to the aromatic ring of the methylphloroacetophenone precursor of the usnic acids is basically different from the orsellinic acid cyclizations leading to precursors of dibenzofuran derivates, but the oxidative coupling and cyclodehydration reactions of pairs of units are similar in the two cases. With the usnic acids, a C-methyl substituent prevents rearomatization of one ring. The structurally isomeric usnic acids are related by the orientation of the A ring about the carbon-carbon bond. In the two types of dibenzofuran derivates, the second unit also remains fixed, but the A ring is reoriented with respect to the ether linkage.

The orientation of the second phenolic unit in the usnic acids resembles that of all the known dibenzofuran derivates in that the position involved in the carbon-carbon coupling is *meta* to a carbonyl substituent and *ortho* and *para* to O-substituted positions. It differs from the dibenzofurans in having a second ortho-hydroxyl substituent instead of an alkyl group. Dibenzofuran derivates could form from the methylphloroacetophenone precursor of usnic acid by a change in the orientation of the C ring with respect to the ether linkage, and such an orientation would still allow a *para*-quinoid radical.

Until recently, it was thought that two usnic acid pigments were produced in lichens, (+)-usnic acid, and its optical isomer. Now structural isomers have also been found in which the carbon-carbon bond remains the same, but the ether linkage forms by dehydrative cyclization at a different hydroxyl group of the A ring. Both optical isomers of these isousnic acids have been found. Since the structures of usnic acid and isousnic acid are very similar, it is probable that the pairs of compounds rotating polarized light in the same direction also have the same configuration about the asymmetric carbon.

The isousnic acids are known only in *Cladonia*, but extensive surveys for the distribution of these compounds have not yet been reported. In *Cladonia*, both (+)- and (−)- usnic acids occur, but in this genus as in others, both optical isomers are never found in a single plant. Some genera produce only one optical isomer. It is probable that the methylphloroacetophenone unit of usnic acid is a common aromatic lichen product for which mechanisms of coupling and cyclodehydration have arisen evolutionarily many times.

Diquinone Only one lichen compound, pyxiferin, has been identified as a diquinone. If pyxiferin were formed by oxidative coupling of acetate-polymalonate-derived units, the first ring would arise by decarboxylation, hydroxylation, and oxidation of orsellinic acids. The second ring would also require oxidative removal of the C_1 substituent. The tentative structure of this compound requires a considerable biosynthetic departure from pathways leading to most of the orsellinic acid derivates in lichens.

1.3.1.2 Aromatic Products Derived from a Single Polyacetyl Chain

Aromatic products derived from a single polyacetyl chain include monoaryl products related to phenolic acids, chromones, xanthones, and anthraquinones.

Lichen Monoaryl Derivates are relatively rare in lichens. All of the lichen monoaryl derivates have lost their free carboxylic acid groups either by esterification or by decarboxylation. These compounds might be synthesized in the lichen from coenzyme A derivates of the initially formed phenolic acid units or by biological decomposition of depsides.

Xanthones are known in free-living fungi, and recent studies indicate that they are rather common in lichens too. Unlike the fungal xanthones, many lichen xanthones have one or more nuclear chlorine substituents. The fundamental structure of the known lichen xanthones could be derived directly by linear condensation of seven acetate and malonate units with one orsellinic acid-type cyclization. The two rings are joined by a ketonic carbon and by an ether-oxygen arising from cyclodehydration. Xanthones include arthothelin, concretin, 2,4-dichloronorlichexanthone, 2,7-dichloronorlichexanthone, lichexanthone, norlichexanthone, thiophanic acid, thiophaninic acid, thuringione, and vinetorin.

The most widespread **anthraquinone** in lichens is parietin (physcion), which is also known in fungi and higher plants. Other known anthraquinones are chiodectonic acid, chrysophanol, 1,3-dihydroxy-8-methoxy-2-chloro-6-methylanthraquinone, emodin, endocrocin, fallacinal, fragilin, 1-hydroxy-3,8-dimethoxy-2-chloro-6-methylanthraquinone, mysaquinone, nephromin, norsolorinic acid, parietin, parietinic acid, rhodocladonic acid, rugulosin, skyrin, solorinic acid, teloschistin, 1,3,8-trihydroxy-2-chloro-6-methylanthraquinone, 1,3,8-trihydroxy-2,4-dichloro-6-methylanthraquinone, and xanthorin.

In addition, three **chromones** are known in lichens: siphulin, sordidone, and lepraric acid.

1.3.1.3 Aliphatic Products Derived From a Single Polyacetyl Chain: Higher Aliphatic Acids

The fatty acids of lichens show some resemblance to those in non-lichen-forming fungi, but none are identical. Aliphatic acids, much rarer in lichens than in other plant groups, appear to be formed in a very different manner through the tricarboxylic acid cycle. Known higher aliphatic acids and related substances are acaranoic acid, acarenoic acid, (+)-aspicilin, (−)-caperatic acid, (−)-lichesterinic acid, linoleic acid, (−)-nephromopsinic acid, (+)-nephrosteranic acid, (+)-nephrosterinic acid, (+)-nonrrangiformic acid, oleic acid, (−)-protolichesterinic acid, (+)-protolichesterinic acid, (+)-*allo*-protolichesterinic acid, (+)-pseudononrrangiformic acid, (+)-rangiformic acid, (+)-roccellaric acid, (+)-roccellic acid, 9,10,12,13-tetrahydroxyheneicosanoic acid, tetrahydroxytricosanoic acid, ventosic acid, and unidentified aliphatic acids.

1.3.2 Secondary Products of the Mevalonic Acid Pathway: Terpenes, Steroids, and Carotenes

All of the acetate-derived aliphatic products described to this point have arisen by head-to-tail condensations leading to unbranched products. Compounds derived by the mevalonic acid pathway involve consideration of an acetyl coenzyme A unit at the methylenic carbon of an acetoacetyl coenzyme A unit to produce a branched precursor, β-hydroxy- β-methylglutarul coenzyme A, which is then reduced to mevalonic acid. Compounds derived by the mevalonic acid pathway resemble those formed by the acetate-polymalonate pathway by requiring acetate and coenzyme A, but malonate units formed by carboxylation of acetate units are not involved, and the mode of condensation differs. Compounds derived by the mevalonic acid pathway includes **carotene** (β-carotene, γ-carotene, violoxanthin, xanthophyll, unspecified carotenoid pigments), **sterols** (ergosterol, fungisterol, β-sitosterol, steroid of unknown structure), and **terpenes** which are divided into diterpenes ((−)-16α-hydroxykaurane, nephrin) and triterpenes (7β-acetoxy-22-hydroxyhopane, 16β-O-acetylleucotylic acid, 6α-O-acetylleucotylin, 6-deoxy-16β-O-acetylleucotylin, 6-deoxyleucotylin, 6-deoxy-16β-di-O-acetylleucotylin, 15α,22-dihydroxyhopane, dolichorrhizin, durvilldiol, durvillonol, eulecanorol, friedelin, *epi*-friedelinol, leucotylic acid, leucotylin, pyxinic acid, retigeranic acid, retigerdiol, taraxerene, triterpene C, triterpene D, triterpene N-1, ursolic acid, zeorin).

1.3.3 Secondary Products of the Shikimic Acid Pathway: Terphenylquinones and Pulvinic Acid Derivates

The **terphenylquinones** and **pulvinic acid** pigments are well-documented examples of secondary lichen products derived by the shikimic acid pathway. The genera of lichens best known for their production of pulvinic acid pigments belong to the *Stictaceae* in which the algal symbionts are frequently blue-green rather than green. Most families of lichens containing only green algae are best known for producing typical acetate-polymalonate-derived aromatic esters and coupled products.

Lichens with green algae generally show a very low nitrogen content unless they grow on high-nitrogen substrates. Although nitrogen metabolisms in lichens are poorly understood, the small number of secondary substances containing nitrogen and the complete absence of alkaloids in lichens suggest that in many species, the supply of available nitrogen to the fungus is severely limited. Some lichens show a strong habitat preference for barnyards, rookeries, or other places high in organic nitrogen. In contrast to the green algae, many blue-green algae show a much enhanced nitrogen content compared to that of their substrates.

Not all lichens with blue-green algae contain pulvinic acid derivates, and many lichens with pulvinic acid pigments or terphenylquinones do not contain nitrogen-fixing algae. Although phenylalanine can serve as a precursor to pulvinic acid derivates in the species studied, the compounds other than rhizocarpic acid and epanorin do not contain nitrogen in the molecule. They might arise from non-nitrogenous, prephenic acid derivates or by some mechanisms allowing a small concentration of nitrogen to be constantly recycled.

Only two terphenylquinones, polyporic acid and thelephoric acid, are known, while pulvinic acid derivates are more numerous (calycin, epanorin, leprapinic acid, leprapinic acid methyl ether, pinastric acid, pulvinic acid, pulvinic dilactone, rhizocarpic acid, stictaurin, vulpinic acid).

1.4 Methods for Determining Lichen Secondary Metabolites

The following methods are used for identification of lichen compounds:

- Classic spot tests
- Microcrystallography
- Paper and thin-layer chromatography
- High-performance liquid chromatography (HPLC)
- Chemical methods
- Gas chromatography and mass spectrometry

Spot Test This test has been used universally as rapid, nonspecific means for detecting the presence of certain unspecified lichen substances. This test is most convenient and simple to perform, even under field conditions. However, this is only a preliminary step in the process of identification of lichens or its substances. In order to identify accurately the secondary metabolite present in the lichen thallus, one has to perform more sensitive test such as TLC or HPLC. Spot test is carried out by placing a small drop of reagent on the lichen thallus, either directly on the upper surface (cortex) or on the medulla. In the latter case, the cortex is scraped or superficially cut with the help of blade. The reagents used are 10% aqueous KOH solution (K), saturated aqueous solution of bleaching powder ($NaOCl_2$) or calcium hypochlorite ($Ca(OCl)_2$) (called as C), and 5% alcoholic *p*-phenylenediamine solution (PD). The color changes at the reagent application point of the thallus are noted as + r −. These color changes take place due to presence of particular secondary metabolite in the thallus, which termed as spotting (Nash 1996; Karunaratne 1999; Karunaratne et al. 2005; Molnár and Farkas 2010; Shukla et al. 2013).

Microcrystallography Some of the secondary metabolites form characteristic crystals when a crystallizing reagent is added and gently warmed. The test is conducted on the glass slide. The lichen compound is extracted using acetone. Glycerol, ethanol, and glacial acetic acids are some of the chemicals used in different combination to make the reagent. Microcrystallography has been largely superseded by the more sensitive and reliable method such as TLC. But the technique is still useful for a number of lichen compounds which are difficult to identify in TLC due to the same Rf class (or value) or spot color (e.g., lecanoric acid and gyrophoric acids, barbatic and diffractaic acids). However, mixture of substances may be difficult to identify with this method, and also minor substances may be undetectable (Nash 1996; Karunaratne 1999; Karunaratne et al. 2005; Molnár and Farkas 2010; Shukla et al. 2013).

Paper Chromatography Paper chromatography is used for the separation of amino acids. Paper is used as the support or adsorbent, but partition probably plays a greater part than adsorption in the separation of the components of mixtures, as the cellulose fibers have a film of moisture round them even in the air-dry state. The technique is therefore closely allied to column partition chromatography, but whereas the latter is capable of dealing with a gram of more material, the former requires micrograms. It is therefore an extremely sensitive technique of enormous value in chemical and biological fields (Nash 1996; Karunaratne 1999; Karunaratne et al. 2005; Molnár and Farkas 2010; Shukla et al. 2013).

Thin-Layer Chromatography (TLC) It is a relatively simple and inexpensive technique which can be performed by anyone with access to basic laboratory facilities. Lichen substances are extracted in acetone, and the extract is spotted on to glass or aluminum plates coated with silica gel. The plate is placed in a sealed tank so that the base of the plate is immersed in a shallow layer of a specific mixture or organic solvents. The different lichen substances present in the sample are separated from each other during the passage of solvent through the silica gel layer and are later made visible by spraying with sulfuric acid. The resulting spots are provisionally identified by their color and relative position in comparison to the control sample (Nash 1996; Karunaratne 1999; Karunaratne et al. 2005; Molnár and Farkas 2010; Shukla et al. 2013).

High-Performance Liquid Chromatography (HPLC) This method was applied to identify and quantify characteristic substances in commercially available oak moss products. This technique provides a powerful complement to the established TLC method. The bonded reverse phase columns are used here, and all the aromatic lichen products are suitable for analysis with this method. Samples are dissolved in methanol and injected into the appropriate portion column, through which an appropriate solvent or sequence of solvents is passed under high pressure. The substances separate and are detected using UV detector. The retention time (Rt or time of passage) and peak intensity are recorded by a chart recorder. HPLC is also used to measure either absolute or relative concentrations of lichen compounds, because the peak intensity (area under curve) is proportional to the concentration. Most workers use HPLC to detect lichen compounds and combine this technique with TLC and/or mass spectrometry to verify the identification of the peaks (Nash 1996; Karunaratne 1999; Karunaratne et al. 2005; Molnár and Farkas 2010; Shukla et al. 2013).

Chemical Methods As with many areas of natural product chemistry, new impetus in the chemistry of lichen substances is provided by the more rapid and improved methods for detecting, isolating, and purifying these compounds and in determining their structure. The techniques of preparative TLC, radial chromatography, and preparative HPLC provide rapid and efficient methods for the purification of lichen substances, and developments in mass spectrometry, proton and carbon-13 NMR (nuclear magnetic resonance) spectroscopy, and X-ray analysis greatly aid structural studies.

The more classical chemical procedures of degradation and total synthesis also developed apace with the use of newer reagents and synthetic methods. For instance, the use of condensing reagents trifluoroacetic anhydride and dicyclohexylcarbodiimide makes the preparation of lichen depsides a relatively straightforward procedure, so that total synthesis is now a common means of structural confirmation.

Gas Chromatography and Mass Spectrometry Xanthones, anthraquinones, dibenzofurans, terpenes, and pulvinic acid derivates which lack thermolabile ester groups can be studied by gas chromatography with mass spectrometry (GCMS). Xanthones in lichens were studied by injecting a lichen extract directly into a mass spectrometer. More recently, the main terpenoid components of the lichens of the family Pyxinaceae have been studied by GMCS (Nash 1996; Karunaratne et al. 2005).

1.5 Application of Lichen Secondary Metabolites in Medicine and Pharmacy

To date, many lichens have proved to be a source of important secondary metabolites for pharmaceutical industries (Huneck 1999; Oksanen 2006) and still hold a considerable interest as alternative treatments in various parts of the world (Richardson 1992). A wide spectrum of biological potential is shown by the lichens, but they have been long neglected by mycologists and overlooked by pharmaceutical industry because of their slow growing nature and difficulties in their artificial cultivation and have scarcely been studied from a biochemical perspective (Crittenden and Porter 1991; Yamamoto et al. 1998; Behera et al. 2003, 2004).

There is a growing interest in the pharmaceutical properties of compounds derived from lichens. However, relatively few lichen substances have been screened in detail for biological activity and therapeutic potential, due principally to difficulties in obtaining them in quantities and purities sufficient for structural elucidation and pharmacological testing. Additionally, precise lichen determination is essential and requires taxonomic expertise (Boustie and Grube 2005).

Lichen secondary metabolites exhibit antimicrobial, antioxidant, anti-inflammatory, cytotoxic, analgesic, antipyretic, and antiviral properties (Table 1.1) and could be potential sources of pharmaceutically useful chemicals. Several examples are described below.

Numerous literature data have supported the antioxidant action of lichen metabolites. The depsidones, physodic, and physodalic acid prevent the formation of reactive metabolites by blocking the oxidation systems present in the hepatic microsomal fraction (Osawa et al. 1991). Hidalgo et al. (1994) reported the antioxidant activity of some depsides, such as atranorin (isolated from *Placopsis* sp.) and divaricatic acid (isolated from *Protousnea malacea*), and depsidones, such as pannarin (isolated from *Psoroma pallidum*) and 1′-chloropannarin (isolated from

Table 1.1 Literature sources mentioning data for some lichen substances with different pharmaceutical activity

Lichen compounds	Studied activities	References
Lecanoric acid	Antitumor, antioxidant, antibacterial, antifungal, antidiabetic, anticancer, anti-inflammatory	Gomes et al. (2003), Lopes et al. (2008), Ranković et al. (2008), Honda et al. (2010), Bogo et al. (2010), Choudhary et al. (2011), Ristić et al. (2016b), Studzinska-Sroka and Dubino (2018)
Atranorin	Antimicrobial, antioxidant, anti-inflammatory, anticancer, antineurodegenerative, antidiabetic	Kumar and Muller (1999), Yilmaz et al. (2004), Turk et al. (2006), Ranković et al. (2008, 2014), Melo et al. (2011), Reddy et al. (2016), Thadhani and Karunaratne (2017), Studzinska-Sroka and Dubino (2018)
Zeorin	Antioxidant, antibacterial, antifungal, antidiabetic	Behera et al. (2008), Kosanić et al. (2010), Karunaratne et al. (2014)
Gyrophoric acid	Antimicrobial, anticancer, antioxidant, antidiabetic	Candan et al. (2006), Ranković et al. (2008), Burlando et al. (2009), Choudhary et al. (2011), Kosanić et al. (2014a, b)
Protocetraric acid	Antibacterial, antifungal, antioxidant, anticancer	Tay et al. (2004), Ranković et al. (2008), Honda et al. (2010), Manojlović et al. (2012)
Fumarprotocetraric acid	Antibacterial, antifungal, antioxidant, anticancer, neuroprotective	Yilmaz et al. (2004), Ranković et al. (2008), Kosanić et al. (2014a, b), Fernández-Moriano et al. (2017)
Stictic acid	Antioxidant, antimicrobial, anticancer	Lohézic-Le Dévéhat et al. (2007), Ranković et al. (2008)
Salazinic acid	Antitumor, antibacterial, antifungal, antioxidant, antidiabetic, antiviral	Candan et al. (2007), Burlando et al. (2009), Manojlović et al. (2012), Verma et al. (2012), Odabasoglu et al. (2006)
Usnic acid	Antiviral, antitumor, antioxidant, antibacterial, antifungal, antipyretic, analgetic, anti-inflammatory, hepatotoxic, antiviral, antidiabetic	Lauterwein et al. (1995), Tay et al. (2004), Ranković et al. (2008, 2014), Paudel et al. (2010), Perry et al. (1999), Odabasoglu et al. (2006), Bazin et al. (2008), Burlando et al. (2009), Ramos and Almeida da Silva (2010), Verma et al. (2012), Thadhani and Karunaratne (2017), Vijayakumar et al. (2000), Okuyama et al. (1995)
Vulpinic acid	Antimicrobial, anticancer	Lauterwein et al. (1995), Burlando et al. (2009)
Evernic acid	Antifungal, antibacterial, antioxidant, anticancer	Halama and Van Haluwin (2004), Kosanić et al. (2013)
Lobaric acid	Antibacterial, antifungal, anticancer, antineurodegenerative	Ingolfsdottir et al. (1998), Piovano et al. (2002), Sundset et al. (2008), Thadhani et al. (2014)
Physodic acid	Antibacterial, antifungal, antioxidant, anticancer, antineurodegenerative, anti-inflammatory	Turk et al. (2006), Ranković et al. (2008, 2014), Kosanić et al. (2013), Emsen et al. (2016), Reddy et al. (2016), Studzinska-Sroka and Dubino (2018)

(continued)

Table 1.1 (continued)

Lichen compounds	Studied activities	References
Protolichesterinic acid	Antitumor, antibacterial, anticancer, anti-inflammatory	Ingolfsdottir et al. (1998), Turk et al. (2003), Bucar et al. (2004), Studzinska-Sroka and Dubino (2018)
Norstictic acid	Antimicrobial, antioxidant, anticancer	Tay et al. (2004), Honda et al. (2010), Ranković et al. (2014)
Ramalin	Antioxidant, antibacterial	Paudel et al. (2008, 2010)
Barbatic acid	Antioxidant, antimicrobial	Verma et al. (2008), Martins et al. (2010)
Divaricatic acid	Antioxidant, antimicrobial, antidiabetic	Hidalgo et al. (1994), Kosanić et al. (2010), Thadhani and Karunaratne (2017), Oh et al. (2018)
Panarin	Antioxidant, anticancer	Hidalgo et al. (1994), Russo et al. (2008)
Umbilicaric acid	Antioxidant, antimicrobial	Buçukoglu et al. (2013).
Variolaric acid	Antioxidant, anticancer	Brisdelli et al. (2013)
Homosekikaic acid	Antioxidant, antibacterial	Sisodia et al. (2013)
Sekikaic acid	Antioxidant, antibacterial, antidiabetic	Verma et al. (2012), Sisodia et al. (2013)
Benzoic acid	Antioxidant	Sisodia et al. (2013)
Diffractaic acid	Analgetic, antiproliferative, antioxidant, antipyretic, analgesic, anti-inflammatory	Brisdelli et al. (2013), Atalay et al. (2011), Kumar and Muller (1999), Okuyama et al. (1995), Studzinska-Sroka and Dubino (2018)
Vicanicin	Antioxidant, anticancer	Brisdelli et al. (2013)
2,4-Dihydroxy-6-propyl	Antioxidant	Sisodia et al. (2013)
1′-Chloropannarin	Antioxidant	Hidalgo et al. (1994)
Biruloquinone	Antineurodegenerative	Luo et al. (2013)
Olivetoric acid	Antineurodegenerative, anti-inflammatory	Emsen et al. (2016), Studzinska-Sroka and Dubino (2018)
Psoromic acid	Antineurodegenerative	Emsen et al. (2016)
Perlatolic acid	Antineurodegenerative, anti-inflammatory	Reddy et al. (2016), Studzinska-Sroka and Dubino (2018)
Methyl β-orcinolcarboxylate	Antidiabetic	Karunaratne et al. (2014)
Methylorsellinate	Antidiabetic	Karunaratne et al. (2014)
Ethyl hematommate	Antidiabetic	Choudhary et al. (2011)
Ethyl orsellinate	Antidiabetic	Choudhary et al. (2011)
Alectorialic acid	Antiviral	Odabasoglu et al. (2006)
Parietin	Antiviral, antimicrobial	Fazio et al. (2007), Basile et al. (2015)
5-Hydroxysekikaic acid	Antiviral	Lai et al. (2013)
2-Dehydro-5-oxysekikaic acid	Antiviral	Lai et al. (2013)
Sekikaic acid	Antiviral	Lai et al. (2013)
Homosekikaic acid	Antiviral	Lai et al. (2013)
4′-O-methylnorhomosekikaic acid	Antiviral	Lai et al. (2013)

(continued)

Table 1.1 (continued)

Lichen compounds	Studied activities	References
4′-O-methylnorsekikaic acid	Antiviral	Lai et al. (2013)
Divaricatinic acid	Antiviral	Lai et al. (2013)
2,3-Dihydroxy-4-methoxy-6-pentylbenzoic acid	Antiviral	Lai et al. (2013)
2,3-Dihydroxy-4-methoxy-6-propylbenzoic acid	Antiviral	Lai et al. (2013)
3,4-Methylenedioxy-3′-methoxybibenzyl	Antiviral	Lai et al. (2013)
Obtusatic acid	Antioxidant, antimicrobial, anticancer	Ristić et al. (2016a)
Methyl evernate	Antioxidant, antimicrobial, anticancer	Ristić et al. (2016a)
O-methyl anziaic acid	Antioxidant, antimicrobial, anticancer	Ristić et al. (2016b)

Erioderma chilense). All of these secondary compounds inhibited rat brain homogenate auto-oxidation and β-carotene oxidation, and depsidones were found to be the most effective. Russo et al. (2008) found that both sphaerophorin (depside) and pannarin (depsidone) inhibited superoxide anion formation in vitro. For other lichen compounds including protocetraric acid (from *Parmelia caperata*), salazinic acid (from *Parmelia saxatilis*), evernic acid (from *Evernia prunastri*), physodic acid (from *Pseudevernia furfuracea*), gyrophoric acid (from *Acarospora fuscata*), obtusatic acid (from *Ramalina fraxinea*), methyl evernate (from *R. fastigiata*), and O-methyl anziaic acid (from *Melanelia fuliginosa*) have been proven to have strong antioxidant activity using various in vitro assays (Manojlović et al. 2012; Kosanić et al. 2013, 2014a, b; Ristić et al. 2016a, b).

There are many studies on the antimicrobial activity of lichen secondary metabolites. For example, atranorin (from *Physcia aipolia*), fumarprotocetraric acid (from *Cladonia furcata*), gyrophoric acid (from *Umbilicaria polyphylla*), lecanoric acid (from *Ochrolechia androgyna*), physodic acid (from *Hypogymnia physodes*), protocetraric acid (from *Flavoparmelia caperata*), stictic acid (from *Xanthoparmelia conspersa*), usnic acid (from *Flavoparmelia caperata*), obtusatic acid (from *Ramalina fraxinea*), methyl evernate (from *R. fastigiata*), O-methyl anziaic acid (from *Melanelia fuliginosa*), divaricatic acid (from *Evernia mesomorpha*), and parietin (from *Xanthoria parietina*) showed relatively strong antimicrobial effects against numerous bacteria and fungi, among which were human pathogens (Ranković and Mišić 2008; Ranković et al. 2008; Basile et al. 2015; Ristić et al. 2016a, b; Oh et al. 2018).

Species of *Usnea* contains high amount of usnic acid, a very active lichen substance used as tumor inhibitor and as analgesic. (+)-Usnic acid was found to be a strong hepatotoxic agent against monogastric murine hepatocytes, due to its ability to uncouple and inhibit the electron transport chain in mitochondria and induce oxidative stress in cells (Han et al. 2004). The (−)-enantiomer of usnic acid (isolated from *Cladonia convoluta*) induced apoptotic cell death in murine lymphocytic leukemia cells and was moderately cytotoxic to various cancer cell lines, such as murine Lewis lung carcinoma, human chronic myelogenous leukemia, human brain metastasis of a prostate carcinoma, human breast adenocarcinoma, and human glioblastoma (Bézivin et al. 2004). Usnic acid also decreased proliferation of human breast cancer cells and human lung cancer cells without any DNA damage (Mayer et al. 2005). Finding cancer therapies that do not have DNA-damaging effects and that do not cause the development of secondary malignancies later in life is of great interest. Accordingly, usnic acid may represent a novel source for a natural non-genotoxic anticancer drug (chemotherapeutic agent).

The para-depside gyrophoric and diffractaic acid also inhibit the proliferation of human keratinocyte cells (Kumar and Muller 1999). Ursolic acid inhibited the growth of human epidermoid carcinoma cells by affecting tyrosine kinase activity (Hollosy et al. 2000). The use of ursolic acid for the manufacture of an anticancer agent for suppressing metastasis has been patented (Ishikawa et al. 1997). Russo et al. (2008) reported that the depside sphaerophorin (isolated from *Sphaerophorus globosus*) and the depsidone pannarin (isolated from *Psoroma pholidotoides*, *P. pulchrum*, and *P. pallidum*) inhibited the growth of M14 human melanoma cells, triggering apoptotic cell death. The anticancer activities of these lichen metabolites are promising in the treatment of this aggressive, therapy-resistant skin tumor. Obtusatic acid isolated from *Ramalina fraxinea* and methyl evernate isolated from *R. fastigiata* were studied for anticancer activity against human epithelial carcinoma Hela cells, human lung carcinoma A549 cells, and human colon carcinoma LS174 cells (Ristić et al. 2016a). Both tested compounds exhibited relatively strong anticancer activity toward all tested cells. The same authors tested anticancer properties for 2′-O-methyl anziaic acid from *Melanelia fuliginosa* and lecanoric acid from *M. subaurifera*. Both tested compounds expressed high anticancer effect (Ristić et al. 2016a). Norstictic acid, physodic acid, evernic acid, usnic acid, salazinic acid, fumarprotocetraric acid, protocetraric acid, atranorin, and zeorin isolated from different lichen species are relatively strong anticancer agents (Kosanić et al. 2010, 2013, 2014a, b; Manojlović et al. 2012; Ranković et al. 2014).

In recent times, lichen metabolites are used in the treatment of nerve diseases such as Alzheimer's disease and Parkinson's disease. For example, Fernández-Moriano et al. (2017) reported the neuroprotective properties of the major secondary metabolite of the *Cetraria islandica*, i.e., the depsidone fumarprotocetraric acid. Also, it has been reported that lobaric acid extracted from the lichen *Heterodermia* sp., biruloquinone from *Cladonia macilenta*, olivetoric acid and physodic acid isolated from *Pseudevernia furfuracea*, and psoromic acid from *Rhizoplaca melanophthalma* showed antineurodegenerative activity (Luo et al. 2013; Thadhani et al. 2014; Emsen et al. 2016). Similar results were obtained for atranorin, perlatolic acid, physodic acid, and usnic acid (Reddy et al. 2016).

A few studies have discovered the positive potential of lichen metabolites as potent antidiabetic agents. Karunaratne et al. (2014) found three potent alpha glucosidase inhibitors (zeorin, methyl *β*-orcinolcarboxylate, and methylorsellinate) from the lichen *Cladonia* sp. Similarly, salazinic acid from *Ramalina celastri*, sekikaic acid from *R. nervulosa*, and usnic acid from *R. pacifica* showed promising antihyperglycemic effect (Verma et al. 2012). Ethyl hematommate, ethyl orsellinate, lecanoric acid, and gyrophoric acid proved to be as antidiabetic agents (Choudhary et al. 2011). In another study, atranorin, divaricatic acid, and usnic acid had shown high antidiabetic activity (Thadhani and Karunaratne 2017).

Lichen metabolites also exhibit antiviral activity. Odabasoglu et al. (2006) found antiviral activities against three viruses, viz., respiratory syncytial virus (RSV), herpes virus 1, and herpes virus 2 for salazinic acid from *Parmelia saxatilis* and alectorialic acid from *Alectoria nigricans*. Fazio et al. (2007) described the antiviral activity of parietin isolated from *Teloschistes chrysophthalmus* against some Junin and Tacaribe arenaviruses. Ethyl-acetate-soluble fraction *(ET4)* from the lichen *Ramalina farinacea* inhibited wild-type viruses, including the herpes simplex virus type 1 and the respiratory syncytial virus (Esimone et al. 2009). Also, from the lichen *Ramalina farinacea*, Lai et al. (2013) isolated 10 compounds including 5-hydroxysekikaic acid, 2-dehydro-5-oxysekikaic acid, sekikaic acid, homosekikaic acid, 4′-O-methylnorhomosekikaic acid, 4′-O-methylnorsekikaic acid, divaricatinic acid, 2,3-dihydroxy-4-methoxy-6-pentylbenzoic acid, 2,3-dihydroxy-4-methoxy-6-propylbenzoic acid, and 3,4-methylenedioxy-3′-methoxybibenzyl. All isolated compounds displayed varying degrees of antiviral activity against the respiratory syncytial virus with one of them sekikaic acid showing the most potent inhibition toward respiratory syncytial virus.

Lichen metabolites are potential candidates for antipyretic and analgesic drugs. Some lichen substances have been shown to relieve pain effectively or reduce fever and inflammation in various mammals, and it is reasonable to assume that these compounds also could be effective in humans. Vijayakumar et al. (2000) reported that (+)-usnic acid, isolated from Roccella montagnei, showed significant, dose-dependent anti-inflammatory activity in rats, reducing carrageenin-induced paw edema. Diffractaic and usnic acids have an analgesic effect in mice in vitro (Okuyama et al. 1995), and usnic acid also is an antipyretic against lipopolysaccharide-induced fever. Except these, another lichen metabolites produced by certain species of lichens have anti-inflammatory properties including atranorin, lecanoric acid, diffractic acid, physodic acid, perlatoric acid, olivetoric acid, and protolichestrinic acid (Studzinska-Sroka and Dubino 2018). The mentioned compounds of a depside, depsidone, and dibenzofuran structure were characterized by the significant anti-inflammatory activity, resulting, among others, from the ability to inhibit enzymes involved in the course of inflammation including cyclooxygenase, lipoxygenase, and prostaglandin synthase.

Several investigators have focused on the possible antigenotoxic potential of lichen metabolites. For example, usnic acid, hypostictic acid, protocetraric acid, salazinic acid, and psoromic acid isolated from different lichen species were found to exhibit antigenotoxic effect (Leandro et al. 2013; Guterres et al. 2017).

Lichens and other natural products remain as an untapped reservoir of potentially useful chemical compounds not only as drugs but also as unique templates that could serve as a starting point for synthetic analogues. Over 50% of all modern clinical drugs are of natural product origin, and natural products play an important role in the drug development programs in the pharmaceutical industries.

Here we want to show that lichen secondary substances exhibit a huge array of remarkable pharmaceutical activities and that these important properties of lichen substances make them possible pharmaceuticals. Lichens and their secondary metabolites exhibit many pharmaceutical roles, primarily including antimicrobial, antioxidant, antiviral, anticancer, antigenotoxic, anti-inflammatory, analgesic, and antipyretic properties. This study suggests that the lichens can be productively used in the pharmaceutical area because of its possible activities reported.

References

Ahmadijan V (1993) The lichen symbiosis. Wiley, New York, pp 1–250

Armaleo D, Zhang Y, Cheung S (2008) Light might regulate divergently depside and depsidone accumulation in the lichen *Parmotrema hypotropum* by affecting thallus temperature and water potential. Mycologia 100:565–576

Atalay F, Halici MB, Mavi AA et al (2011) Antioxidant phenolics from *Lobaria pulmonaria* (L.) Hoffm. and *Usnea longissima* Ach. Lichen species. Turk J Chem 35:647–661

Bačkor M, Fahselt D (2008) Lichen photobionts and metal toxicity (review article). Symbiosis 46:1–10

Bačkorová M, Jendželovský R, Kello M et al (2012) Lichen secondary metabolites are responsible for induction of apoptosis in HT-29 and A2780 human cancer cell lines. Toxicol In Vitro 26:462–468

Basile A, Rigano D, Loppi S et al (2015) Antiproliferative, antibacterial and antifungal activity of the lichen *Xanthoria parietina* and its secondary metabolite parietin. Int J Mol Sci 16:7861–7875

Bazin MA, Le Lamer AC, Delcros JG et al (2008) Synthesis and cytotoxic activities of usnic acid derivatives. Bioorg Med Chem Lett 16:6860–6866

BeGora MD, Fahselt D (2001) Usnic acid and atranorin concentrations in lichens in relation to bands of UV irradiance. Bryologist 104:134–140

Behera BC, Adawadkar B, Makhija U (2003) Inhibitory activity of xanthine oxidase and superoxide-scavenging activity in some taxa of the lichen family *Graphidaceae*. Phytomedicine 10:536–543

Behera BC, Adawwadkar B, Makhija U (2004) Capacity of some Graphidaceous lichens to scavenge superoxide and inhibition of tyrosinase and xanthine oxidase activities. Curr Sci 87:83–87

Behera BC, Verma N, Sonone A et al (2008) Antioxidant and antibacterial properties of some cultured lichens. Bioresour Technol 99:776–784

Bentley R (1999) Secondary metabolite biosynthesis: the first century. Crit Rev Biotechnol 19:1–40

Bézivin C, Tomasi S, Rouaud I et al (2004) Cytotoxic activity of compounds from the lichen: *Cladonia convoluta*. Planta Med 70:874–877

Bjerke JW, Dahl T (2002) Distribution patterns of usnic acid producing lichens along local radiation gradients in West Greenland. Nova Hedwigia 75:487–506

Bjerke JW, Zielke M, Solheim B (2003) Long-term impacts of simulated climatic change on secondary metabolism, thallus structure and nitrogen fixation activity in two cyanolichens from the Arctic. New Phytol 159:361–367

Bjerke JW, Joly D, Nilsen L et al (2004) Spatial trends in usnic acid concentrations of the lichen *Flavocetraria nivalis* along local climatic gradients in the Arctic (Kongsfjorden, Svalbard). Polar Biol 27:409–417

Bjerke JW, Gwynn-Jones D, Callaghan TV (2005) Effects of enhanced UV-B radiation in the field on the concentration of phenolics and chlorophyll fluorescence in two boreal and arctic-alpine lichens. Environ Exp Bot 53:139–149

Bogo D, de Fatima Cepa Matos M, Honda NK et al (2010) In vitro antitumour activity of orsellinates. Z Naturforsch C 65:43–48

Boustie J, Grube M (2005) Lichens – a promising source of bioactive secondary metabolites. Plant Genet Resour 3:273–287

Brisdelli F, Perilli M, Sellitri D et al (2013) Cytotoxic activity and antioxidant capacity of purified lichen metabolites: an in vitro study. Phytother Res 27:431–437

Brunauer G, Hager A, Grube M et al (2007) Alterations in secondary metabolism of aposymbiotically grown mycobionts of *Xanthoria elegans* and cultured resynthesis stages. Plant Physiol Biochem 45:146–151

Bucar F, Schneider I, Ogmundsdottir H et al (2004) Antiproliferative lichen compounds with inhibitory activity on 12(S)-HETE production in human platelets. Phytomedicine 11:602–606

Buçukoglu TZ, Albayrak S, Halici MG et al (2013) Antimicrobial and antioxidant activities of extracts and lichen acids obtained from some *Umbilicaria* species from Central Anatolia, Turkey. J Food Process Preserv 37:1103–1110

Burlando B, Ranzato E, Volante A et al (2009) Antiproliferative effects on tumour cells and promotion of keratinocyte wound healing by different lichen compounds. Planta Med 75:607–613

Candan M, Yilmaz M, Tay T et al (2006) Antimicrobial activity of extracts of the lichen *Xanthoparmelia pokornyi* and its gyrophoric and stenosporic acid constituents. Z Naturforsch 61:319–323

Candan M, Yilmaz M, Tay T et al (2007) Antimicrobial activity of exracts of the lichen *Parmelia sulcata* and its salazinic acid constituent. Z Naturforsch 62:619–621

Choudhary MI, Ali M, Wahab AT et al (2011) New antiglycation and enzyme inhibitors from *Parmotrema cooperi*. Sci China Chem 54:1926–1931

Cox DD (2003) A naturalist's guide to forest plants: an ecology for Eastern North America. Syracuse University Press, Syracuse, NY

Crittenden D, Porter N (1991) Lichen-forming fungi: potential sources of novel metabolites. Trends Biotechnol 9:409–414

Culberson WL (1970) Chemosystematics and ecology of lichen-forming fungi. Annu Rev Ecol Syst 1:153–170

Culberson CF, Armaleo D (1992) Induction of a complete secondary-product pathway in a cultured lichen fungus. Exp Mycol 16:52–63

Culberson CF, Culberson WL (2001) Future directions in lichen chemistry. Bryologist 104(2):230–234

Culberson WL, Culberson CF, Johnson A (1983) Genetic and environmental effects on growth and production of secondary compounds in *Cladonia cristatella*. Biochem Syst Ecol 11:77–84

Egan RS (1986) Correlations and non-correlations of chemical variation patterns with lichen morphology and geography. Bryologist 89(2):99–110

Emsen B, Aslan A, Togar B et al (2016) In vitro antitumor activities of the lichen compounds olivetoric, physodic and psoromic acid in rat neuron and glioblastoma cells. Pharm Biol 54:1748–1762

Ernst-Russell MA, Chai CLL, Hurne AM et al (1999) Revision and cytotoxic activity of the scabrosin esters, epidithiopiperazinediones from the lichen *Xanthoparmelia scabrosa*. Aust J Chem 52:279–283

Esimone CO, Grunwald T, Nworu CS et al (2009) Broad spectrum antiviral fractions from the lichen *Ramalina farinacea* (L.) Ach. Chemotherapy 55:119–126

Fazio AT, Adler MT, Bertoni MD et al (2007) Lichen secondary metabolites from the cultured lichen mycobionts of *Teloschistes chrysophthalmus* and *Ramalina celastri* and their antiviral activities. Z Naturforsch 62c:543–549

Fehrer J, Slavíková-Bayerová Š, Orange A (2008) Large genetic divergence of new, morphologically similar species of sterile lichens from Europe (*Lepraria, Stereocaulaceae, Ascomycota*): concordance of DNA sequence data with secondary metabolites. Cladistics 24:443–458

Fernández-Moriano C, Divakar PK, Crespo A et al (2017) In vitro neuroprotective potential of lichen metabolite fumarprotocetraric acid via intracellular redox modulation. Toxicol Appl Pharmacol 316:83–94

Fessenden RJ, Fessenden JS (1986) Organic chemistry, 3rd edn. Brooks/Cole, Monterey, CA, p 1129

Galun M, Shomer-Ilan A (1988) Secondary metabolic products. In: Galun M (ed) CRC handbook of lichenology, vol III. CRC Press, Boca Raton, FL, pp 3–8

Gilbert O (2000) Lichens. Harper Collins, London

Gomes AT, Smania AJ, Seidel C et al (2003) Antibacterial activity of orsellinates. Braz J Microbiol 34:194–196

Guterres ZR, Honda NK, Coelho RG et al (2017) Antigenotoxicity of depsidones isolated from Brazilian lichens. Orbital Electron J Chem 9:50–54

Hager A, Brunauer G, Türk R et al (2008) Production and bioactivity of common lichen metabolites as exemplified by *Heterodea muelleri* (Hampe) Nyl. J Chem Ecol 34:113–120

Halama P, Van Haluwin C (2004) Antifungal activity of lichen extracts and lichenic acids. Biol Control 49:95–107

Hale ME (1983) The biology of lichens. Edward Arnold, London

Hamada N (1982) The effect of temperature on the content of the medullary depsidone salazinic acid in *Ramalina siliquosa*. Can J Bot 60:383–385

Hamada N (1984) The content of lichen substances in *Ramalina Siliquosa* cultured at various temperatures in growth cabinets. Lichenologist 16:96–98

Hamada N (1991) Environmental factors affecting the content of usnic acid in the lichen mycobiont of *Ramalina siliquosa*. Bryologist 94:57–59

Han D, Matsumaru K, Rettori D et al (2004) Usnic acid-induced necrosis of cultured mouse hepatocytes: inhibition of mitochondrial function and oxidative stress. Biochem Pharmacol 67:439–451

Hidalgo ME, Fernández E, Quilhot W et al (1994) Antioxidant activity of depsides and depsidones. Phytochemistry 37:1585–1587

Hollosy F, Meszaros G, Bokonyi G et al (2000) Cytostatic, cytotoxic and protein tyrosine kinase inhibitory activity of ursolic acid in human tumor cells. Anticancer Res 20:4563–4570

Honda NK, Pavan FR, Coelho RG et al (2010) Antimycobacterial activity of lichen substances. Phytomedicine 17:328–332

Honegger R (1991) Functional aspects of the lichen symbioses. Annu Rev Plant Physiol Plant Mol Biol 42:553–578

Huneck S (1999) The significance of lichens and their metabolites. Naturwissenschaften 86:559–570

Huneck S, Himmelreich U (1995) Arthogalin, a cyclic depsipeptide from the lichen *Arthothelium galapagoense*. Z Naturforsch 50B:1101–1103

Ingolfsdottir K, Chung GAC, Skulason VG et al (1998) Antimycobacterial activity of lichens metabolites in vitro. Eur J Pharm Sci 6:141–144

Ishikawa H, Nishimuro S, Watanbe T et al (1997) Use of ursolic acid for the manufacture of a medicament for suppressing metastasis. Eur Pat Appl EP774255

Karunaratne V (1999) Lichen substances: biochemistry, ecological role and economic uses. Cey J Sci (Phys Sci) 6:13–28

Karunaratne V, Bombuwela K, Kathirgamanathar S et al (2005) Lichens: a chemically important biota. J Nat Sci Found Sri Lanka 33:169–186

Karunaratne V, Thadhani VM, Khan SN et al (2014) Potent FC;-glucosidase inhibitors from the lichen Cladonia species from Sri Lanka. J Nat Sci Found Sri Lanka 42:95–98
Kinoshita Y (1993) The production of lichen substances for pharmaceutical use by lichen tissue culture. Nippon Paint, Osaka
Kirk PM, Cannon PF, Minter DW et al (eds) (2008) Dictionary of the fungi, 10th edn. CAB International, Wallingford
Kosanić M, Ranković B, Sukdolak S (2010) Antimicrobial activity of the lichen *Lecanora frustulosa* and *Parmeliopsis hyperopta* and their divaricatic acid and zeorin constituents. Afr J Microbiol Res 4:885–890
Kosanić M, Manojlović N, Janković S et al (2013) *Evernia prunastri* and *Pseudoevernia furfuraceae* lichens and their major metabolites as antioxidant, antimicrobial and anticancer agents. Food Chem Toxicol 53:112–118
Kosanić M, Ranković B, Stanojković T et al (2014a) *Cladonia* lichens and their major metabolites as possible natural antioxidant, antimicrobial and anticancer agents. LWT Food Sci Technol 58:518–525
Kosanić M, Ranković B, Stanojković T et al (2014b) Biological activities and chemical composition of lichens from Serbia. EXCLI J 13:1226–1238
Kumar KC, Muller K (1999) Lichen metabolites, 2: Antiproliferative and cytotoxic activity of gyrophoric, usnic, and diffractaic acid on human keratinocyte growth. J Nat Prod 62:821–823
Lai D, Odimegwu DC, Esimone C et al (2013) Phenolic compounds with in vitro activity against respiratory syncytial virus from the Nigerian lichen *Ramalina farinacea*. Planta Med 79:1440–1446
Lauterwein M, Oethinger M, Belsner K et al (1995) In vitro activities of the lichen secondary metabolites vulpinic acid, (+)-usnic acid, and (−)- usnic acid and against aerobic and anaerobic microorganisms. Antimicrob Agents Chemother 39:2541–2543
Lawrey JD (1986) A biological review of lichen substances. Bryologist 89:111–122
Leandro LF, Munari CC, Sato VL et al (2013) Assessment of the genotoxicity and antigenotoxicity of (+)-usnic acid in V79 cells and Swiss mice by the micronucleus and comet assays. Mutat Res 753:101–106
Leuckert C, Ahmadjian V, Culberson CF et al (1990) Xanthones and depsidones of the lichen *Lecanora dispersa* in nature and of its mycobiont in culture. Mycologia 82:370–378
Lohézic-Le Dévéhat F, Tomasi S, Elix JA et al (2007) Stictic acid derivatives from the lichen *Usnea articulata* and their antioxidant activities. J Nat Prod 70:1218–1220
Lopes TIB, Coelho RG, Yoshida NC et al (2008) Radical-scavenging activity of orsellinates. Chem Pharm Bull 56:1551–1554
Lumbsch HT (1998) Taxonomic use of metabolic data in lichen-forming fungi. In: Frisvad JC, Bridge PD, Arora DK (eds) Chemical fungal taxonomy. Marcel Dekker, New York, pp 345–387
Luo H, Li C, Kim JC et al (2013) Biruloquinone, an acetylcholinesterase inhibitor produced by lichen-forming fungus *Cladonia macilenta*. J Microbiol Biotechnol 23:161–166
Manojlović N, Ranković B, Kosanić M et al (2012) Chemical composition of three Parmelia lichens and antioxidant, antimicrobial and cytotoxic activities of some their major metabolites. Phytomedicine 19:1166–1172
Marques J (2013) A framework for assessing the vulnerability of schist surfaces to lichen-induced weathering in the Upper Douro region (NE Portugal). Directores: Rubim Almeida y Graciela Paz. Universidad, Universidade de Porto, Fecha de lectura, 2013
Martins MCB, Gonçalves de Lima MJ, Silva FP et al (2010) *Cladia aggregata* (lichen) from Brazilian Northeast, chemical characterization and antimicrobial activity. Braz Arch Biol Technol 53:115–122
Mattsson JE (1994) Lichen proteins, secondary products and morphology: a review of protein studies in lichens with special emphasis on taxonomy. J Hattori Bot Lab 76:235–248
Mayer M, O'Neill MA, Murray KE et al (2005) Usnic acid: a non-genotoxic compound with anticancer properties. Anti-Cancer Drugs 16:805–809

Melo MG, Dos Santos JP, Serafini MR et al (2011) Redox properties and cytoprotective actions of atranorin, a lichen secondary metabolite. Toxicol In Vitro 25:462–468

Moe R (1997) *Verrucaria tavaresiae* sp. nov., a marine lichen with a brown algal photobiont. Bull Calif Lichen Soc 4:7–11

Molnár K, Farkas E (2010) Current results on biological activities of lichen secondary metabolites. Z Naturforsch C 65:157–173

Nash TH (1996) Lichen biology. Cambridge University Press, Cambridge

Nash TH (2008) Lichen biology, 2nd edn. Cambridge University Press, Cambridge

Nelsen MP, Gargas A (2008) Phylogenetic distribution and evolution of secondary metabolites in the lichenized fungal genus *Lepraria* (*Lecanorales: Stereocaulaceae*). Nova Hedwigia 86:115–131

Nordin A, Tibell L, Owe-Larsson B (2007) A preliminary phylogeny of *Aspicilia* in relation to morphological and secondary product variation. Bibl Lichenol 96:247–266

Odabasoglu F, Cakir A, Suleyman H et al (2006) Gastroprotective and antioxidant effects of usnic acid on indomethacine-induced gastric ulcer in rats. J Ethnopharmacol 1:59–65

Oh JM, Kim YJ, Gang HS et al (2018) Antimicrobial activity of divaricatic acid isolated from the lichen *Evernia mesomorpha* against Methicillin-Resistant *Staphylococcus aureus*. Molecules 23:3068

Oksanen I (2006) Ecological and biotechnological aspects of lichens. Appl Microbiol Biotechnol 73:723–734

Okuyama E, Umeyama K, Yamazaki M et al (1995) Usnic acid and diffractaic acid as analgesic and antipyretic components of *Usnea diffracta*. Planta Med 61:113–115

Osawa K, Marsumoto T, Yasuda H et al (1991) The inhibitory effect of plant extracts on the collagenolytic activity and cytotoxicity of human gingival fibroblasts by *Porphyromonas gingivalis* crude enzyme. Bull Tokyo Dent Coll 32:1–7

Paracer S, Ahmadijan V (2000) Symbiosis: an introduction to biological associations. Oxford University Press, Oxford, 291p

Paudel B, Bhattarai HD, Lee JS et al (2008) Antibacterial potential of Antarctic lichens against human pathogenic Gram-positive bacteria. Phytother Res 22:1269–1271

Paudel B, Bhattarai HD, Lee HK et al (2010) Antibacterial activities of ramalin, usnic acid and its three derivatives isolated from the Antarctic lichen *Ramalina terebrata*. Z Naturforsch C 65:34–38

Perry NB, Benn MH, Brennan NJ et al (1999) Antimicrobial, antiviral and cytotoxic activity of New Zeland lichens. Lichenologist 31:627–636

Piovano M, Garbarino JA, Giannini FA et al (2002) Evaluation of antifungal and antibacterial activities of aromatic metabolites from lichens. Bol Soc Chil Quím 47:235–240

Ramos DF, Almeida da Silva PE (2010) Antimycobacterial activity of usnic acid against resistant and susceptible strains of *Mycobacterium tuberculosis* and non-tuberculous mycobacteria. Pharm Biol 48:260–263

Ranković B, Mišić M (2008) The antimicrobial activity of the lichen substances of the lichens *Cladonia furcata, Ochrolechia androgyna, Parmelia caperata* and *Parmelia conspresa*. Biotechnol Biotechnl Eq 22:1013–1016

Ranković B, Mišić M, Sukdolak S (2008) The antimicrobial activity of substances derived from the lichens *Physcia aipolia, Umbilicaria polyphylla, Parmelia caperata* and *Hypogymnia physodes*. World J Microbiol Biotechnol 24:1239–1242

Ranković B, Kosanić M, Manojlovic N et al (2014) Chemical composition of *Hypogymnia physodes* lichen and biological activities of some its major metabolites. Med Chem Res 23:408–416

Reddy RG, Veeraval L, Maitra S et al (2016) Lichen-derived compounds show potential for central nervous system therapeutics. Phytomedicine 23:1527–1534

Řezanka T, Dembitsky V (1999) Novel brominated lipidic compounds from lichens of Central Asia. Phytochemistry 51:963–968

Řezanka T, Guschina IA (1999) Brominated depsidones from *Acarospora gobiensis*, a lichen of Central Asia. J Nat Prod 62:1675–1677

Řezanka T, Guschina IA (2000) Glycosidic compounds of murolic, protocontipatic and allo-murolic acids from lichens of Central Asia. Phytochemistry 54:635–645

Řezanka T, Guschina IA (2001a) Glycosides esters from lichens of Central Asia. Phytochemistry 58:509–516

Řezanka T, Guschina IA (2001b) Further glucosides of lichens' acids from Central Asian lichens. Phytochemistry 56:181–188

Řezanka T, Guschina IA (2001c) Macrolactones glycosides of three lichen acids from *Acarospora gobiensis*, a lichen of Central Asia. Phytochemistry 58:1281–1287

Řezanka T, Jáchymová J, Dembitsky VM (2003) Prenylated xanthone glucosides from Ural's lichen *Umbilicaria proboscidea*. Phytochemistry 62:607–612

Řezanka T, Temina M, Hanuš L et al (2004) The tornabeatins, four tetrahydro-2-furanone derivatives from the lichenized ascomycete *Tornabea scutellifera* (With.) J.R. Laundon. Phytochemistry 65:2605–2612

Richardson DHS (1992) Pollution monitoring with lichens. Naturalist's handbook 19. Richmond, Richmond

Ristić S, Ranković B, Kosanić M et al (2016a) Biopharmaceutical potential of two Ramalina lichens and their metabolites. Curr Pharm Biotechnol 17:651–658

Ristić S, Ranković B, Kosanić M et al (2016b) Phytochemical study and antioxidant, antimicrobial and anticancer activities of *Melanelia subaurifera* and *Melanelia fuliginosa* lichens. J Food Sci Technol 53:2804–2816

Rubio C, Fernández E, Hidalgo ME et al (2002) Effects of solar UV-B radiation in the accumulation of rhizocarpic acid in a lichen species from alpine zones of Chile. Bol Soc Chil Quim 47:67–72

Rundel PW (1978) The ecological role of secondary lichen substances. Biochem Syst Ecol 6:157–170

Russo A, Piovano M, Lombardo L et al (2008) Lichen metabolites prevent UV light and nitric oxide-mediated plasmid DNA damage and induce apoptosis in human melanoma cells. Life Sci 83:468–474

Shukla V, Upreti DK, Rajesh B (2013) Lichens to biomonitor the environment. Springer, Delhi

Sisodia R, Geol M, Verma S et al (2013) Antibacterial and antioxidant activity of lichen species *Ramalina roesleri*. Nat Prod Res 27:2235–2239

Solhaug KA, Lind M, Nybakken L et al (2009) Possible functional roles of cortical depsides and medullary depsidones in the foliose lichen *Hypogymnia physodes*. Flora 204:40–48

Stephenson NL, Rundel PW (1979) Quantitative variation and the ecological role of vulpinic acid and atranorin in the thallus of *Letharia vulpina*. Biochem Syst Ecol 7:263–267

Stocker-Wörgötter E (2008) Metabolic diversity of lichen-forming ascomycetous fungi: culturing, polyketide and shikimate metabolite production, and PKS genes. Nat Prod Rep 25:188–200

Stocker-Wörgötter E, Elix JA (2002) Secondary chemistry of cultured mycobionts: formation of a complete chemosyndrome by the lichen fungus of *Lobaria spathulata*. Lichenologist 34:351–359

Studzinska-Sroka E, Dubino A (2018) Lichens as a source of chemical compounds with antiinflammatory activity. Herba Pol 64:56–64

Sundset MA, Kohn A, Mathiesen SD et al (2008) *Eubacterium rangiferina*, a novel usnic acid resistant bacterium from the reindeer rumen. Naturwissensch 95:741–749

Swanson A, Fahselt D, Smith D (1996) Phenoli levels in *Umbilicaria americana* in relation to enzyme polymorphism, altitude and sampling date. Lichenologist 28:331–339

Tay T, Türk AO, Yılmaz M et al (2004) Evaluation of the antimicrobial activity of the acetone extract of the lichen *Ramalina farinacea* and its (+)-usnic acid, norstictic acid, and protocetraric acid constituents. Z Naturforsch 59:384–388

Thadhani VM, Karunaratne V (2017) Potential of lichen compounds as antidiabetic agents with antioxidative properties: a review. Oxid Med Cell Longev:2079697

Thadhani VM, Naaz Q, Choudhag MI et al (2014) Enzyme inhibitory and immunomodulatory activities of the depsidone lobaric acid extracted from the lichen Heterodermia sp. J Nat Sci Found Sri Lanka 42:193–196

Torres A, Hochberg M, Pergament I et al (2004) A new UV-B absorbing mycosporine with photoprotective activity from the lichenized ascomycetes *Collema cristatum*. Eur J Biochem 271:780–784

Turk AO, Yilmaz M, Kivanc M et al (2003) The antimicrobial activity of extracts of the lichen Cetraria aculeata and its protolichesterinic acid constituent. Z Naturforsch 58:850–854

Turk H, Yilmaz M, Tay T et al (2006) Antimicrobial activity of extracts of chemical races of the lichen *Pseudevernia furfuracea* and their physodic acid, chloroatranorin, atranorin, and olivetoric acid constituents. Z Naturforsch C 61:499–507

Verma N, Behera BC, Sonone A et al (2008) Cell aggregates derived from natural lichen thallus fragments: antioxidant activities of lichen metabolites developed. Nat Prod Commun 3(11):1911–1918

Verma N, Behera BC, Sharma BO (2012) Glucosidase inhibitory and radical scavenging properties of lichen metabolites salazinic acid, sekikaic acid and usnic acid. Hacettepe J Biol Chem 40:7–21

Vijayakumar CS, Viswanathan S, Reddy MK et al (2000) Anti-inflammatory activity of (+) usnic acid. Fitoterapia 71:564–566

Waring B (2008) Light exposure affects secondary compound diversity in lichen communities in Monteverde, Costa Rica. PennScience 6:11–13

Yamamoto Y, Mizuguchi R, Yamada Y (1985) Tissue cultures of *Usnea rubescens* and *Ramalina yasudae* and production of usnic acid in their cultures. Agric Biol Chem 49:3347–3348

Yamamoto Y, Miura Y, Higuchi M et al (1993) Using lichen tissue cultures in modern biology. Bryologist 96:384–393

Yamamoto Y, Kinoshita Y, Matsubara H et al (1998) Screening of biological activities and isolation of biological-active compounds from lichens. Recent Res Dev Phytochem 2:23–34

Yilmaz M, Türk AO, Tay T et al (2004) The antimicrobial activity of extracts of the lichen *Cladonia foliacea* and its (−)-usnic acid, atranorin, and fumarprotocetraric acid constituents. Z Naturforsch C 59:249–254

Yoshimura I, Kurokawa T, Kinoshita Y et al (1994) Lichen substances in cultured lichens. J Hattori Bot Lab 7:249–261

Zhou QM, Guo SY, Huang MR et al (2006) A study of the genetic variability of *Rhizoplaca chrysoleuca* using DNA sequences and secondary metabolic substances. Mycologia 98(1):57–67

Lichens Used in Traditional Medicine

2

Stuart D. Crawford

Abstract

Lichens are used in traditional medicine by cultures across the world, particularly in temperate and arctic regions. Knowledge of these medicinal uses is available to us because of the contributions of traditional knowledge holders in these cultures.

The traditional medicinal uses of 60 lichen genera are summarized in this paper. Cultures in different regions of the world tend to emphasize different lichen genera in their traditional medicines, with *Usnea* being the most widely used genus. The folk taxonomy of lichens within a given culture is not synonymous with the scientific taxonomy and reflects the cultural value of those lichens and the traditional method of their identification. Even within Western science the identity and taxonomy of lichens has not remained constant throughout history.

Lichens in traditional medicine are most commonly used for treating wounds, skin disorders, respiratory and digestive issues, and obstetric and gynecological concerns. They have been used for both their secondary metabolites and their storage carbohydrates. The European uses of lichens have been exported world wide and sometimes influence the use of lichens by other cultures. These European uses started in the fifteenth and sixteenth centuries and arose from interpretations of Ancient Greek uses, as well as the application of the Doctrine of Signatures.

2.1 Introduction

Lichens are important traditional medicine in many different cultures. This information has been made available to us from the contributions of hundreds of traditional knowledge holders in communities across the world. It is our responsibility to

S. D. Crawford (✉)
Crawford Ecological Consulting, Masset, BC, Canada

B. Ranković (ed.), *Lichen Secondary Metabolites*,
https://doi.org/10.1007/978-3-030-16814-8_2

respect and value the knowledge that has been given to us. This chapter is a tribute to the wealth of traditional knowledge that exists about lichens.

There have been a few previous reviews on the traditional uses of lichens for medicine. The traditional uses of lichens in Europe were reviewed by Smith (1921), with later contributions by Llano (1948) and Richardson (1974). Sharnoff (1997) compiled the first global review of lichen uses, which was added to by Crawford (2007). Upreti and Chatterjee (2007) reviewed the medicinal uses of lichens in India and republished Sharnoff's (1997) database on medicinal uses elsewhere in the world. Wang and Qian (2013) reviewed the medicinal uses of lichens in China. The first edition of this paper (Crawford 2015) compiled all the medicinal uses recorded by these previous authors along with many additional records and was republished by Upreti et al. (2016). The current paper updates and expands the first edition and is the most comprehensive review to date, but it is still far from complete.

2.2 Cultures That Use Lichens

There are records of medicinal uses of lichens by cultures in Africa, Europe, Asia, Oceania, North America, and South America. The majority of these uses are in North America, Europe, India, and China, but this is most likely because that is where the majority of the ethnographic work has been done. Interestingly, no records have been found for any traditional use of lichens in Australia.

It is difficult to determine the prevalence of lichens in traditional medicine across the world. Most ethnobotanists and ethnographers have ignored cryptogams, both historically and currently.

If the ethnographic literature on a culture does not mention lichens, it might be because that culture does not utilize lichens. However, it might also be because the ethnographer's culture does not value lichens, and the ethnographer therefore did not notice and record the value of lichens in the culture that they were documenting. In the cultures for which traditional uses of lichens have been recorded, there are usually between one and three medicinal lichens. There are more records of lichen use among cultures in temperate and arctic areas and less in the tropics. This probably represents the relative dominance of lichens in these zones.

A few ethnobotanists have recognized the cultural value of lichens, and their work has been invaluable in documenting lichens in traditional medicines. These workers include, among others, N. J. Turner (Canada), M. R. González-Tejero (Spain), L. S. Wang (China), and D. K. Upreti (India). As a result, there is an overrepresentation of these geographic areas in this current analysis.

2.3 The Lichens That Are Used in Traditional Medicine

This chapter documents a total of 60 different genera of lichens that are used in traditional medicines. The most commonly used genus of lichen is *Usnea*, which is used across the world for medicine, although it is often used synonymously with

Table 2.1 Lichen genera commonly used in traditional medicine

Lichen genus	Main area of use
Usnea	Worldwide (except Australia)
Evernia and *Pseudevernia*	Europe and North Africa
Letharia	North America
Lethariella	China
Cetraria	Europe
Parmotrema and *Hypotrachyna*	India
Xanthoparmelia	North America and Africa
Cladonia and *Cladina*	N. America, Europe, and Asia
Thamnolia	Asia
Ramalina	N. America, Europe, and Asia
Lobaria and *Peltigera*	N. America, Europe, and Asia
Umbilicaria	North America and Asia

other arboreal hair lichens. Despite its worldwide importance, *Usnea* is not traditionally one of the dominant medicinal lichens in Europe. Numerous other genera of lichens have particular importance in certain parts of the world, as is shown in Table 2.1.

2.3.1 The Folk Taxonomy of Lichens

All cultures develop a folk taxonomy of living organisms that allows people to make sense of the world around them. Folk taxonomies are unique to a specific culture and usually reflect its particular environment and values. Some cultures have a very detailed folk taxonomy for lichens. The traditional taxonomy of the Saami recognizes lichens as being a distinct life form from mosses and divides lichens into three different generic taxa and numerous specific taxa (Nissen 1921). Other cultures placed less value on lichens, which is reflected in a much more simplistic folk taxonomy for lichens. European botanists in the fifteenth century lumped all lichens, and many other cryptogams, into a single life form category of ***moss***.

Folk taxonomies can be very accurate, but they are often different than the scientific taxonomy. This mismatch between folk and scientific taxonomies is particularly prevalent in lichens. For instance, the Saami folk taxon ***jægel*** includes *Cetraria*, *Cladina*, and *Stereocaulon* but excludes *Parmelia*, which is placed in the folk taxon ***gadna***. The scientific taxonomy would lump *Parmelia* and *Cetraria* together in Parmeliaceae and exclude *Cladina* and *Stereocaulon*. Another example is the common practice within folk taxonomies of classifying lichens according to their substrate. There is often a folk genera that includes all arboreal hair lichens (and sometimes mosses), which are then divided into different species depending on what type of tree they are growing on.

One of the biggest challenges in ethnolichenology is that a folk taxon of lichens that has cultural significance may not be synonymous with any scientific taxon. This means that if a culturally important lichen is identified according to the scientific taxonomy without understanding the folk taxonomy, it may be recorded as the wrong lichen. For example, a botanist recorded that the Saami used *Usnea plicata* for blisters, but maybe the lichen that they saw only happened to be *U. plicata*, and the Saami actually used any species of *Alectoria*, *Bryoria*, or *Usnea* as long as it is growing on a birch tree.

Folk taxonomies of lichens are intrinsically linked with the traditional methods of identifying lichens. It is very common to identify lichens based on where they are found. Lichens are often thought to imbibe their desirable properties from the substrate on which they are growing. For example, Nuxalk consider alectoroid lichens to be better medicine if growing on alder, the Gitga'at consider *Lobaria oregana* to be better if on fir, and the Ancient Greeks thought that *Evernia* was better if growing on cedar. The medicinal properties of a lichen species may change depending on where it is growing. However, this may also be a clever aid for identification. Many lichens have specific microhabitat preferences, and selecting lichens from only a specific substrate will result in preferentially selecting certain species.

Another interesting identification method is employed by the Quichua of Saraguro, Ecuador, who have determined that an effective medicine requires seven different colours of rock lichens. It is possible that there is a synergistic effect between the different lichen species. It is also possible that collecting seven different species makes it much more likely to collect the correct one.

2.3.2 Development of Lichen Taxa in Western Science

The meaning of the word lichen has changed over time, which can make it complicated to identify culturally important lichens in old documents. ***Lichen*** comes from the Ancient Greek Λειχήν (***leikhēn***), the first record of which is from Theophrastus in 300 BCE (Richardson 1974). Theophrastus was probably referring to thalloid liverworts, but subsequent Ancient Greek authors may have used that name for a lichen (see Ancient Greek use of *Ramalina* spp.). Early European botanists lumped together a variety of cryptogams into the same taxon, usually including lichens, mosses, liverworts, fungi, seaweed, and sometimes even coral. de Tournefort (1694) was the first European author to distinguish lichens by the name ***lichen***, but he also included some thalloid liverworts in his taxon and excluded some lichens. It was Dillenius (1742) who reorganized the ***lichen*** taxon to make it synonymous with our modern concept.

The taxonomy and names of lichens have changed radically since Dillenius and are continuing to change in contemporary times. This can make it difficult to determine what lichen is being discussed in ethnographic literature. To add further complications, most authors know very little about lichens and thus frequently use names that are outdated or even just completely wrong.

The genus *Usnea* was created by Dillenius (1742). Linnaeus (1753) described five *Usnea* species, but lumped them all together in his all-encompassing genus *Lichen*. They were moved to the *Usnea* genus by Weber and Wiggers (1780). Four of the original species are often mentioned in ethnographic literature: *Usnea barbata*, *U. florida*, *U. hirta*, and *U. plicata*. The number of *Usnea* species has now increased to around 350 species (Thell et al. 2012), so any reference to one of the original *Usnea* species in old herbals or ethnographies is suspect. Of the original five, only *Usnea hirta* occurs in North America (Esslinger 2018). References to *Usnea barbata* are particularly ambiguous, as the taxonomy of this species is confusing and still being determined (Articus 2004).

The *Usnea*-like lichens (those with a cartilaginous central cord) form a large clade which has been alternately lumped and split by various authors. This chapter follows Truong et al. (2013), where *Lethariella* and *Protousnea* are considered separate genera, while *Dolichousnea*, *Eumitria*, and *Neuropogon* are considered to be subgenera and lumped within *Usnea*.

The pendant *Bryoria* species were originally all lumped together as *Lichen jubatus* (Linnaeus 1753), which became *Alectoria jubata* (Acharius 1810). The taxonomy of *Bryoria* was not well understood until Brodo and Hawksworth (1977) created the genus *Bryoria*, so references to specific *Bryoria* species prior to that are ambiguous.

The Parmeliaceae is a large and diverse family of lichens that includes many culturally significant lichens. This family currently contains around 80 genera and over 2000 species (Thell et al. 2012). Five culturally significant genera of Parmeliaceae were described before 1810: *Usnea*, *Parmelia*, *Cetraria*, *Alectoria*, and *Evernia*. By 1903, *Letharia* and *Pseudevernia* had been split from *Evernia*, and *Parmotrema* and *Hypogymnia* had been split from *Parmelia*, although historically not all authors have recognized these genera. The taxonomy of Parmeliaceae remained relatively constant until 1965, when the genus *Cetraria* began to be split into numerous different genera. The genus *Parmelia* was also split up starting in 1974. This splitting was mostly completed by the early 1990s, by which time there were over 80 genera in the family (Thell et al. 2004). Recent molecular work has resulted in some genera being lumped and others split, such that Thell et al. (2012) recognize 79 genera. Currently, the original genus *Parmelia* is divided into 32 genera and *Cetraria* into 22 genera.

For practical reasons, lichenologists sometimes lump the morphologically similar genera that were previously included in *Parmelia* and *Cetraria* back together into the categories of parmelioid (Hale and DePriest 1999) and cetrarioid lichens (Randlane et al. 2013). These morphological groupings are not entirely monophyletic (Thell et al. 2012), but they can still be useful. A third morphological grouping of Parmeliaceae lichens that is often used is the alectorioid lichens, which include several similar-looking genera of hair lichens that were previously lumped together in the genus *Alectoria*. The genus *Usnea* is sometimes included in this category.

One result of the profusion of genera within Parmeliaceae is that any reference to an unidentified species of *Parmelia* or *Cetraria* in an older ethnographic work is very ambiguous. The categories of parmelioid, cetrarioid, and alectorioid lichens are very useful when dealing with folk taxonomies of lichens, so they will be utilized in the current work.

2.4 The Medicinal Uses of Lichens

Lichens are used for many different medicinal purposes, but there are some general categories of use that reoccur across the world. Lichens are often used externally for dressing wounds, either as a disinfectant or to stop bleeding. Other common topical uses are for skin infections and sores, including sores in the mouth. The importance of this use is apparent in the name ***lichen*** (from ***leikhēn***, "what eats around itself"), which comes from the Ancient Greek practice of using a cryptogam to cure a skin disease.

Lichens are often drunk as a decoction to treat ailments relating to either the lungs or the digestive system. This is particularly common in Europe, but is also found across the world. Many other uses of lichens are related to obstetrics or treating gynecological issues. This may be related to the common use of lichens for treating sexually transmitted infections and ailments of the urinary system. Three other uses of lichens that are less common, but reoccur in several different cultures, are for treating eye afflictions, treating issues with feet, and for use in smoking mixtures.

Many of the traditional medicinal uses of lichens are probably related to their secondary metabolites, many of which are known to both be physiologically active and to act as antibiotics. However, some of the traditional uses of lichens also rely on the qualities of the lichen carbohydrates. In particular, the lichenins [β-(1→3)-(1→4)-linked D-glucans] are common in the Parmeliaceae and have a remarkable ability to absorb water and form a gel (Crawford 2007). Many of the traditional uses of lichens involve boiling the lichen to create a mucilage which is drunk for lung or digestive ailments or applied topically for other issues. Other lichen carbohydrates which may be important are the isolichenins and galactomannans, which are taxonomically widespread, and the pustulins that are found in Umbilicariaceae.

2.4.1 Medicinal Lichens of Europe

Lichens are used in traditional medicine across the world. As is to be expected, many of these uses of lichens are completely unrelated to how these lichens are used in Europe. However, European uses of lichens have been exported worldwide, and there are numerous instances where the European use for a lichen appears to be associated with its traditional use in a different culture. This dispersal of European uses of lichens is related to the general dispersal of other aspects of European culture across the world. One specific source of this bias may be that most ethnographers who recorded traditional uses of lichens are from a European background, and their personal cultural bias can affect what they have documented. Another source is that most literature on lichens is from a European background, and if it features any uses of lichens, those uses are generally European.

An understanding of the traditional use of lichens in Europe can therefore be important for understanding traditional uses elsewhere. The origins of the medicinal use of lichens in Europe date back to the fourth and third century BCE, when medicinal lichens were recorded by the Ancient Greek scholars Hippocrates and Theophrastus (Lebail 1853). The use of lichens continued to be recorded by various scholars throughout the rest of the classical era, including Pedanius Dioscorides and

Pliny the Elder (Rome, first century CE), Galen of Pergamon (Greece, second century CE), Paul of Aegina (Greece, seventh century CE), and Serapion the Younger (a twelfth- or thirteenth-century compilation). These authors discuss at least three different cryptogams that might be lichens, but the most important for subsequent pharmacopoeias was an arboreal fruticose lichen called ***splanchnon*** ("intestine"). According to the original writings of Dioscorides, ***splanchnon*** was not only a powerful medicine, it was also sweet-smelling and used as a perfume (López Eire et al. 2006).

In the Middle Ages, various Persian scholars like Rhazes (tenth century) and Avicenna (eleventh century) wrote about the medicinal properties of ***splanchnon***, and it was adopted into Unani medicine under the name ***ushna***. This lichen is currently interpreted as being *Usnea* spp.

At the start of the modern era (~fifteenth century), herbalism flourished in Western Europe, with many authors adopting Greek herbal knowledge. These Europeans lumped together all fruticose arboreal lichens into one taxon, which they called ***usnea*** (borrowing from the Arabic ***ushna***), ***tree moss***, or ***oak moss*** (Dorstenius 1540; de L'Obel 1576; Gerarde 1597; Ray 1686; Quincy 1724; Culpeper 1788). This taxon was considered to be synonymous with the Ancient Greek ***splanchnon***, with all of its medicinal and perfume qualities. Parkinson (1640) accurately distinguished between numerous genera, but considered them all types of ***oak moss***, and attributed the same medicinal values to all of them.

It was not until the late 1700s that a distinction was made between the different genera of ***oak moss***, at which time the name *Usnea* was only applied to our modern genus. From this time onward, most authors decided that the medicinal values of ***splanchnon*** were referring to *Usnea* (Lightfoot 1777; Willemet 1787; Adams 1847; Lebail 1853), although the same medicinal properties were sometimes applied to *Evernia prunastri* (Willemet 1787; Lebail 1853).

Oak moss was used to make a popular scented hair powder called Cyprus powder in Europe in the late 1600s (Bauhin and Cherler 1650; Zwelfer 1672). By the time European botanists could distinguish different genera, Cyprus powder was found to contain a variety of lichen genera, including *Usnea*, *Pseudevernia*, and other arboreal lichens (Amoreux 1787). At this time, *Evernia prunastri* was the preferred lichen to use for perfumes in France (Amoreux 1787). In more recent times, ***oak moss*** refers to only *Evernia prunastri* and ***tree moss*** to *Pseudevernia furfuracea*, and these are the two lichen species harvested for perfume (Moxham 1986).

When Europeans first adopted Ancient Greek herbal knowledge, they were confused as to the identity of ***splanchnon***, but eventually decided that it was *Evernia/Pseudevernia* when used for perfume and *Usnea* when used for medicine. Dioscorides' description of ***splanchnon*** is ambiguous and its identity cannot be determined with certainty, but Richardson (1974) suggests that it is referring to *Evernia prunastri* and *Pseudevernia furfuracea*. He may be correct, as these lichens were used medicinally in Europe and North Africa from ancient times to present. Europeans have added medicinal properties to *Usnea* that were not originally associated with ***splanchnon*** by the Ancient Greeks. Perhaps these medicinal uses for *Usnea* existed in Europe independent of the Ancient Greek writings, and this caused Europeans to wrongly associate the Ancient Greek medicinal uses of *Evernia/Pseudevernia* with *Usnea*.

Starting in the 1500s, the Doctrine of Signatures was a ubiquitous concept in European medicine. It was thought that plants looked like the organ or ailment that they cured and various lichens were adopted into the European pharmacopoeia as a result. The main medicinal lichens in early modern era Europe were *Cetraria islandica*, *Cladonia pyxidata*, *Peltigera canina*, *Peltigera aphthosa*, *Usnea* spp., *Lobaria pulmonaria*, *Xanthoria parietina*, and *Evernia prunastri.* For more details, refer to these lichens in the tables below. The widespread use of these lichens had been mostly abandoned by 1800, with the exception of *Cetraria islandica*, which has persisted as a medicinal lichen in parts of Europe until today. For a more detailed description of the rise and fall of the use of lichens for medicine in early modern Europe, see Mitchell (2015).

2.5 Known Records of Lichens Used in Traditional Medicine

The following tables document all of the traditional medicinal uses of lichens for which the author has found records. Tables 2.2 and 2.3 provide a list of the different genera and an index to the table where they can be found. Tables 2.4, 2.5, 2.6, 2.7, 2.8, 2.9, 2.10, 2.11, 2.12, 2.13, 2.14, 2.15, 2.16, 2.17, 2.18, 2.19, 2.20, 2.21, 2.22, 2.23, 2.24, 2.25, 2.26, and 2.27 are organized taxonomically by lichen family and provide the details on each traditional use.

Table 2.2 Lichen genera used in traditional medicine

Alectoria, Alectorioid	*Lasallia*, Umbilicariaceae	*Pseudevernia*, Parmeliaceae
Anaptychia, Physciaceae	*Lepra*, Pertusariaceae	*Punctelia*, Parmelioid
Anzia, Parmeliaceae	*Leptogium*, Collemataceae	*Ramalina*, Ramalinaceae
Aspicilia, Megasporaceae	*Letharia*, Parmeliaceae	*Rhizoplaca*, Lecanoraceae
Bryoria, Alectorioid	*Lethariella*, Alectorioid	*Roccella*, Roccellaceae
Canoparmelia, Parmeliaceae	*Lobaria*, Lobariaceae	*Rusavakia*, Teloschistaceae
Cetraria, Cetrarioid	*Masonhalea*, Cetrarioid	*Siphula*, Icmadophilaceae
Cetrelia, Cetrarioid	*Mycoblastus*, Mycoblastaceae	*Stereocaulon*, Stereocaulonaceae
Cladina, Cladoniaceae	*Nephroma*, Nephromataceae	*Sticta*, Lobariaceae
Cladonia, Cladoniaceae	*Nephromopsis*, Cetrarioid	*Sulcaria*, Alectorioid
Crocodia, Lobariaceae	*Ophioparma*, Ophioparmaceae	*Teloschistes*, Teloschistaceae
Dendriscosticta, Lobariaceae	*Parmelia*, Parmelioid	*Thamnolia*, Icmadophilaceae
Dermatocarpon, Verrucariaceae	*Parmelinella*, Parmelioid	*Umbilicaria*, Umbilicariaceae
Dictyonema, Hygrophoraceae	*Parmotrema*, Parmelioid	*Usnea*, Alectorioid
Evernia, Parmeliaceae	*Peltigera*, Peltigeraceae	*Varicellaria*, Pertusariaceae
Flavocetraria, Cetrarioid	*Pertusaria*, Pertusariaceae	*Vermilacinia*, Ramalinaceae
Flavoparmelia, Parmelioid	*Physcia*, Physciaceae	*Vulpicida*, Cetrarioid
Heterodermia, Physciaceae	*Polycauliona*, Teloschistaceae	*Xanthoparmelia*, Parmelioid
Hypogymnia, Parmeliaceae	*Protoparmeliopsis*, Lecanoraceae	*Xanthomendoza*, Teloschistaceae
Hypotrachyna, Parmelioid	*Protousnea*, Alectorioid	*Xanthoria*, Teloschistaceae

Table 2.3 Index to tables of lichen families used in traditional medicine

Ascomycota		**Ascomycota**		**Ascomycota**	
Lecanorales		Peltigerales		Pertusariales	
Cladoniaceae	10	Collemataceae	40	Icmadophilaceae	48
Lecanoraceae	13	Lobariaceae	40	Megasporaceae	50
Mycoblastaceae	14	Nephromataceae	43	Pertusariaceae	50
Parmeliaceae		Peltigeraceae	43	Verrucariales	
Alectorioid	14	Teloschistales		Verrucariaceae	50
Cetrarioid	25	Teloschistaceae	45		
Parmelioid	27	Arthoniales		**Basidiomycota**	
Other	34	Roccellaceae	46	Agaricales	
Physciaceae	37	Umbilicariales		Hygrophoraceae	50
Ramalinaceae	37	Ophioparmaceae	47		
Stereocaulaceae	39	Umbilicariaceae	47	**Unidentified lichens**	51

Table 2.4 Cladoniaceae used in traditional medicines around the world

Culture and *folk name*	Traditional use
***Cladina* spp. Nyl.**	
Den'ina (Alaska, USA) *k'udyi*	Decoction used for diarrhea (Kari 1987)
Upper Tanana (AK, USA)	A "liquor" prepared from plant was drunk for colds (McKennan 1959)
Aleut (Alaska, USA) *kinadam aiyukax*	Drunk as a tea for chest pains. Hunters who are climbing hills chew the lichen to maintain their wind (Bank 1953; Smith 1973)
Nganasans (Siberia)	Remedy for scurvy (Eidlitz 1969)
Saami (Scandinavia) *ullo-jægel* ("wool lichen")	Decoction for unspecified medicine (Nissen 1921; Eidlitz 1969)
***Cladina arbuscula* (Wallr.) Burgaz**	
China 林石蕊 (*lín shí ruǐ*, "forest stone bud")	Used for dizziness, hypertension, pulmonary tuberculosis, fever, trauma with pus formation, and skin infections due to external injury (Wang and Qian 2013)
***Cladina mitis* (Sandst.) Mong.**	
Himalayas (Sikkim, India)	Medicine for respiratory health and hepatic health (O'Neill et al. 2017)
***Cladina laevigata* (Vain.) C.W. Dodge**	
Selk'nam (Tierra del Fuego, Argentina) *chepl, chispl, shûj*	Used for personal hygiene: Lichen used to wash body before drying with *Protousnea magellanica* (Berihuete-Azorín 2013)
***Cladina rangiferina* (L.) Nyl.**	
Ojibwe (MN and WI, USA) *asa' gûniñk'*	Boil and use water to wash a newborn baby (Smith 1932)

(continued)

Table 2.4 (continued)

Culture and *folk name*	Traditional use
Whapmagoostui Cree (Quebec, Canada) *whapskumuk; epshatuk*	Used to treat inflammation associated with diabetes (Fraser 2006).
Finland	Remedy for coughs and tuberculosis. Boil in water and drink (Richardson 1974)
China 鹿石蕊 (*lù shí ruǐ*, "deer stone bud"), 石蕊 (*shí ruǐ*, "stone bud"), 太白树 (*tàibái shù*, "white tree"), 石湍 (*shí tuān*), 石芥 (*shí jiè*), 云茶 (*yún chá*), 蒙顶茶 (*méng dǐng chá*), 石蕊花 (*shí ruǐhuā*, "stone flower bud"), 是云茶 (*shì yún chá*), 云芝茶 (*yún zhī chá*), 蒙山茶 (*méng shān chá*), 酶苔 (*méi tái*), 石花 (*shí huā*, "stone flower"), 刀伤药 (*dāo shāng yào*, "knife wound medicine"), 地花菜 (*de huācài*, "earth cauliflower")	Used for fever, headaches, cuts, coughing up blood, jaundice, blurred vision, cloudy cornea, difficulty urinating, urinary tract infection, irritable depression, rheumatism, phlegm due to dry throat. Drink decoction; or apply decoction or powdered lichen to the affected area (Wang and Qian 2013)
Himalayas (Sikkim, India)	Medicine for respiratory health (O'Neill et al. 2017)
Monpa (Arunachal Pradesh, India)	Remedy for kidney stones. Half teaspoon of sun-dried, ground lichen added to one cup boiling water. Drunk in morning on empty stomach for 1 month or until cured (Rout et al. 2005)
***Cladina stellaris* (Opiz) Brodo [*Cladina alpestris*]**	
Nihithawak (SK, Canada) *wāpiskastastkamihk*; or *atikōmīciwin*	Drink to expel intestinal worms: either decoction or powdered lichen added to water (Leighton 1985)
Inuit (Nunavut, Canada) *nirait*	Broth used for sickness and eye infections (Black et al. 2008)
Primorsky and Sakhalin (Russian Far East)	Powdered form used to treat wounds and some infections (Moskalenko 1986)
China 雀石蕊 (*què shí ruǐ*, "bird stone bud"), 太白花 (*tàibái huā*, "white flower"), 高山石蕊 (*gāoshān shí ruǐ*, "high mountain stone bud"), 山岭石蕊 (*shānlǐng shí ruǐ*, "mountain stone bud"), 岭石蕊 (*lǐng shí ruǐ*, "mountain stone bud"), 雀儿石蕊 (*qiǎo er shí ruǐ*)	Used for hypertension, headaches, nosebleeds, eye diseases, tuberculosis, menstrual disorders, and vaginal discharge. Drink decoction (Hu et al. 1980; Wang and Qian 2013)
***Cladina subtenuis* (Abbayes) Mattick**	
Cherokee (NC, USA)	Used to relieve the pain of insect stings. Lichen chewed and put on sting, sometimes mixed with tobacco (Garrett 2003)
***Cladonia* sp. P. Brown**	
Santiago Camotlán (Oaxaca, Mexico) *esponjita*	Used to treat "culture-bound syndromes" (Pérez-Nicolás et al. 2017)

(continued)

Table 2.4 (continued)

Culture and *folk name*	Traditional use
Cladonia amaurocraea **(Flörke) Schaer.**	
China 黑穗石蕊 (*hēi suì shí ruǐ*), 太白树 (*tàibái shù*)	Used for headaches and dizziness (Wang and Qian 2013)
Cladonia bellidiflora **(Ach.) Schaerer**	
Tlingit (Alaska, USA)	Treatment for eye disease when mixed with mother's milk (Garibaldi 1999)
Haida (BC, Canada)	Red ends dipped in mother's milk and applied to sore eyes (Turner 2004a)
Cladonia cervicornis **(Ach.) Flot.**	
China 千层石蕊 (*qiān céng shí ruǐ*, "thousand levels stone bud"), 小喇叭 (*xiǎo lǎbā*), 树格达 (*shù gé dá*), 地喇叭 (*de lǎbā*), 小叶下层石蕊 (*xiǎoyè xiàcéng shí ruǐ*), 干层石蕊 (*gàn céng shí ruǐ*), 小叶亚种 (*xiǎoyè yà zhǒng*)	Used for scalds, cuts, and coughing up blood. Drink decoction; or apply decoction or powdered lichen to the affected area (Wang and Qian 2013)
Cladonia chlorophaea **(Flörke ex Sommerf.) Sprengel**	
Syilx (BC, Canada) *penp̓en̓emekxísxn̓* ("liver on rock")	Decoction used to wash sores which were slow to heal (Turner et al. 1980)
Britain *chalice-moss*; *cup-moss*; or *Our Lady's chalice*; *cwpanau pas* (Welsh)	Used like *C. pyxidata* for whooping cough, use has continued to contemporary times in Welsh counties of Merionethshire and Denbighshire. In Waterford (Ireland), used for the same purpose boiled in new milk (Allen and Hatfield 2004)
Cladonia coccifera **(L.) Willd. [syn.** ***Scyphophorus cocciferus*****]**	
Europe (early modern era) *cup moss*	Decoction used for fever and whooping cough in children, like *C. pyxidata* (Willemet 1787; Luyken 1809; Lindley 1838; Pereira 1850)
Cladonia cornuta **(L.) Hoffm.**	
Europe (early modern era) *horn moss*	Used with *C. pyxidata* against persistent coughs in children (Watson 1756)
Cladonia fenestralis **Nuno**	
Tibetans (Sichuan, China) 繁鳞石蕊 (*fán lín shí ruǐ*)	Medicinal tea (Wang and Qian 2013)
Cladonia fimbriata **(L.) Fr. [syn.** ***Scyphophorus fimbriatus*****]**	
Europe (early modern era) *cup moss*	Used with *C. pyxidata* to treat whooping cough (Pereira 1850)
Cladonia fruticulosa **Kremp.**	
China 亚麸石蕊 (*yà fū shí ruǐ*)	Extract used for bacterial infections on skin (Wang and Qian 2013)
Cladonia gracilis (L.) Willd.	
China 细石蕊 (*xì shí ruǐ*, "slender stone bud"), 太白鹿角 (*tàibái lùjiǎo*, "white antlers"), 匙石蕊 (*shi shí ruǐ*), 地蓬草 (*de péng cǎo*), 太白针 (*tàibái zhēn*, "white pins"), 细石蕊 (*xì shí ruǐ*)	Used for dizziness, difficult or painful urination, nose bleeding, impetigo, and pink eye. Drink decoction; or apply decoction or powdered lichen to the affected area (Hu et al. 1980; Wang and Qian 2013)

(continued)

Table 2.4 (continued)

Culture and *folk name*	Traditional use
***Cladonia macroceras* (Delise) Ahti**	
China 硬柄石蕊 (*yìng bǐng shí ruǐ*), 太白树 (*tàibái shù*, "white tree"), 石蕊 (*shí ruǐ*, "stone bud"), 太白鹿角 (*tàibái lùjiǎo*, "white antlers"), 长柄石蕊 (*cháng bǐng shí ruǐ*)	Drunk as decoction to relieve blockage of urination, bring down swelling, and remove toxic substances (Wang and Qian 2013)
***Cladonia miniata* G. Meyer [*Cladonia sanguinea*]**	
Brazil	Rubbed down with sugar and water, used as remedy for mouth ulcers (Lindley 1838)
***Cladonia pleurota* (Flörke) Schaer.**	
China 粉杯红石蕊 (*fěn bēi hóng shí ruǐ*)	To clear *heat*, cool liver, dissolve *phlegm*, and eliminate *dampness* (Wang and Qian 2013)
***Cladonia pyxidata* (L.) Hoffm. [syn. *Scyphophorus pyxidata*]**	
Europe (early modern era) *cup moss*	Widely used for whooping cough in children as a powder, decoction, or syrup (Quincy 1724; Gedner 1756; Lightfoot 1777; Willemet 1787; Pereira 1850). Also for fevers and kidney stones (Luyken 1809; Lindley 1838; Lebail 1853). In Finland taken with milk for pulmonary tuberculosis (Vartia 1973).
***Cladonia scabriuscula* (Delise) Nyl.**	
Keyagana (Papua New Guinea) *lanefa-kikinofa*	Heated and taken orally for vaginal discharge/bleeding (Jorim et al. 2012)

Table 2.5 Lecanoraceae used in traditional medicines around the world

Culture and *folk name*	Traditional use
***Protoparmeliopsis muralis* (Schreb.) M. Choisy [*Parmelia saxicola*]**	
Nishinam (CA, USA)	Made into a tea and used to treat colic (Powers 1877)
***Rhizoplaca chrysoleuca* (Sm.) Zopf.**	
China 红脐鳞 (*hóng qí lín*)	Used for tuberculosis, intestinal obstruction, trauma with pus formation, burns and scalds, skin infections, cancer, and pain relief. Used externally or orally (Wang and Qian 2013)

Table 2.6 Mycoblastaceae used in traditional medicines around the world

Culture and *folk name*	Traditional use
***Mycoblastus alpinus* (Fr.) Kernst.**	
China 石霜 (*shí shuāng*, "stone frost")	Used for stopping bleeding from external injury, draining pus, burns, and nocturnal seminal emission. Drink decoction or apply powder to the affected area (Wang and Qian 2013)

Table 2.7 Alectorioid lichens (Parmeliaceae) used in traditional medicines

Culture and *folk name*	Traditional use
***Alectoria* Ach. spp.**	
Scandinavia	Decoction for bathing chapped skin on babies or the feet of adults. Same use for *Lobaria pulmonaria*, *Usnea* sp., and *Peltigera aphthosa* (Richardson 1974)
***Alectoria ochroleuca* (Hoffm.) A. Massal.**	
Chugach (Alaska, USA)	Possibly same as Chugach use of *Bryoria trichodes* (Wennekens 1985)
***Alectoria sarmentosa* Ach.**	
Haida (BC, Canada) *k'aalts'idaa liisɢa* or *k'aalts'adaa liijaa* ("crow whiskers/wool")	Used to strain impurities out of hot pitch when making medicine, and for other unspecified medicines. Also used *Usnea longissima* (Turner 1998, 2004a)
Nuxalk (BC, Canada) *suts'wakt* or *ipts-aak* ("limb moss")	Warmed and applied to a broken boil or festering sore (if growing on red alder). Possibly *Usnea* spp. (Smith 1929; Turner 1973)
Ditidaht (BC, Canada) *p'u7up*	Used for wound dressing, baby diapers, and sanitary napkins. Also used *Usnea* spp. (Turner et al. 1983)
Flathead (Montana, USA) *sqalīō*	Mother drinks tea of *sqalīō* and *Matricaria discoidea* to make her deliver her placenta (Stubbs 1966). Possibly *Usnea* spp.
Umatilla, Cayuse (OR, USA) *laxpt* or *mak'hl*	Boiled and applied as compress for open sores, arthritis, and *achash-pama* (an eye problem) (Hunn 2005). Possibly *Usnea* spp.
Pallars (Spain) *cabellera de pi*	Drunk as tea for asthma and catarrh (Agelet and Vallès 2003)
***Bryoria* spp. Brodo and D. Hawksw.**	
Atsugewi (California, USA)	Applied as poultice to reduce swellings. Either boiled or used dry (Garth 1953)
Tsilhqot'in (BC, Canada) *texa*; *taxa*	Burn *texa* with own hair and rub ashes on hair and scalp to stop hair from going grey (Kay 1995; Turner 2004b)
France (eighteenth century)	Used for healing skin abrasions, diarrhea, and vaginal discharge (Gedner 1756; Willemet 1787)
***Bryoria asiatica* (Du Rietz) Brodo and D. Hawksw.**	
China 亚洲小孢发 (*Yàzhōu xiǎo bāo fā*), 头发七 (*tóufǎ qī*, "seven hair"), 亚洲树发 (*Yàzhōu shù fā*, "Asian tree hair"), 黑丝草 (*hēi sī cǎo*), 人头七 (*réntóu qī*), 黑丝弗 (*hēi sī fú*)	Used for kidney deficiency and general weakness, dizziness, heart palpitation, involuntary ejaculation, night sweats, difficulty urinating, edema, impetigo, draining pus, and improving eyesight. Drink decoction; or apply decoction or powdered lichen to the affected area (Wang and Qian 2013)

(continued)

Table 2.7 (continued)

Culture and *folk name*	Traditional use
***Bryoria bicolor* (Ehrh.) Brodo and D. Hawksw.**	
China 双色小孢发 (*shuāngsè xiǎo bāo fā*)	Same as Chinese use of *B. asiatica* (Wang and Qian 2013)
***Bryoria fremontii* (Tuck.) Brodo and D. Hawksw.**	
Sahaptin (OR and WA, USA) *k̓unč*	Boiled and used as poultice for arthritis (Hunn 1990)
Nimi'ipuu (Montana, USA) *ho.póp*	Good for upset stomach, indigestion, and diarrhea (Hart 1976; Marshall 1977)
Flathead (Montana, USA) *caúmtemkan*, *st'telu*; *skolápkan*; *skolkéin*; *sqatlo*; or *šáwtəmqən*	Important food when baked with root vegetables; when baked alone it is more a tonic for the sick than a food (Turney-High 1937; Stubbs 1966; Hart 1974)
Syilx (BC, Canada) *sk̲w̲elíp*	Mixed with berry juices and melted into syrup: given to newly weaned babies for their health (Gabriel and White 1954). Dried, powdered, and mixed with grease: rubbed on the navel of newborn babies to protect against infection (Turner et al. 1980).
Nlaka'pamux (BC, Canada) *wí7e*	Warts removed by cutting them off and covering the fresh wound with *wí7e* that had been heated on the fire (Teit and Boas 1900; Turner et al. 1990)
***Bryoria trichodes* (Michaux) Brodo and D. Hawksw. [*Alectoria americana*]**	
Sugpiaq (Alaska, USA) *nakuraartum nuyii*; or *napam ungagua'i*	Piled on sick person in the steam bath to hold the heat on his body, also used to staunch blood from wounds. Might also use *Alectoria ochrolechia* (Wennekens 1985)
***Lethariella cladonioides* (Nyl.) Krog [syn. *Usnea cladonioides*, *Lethariella cashmeriana*, and *Lethariella sernanderi*]**	
Tibet and Nakhi (nw Yunnan, China) *gangge* (Tibet); 金丝刷 (*jīn sī shuā*, "gold thread brush"), 金刷把 (*jīn shuā bǎ*, "golden brush"), 鹿心茶 (*lù xīn chá*, "deer heart tea"), 红雪茶 (*hóng xuě chá*, "red snow tea"), 松石蕊 (*sōng shí ruǐ*, "bushy stone bud") (China)	Used for health-promoting tea to tranquilize the mind and treat a decrease in vital energy, schizophrenia, and epilepsy; to reduce blood pressure, body fat, and inflammation; and also drunk non-medicinally. Boiling water is added to dry thalli in a cup, and the infusion is drunk after 3–5 min. Also drunk as an infusion or decoction, or used as a powder applied to the affected area, for relieving pain and treating burns and scalds (Zhang and Hu 1981; Wang et al. 2001; Fu et al. 2005; Wang and Qian 2013; Ju et al. 2013)
***Lethariella sinensis* Wei and Jiang**	
Nakhi (nw Yunnan, China)	Same as Nakhi use of *L. cladonioides* (Wang et al. 2001)

(continued)

Table 2.7 (continued)

Culture and *folk name*	Traditional use
***Lethariella zahlbruckneri* (Du Rietz) Krog [syn. *Usnea zahlbruckneri*]**	
China 金丝带 (*jīn sīdài*, "gold ribbon"), 红雪茶 (*hóng xuě chá*, "red snow tea"), 绿树发 (*lǜ shù fā*, "green tree hair"), 金腰带 (*jīn yāodài*, "gold belt")	Used for aching back and weak legs, paralysis, menstrual disorders, vaginal discharge, dizziness, impotency, and epilepsy. Drink decoction or make tea or wine; or apply powder to the affected area (Fu et al. 2005; Wang and Qian 2013)
***Protousnea magellanica* (Mont.) Krog [syn. *Usnea magellanica*]**	
Selk'nam (Tierra del Fuego, Argentina) *ánhuel, anhól, ánjôl*	Used for personal hygiene: Lichen used to dry body after washing with *Cladina laevigata* (Berihuete-Azorín 2013)
***Protousnea poeppigii* (Nees and Flot) Krog [syn. *Usnea campestris*, *Usnea malacea*]**	
Mendocina (Argentina) *barba de piedra*	Unspecified medicine (Garcia et al. 1990)
***Sulcaria sulcata* (Lév.) Bystrek ex Brodo and D. Hawksw.**	
China 槽枝衣 (*cáo zhī yī*), 头发七 (*tóufǎ qī*, "hair seven"), 沟树发 (*gōu shù fā*)	Used for dizziness, kidney deficiency, general weakness, heart palpitation, involuntary excessive ejaculation, night sweating, edema, impetigo, and sores. Drink decoction or apply to the affected area (Wang and Qian 2013)
***Sulcaria virens* (Tayl.) Bystr.**	
China 绿丝槽枝衣 (*lǜ sī cáo zhī yī*), 金丝带 (*jīn sīdài*, "gold silk ribbon"), 绿树发 (*lǜ shù fā*, "green tree hair"), 金腰带 (*jīn yāodài*, "gold belt")	Used for aching back and legs, traumatic bleeding, menstrual irregularities, uterine prolapse, vaginal discharge, epilepsy, paralysis, impotence, and dizziness. Drink decoction or apply to the affected area (Wang and Qian 2013)
***Usnea* spp. Dill. ex Adans. [includes *Dolichousnea*, *Eumitria*, and *Neuropogon*]**	
Maasai (Kenya) *intanasoito* or *ntana oosoito*	Used for stomachache, heartburn, malaria, backache, fever, loss of appetite, and typhoid. Crush, boil in water, and sieve (Kiringe 2008; Muthee et al. 2011)
Mt. Kilimanjaro (Tanzania)	Ingredient in herbal tea to relieve altitude sickness (Sharnoff 1997)
Iran, Iraq *lihayat-as-shāyib* ("old man's beard")	Taken to correct bad breath (Hooper 1937)
Unani medicine (Muslim culture in India and south-central Asia) *ushna*; *shaibat-al-ajooz* ("old women's hair," Arabic)	An important medicine used from ~1000 CE to the present. Used for heart troubles, for reducing inflammation, for promoting digestion, and for improving appetite, as an antidote, as an astringent, and as an analgesic. Helps wounds heal, and helps lactation in women if applied as a paste on breast. Taken orally and inserted into vagina to stimulate

(continued)

Table 2.7 (continued)

Culture and *folk name*	Traditional use
	menstruation or induce abortion. *Parmotrema* spp. is sometimes included as *ushna*, perhaps resulting from confusion with *shaileya* of Ayurvedic medicine (Razzack and Fazal 1993; Rauf et al. 2006, 2011; Yavuz and Çobanoğlu 2010). Sometimes identified as *U. longissima*
Bhutan *dngul-skud*	Used in *Sowa Rigpa* traditional medicine to heal lung, liver, nerve, and poison-related diseases (Wangchuk et al. 2016)
Taplejung (Nepal) *jhyau*	Fired powder of *jhyau* is mixed with water and taken for tonic, fever, and throat pain (Poudel 2008)
Tibet *bi ba beng suo* (Lhoba)	Soaked in water, used to wash feet for treating beriberi and ulcers (Lhoba). Used for ulcers, toxic fever, pneumonia, tracheitis, and mastitis (Tibetan medicine) (Li et al. 2015)
Ancient China 松萝 (*sōng luó*), 女萝 (*nǚ luó*, "female vine"), "Lao-Tzu's beard"	Recorded in the oldest Chinese herbals (200–250 CE). Treats indignation and anger, *xie qi* ("evil energy"), and relieves vacuity sweating and head *feng* ("wind"). In females, it relieves genital *ban* ("cold"), swelling, and pain. Picked in the fifth lunar month and dried in shade (Shou-Zhong 1998; Sharnoff 1997). See also Chinese use of *U. diffracta*, *U. longissima*, and other *Usnea* species
Muslims, Yuan Dynasty, China *ushnat* (Arabic)	Listed in *Hui Hui Yao Fang*, the official list of Islamic medicines for the Yuan Dynasty (1250–1350 CE) (Kong and Chen 1996)
New Ireland (Papua New Guinea)	Used to induce menstruation (Lee et al. 1977)
Doi Inthanon (Chiang Mai, Thailand)	Used in a bath for women following childbirth, to aid parturition and prevent infection (Sharnoff 1997)
Maori (New Zealand) *angiangi* or *kohukohu*	Steeped in water and placed on affected parts for venereal disease (Best 1905). Dried, powdered, and rubbed on skin for various skin afflictions (Kerry-Nicholls 1886; Goldie 1904). Crushed with hand and lightly bandaged onto wound to stop bleeding (Brooker and Cooper 1962; Macdonald 1974). Along with moss, used as sanitary napkin, as a diaper, and to keep newborn babies warm (Goldie 1904)

(continued)

Table 2.7 (continued)

Culture and *folk name*	Traditional use
Europe (early modern era) *oak moss*; *tree moss*; *usnea*	The Ancient Greeks had important medicinal uses for a fruticose arboreal lichen called *splanchnon*, which was likely *Evernia prunastri* or *Pseudevernia furfuracea* (see Ancient Greek use of *E. prunastri*). This lichen entered European pharmacopoeias in the early 1500s and included all fruticose arboreal lichens. By the late 1700s, it was only *Usnea* spp. Europeans added to the Ancient Greek uses of *splanchnon* and used a decoction of *Usnea* spp. for a styptic, for drying skin lesions, as an anti-inflammatory, as a skin moisturizer, and for nausea, diarrhea, whooping cough, smallpox, insomnia, umbilical hernias, and uterine medicine (Lebail 1853). It was also used for diseases of the scalp and to cure dandruff (Allen and Hatfield 2004) and as *usnea cranii humani* (see *Parmelia saxatilis*).
Jewish Community, eleventh to fourteenth century, in Cairo	Used to treat cold, coughs, heart diseases, and eye diseases. Also used as an anesthetic and to improve eyesight (Lev 2007)
Kartitsch (Austria)	Gathered as a medicinal plant (Christanell et al. 2010)
Aragon (Spain)	Used for respiratory ailments (González-Tejero et al. 1995)
Valsugana Valley (Italy)	Shepherds put it in their shoes to prevent or treat blisters (Sharnoff 1997)
Scandinavia	Decoction for bathing chapped skin on babies and the feet of adults. *Alectoria* sp., *Lobaria pulmonaria*, and *Peltigera aphthosa* also used (Richardson 1974)
Saami (Scandinavia) *lappo*	Powdered and sprinkled on external wounds, and on sores from long journeys. Also used for curing ringworm and scabies (Lebail 1853; Nissen 1921)
Finland	Put fresh lichen on infected wounds, athlete's foot, and other skin eruptions. Taken orally for sore throat and toothache. *Alectoria* spp. also used (Vartia 1973)
Dalarna (Sweden)	Used to treat foot blisters (Ahmadjian and Nilsson 1963)
Nuxalk (BC, Canada)	Probably same as Nuxalk use of *Alectoria sarmentosa* (Turner 1973)
Ditidaht (BC, Canada)	Same as Ditidaht use of *Alectoria sarmentosa* (Turner et al. 1983)
Makah (WA, USA)	Used for boils (Gill 1983)

(continued)

Table 2.7 (continued)

Culture and *folk name*	Traditional use
Nihitahawak (SK, Canada) *mithāpākwan*	Fresh lichen inserted into the nostril to stop a nose bleed (Leighton 1985)
Wabasca (AB, Canada) *miyapakwan*	Decoction used to wash sore or infected eyes. Possibly *U. hirta* (Siegfried 1994; Marles et al. 2000)
Flathead (Montana, USA)	Probably same as Flathead use of *Alectoria sarmentosa* (Stubbs 1966)
Umatilla, Cayuse (OR, USA)	Probably same as Umatilla and Cayuse use of *Alectoria sarmentosa* (Hunn 2005)
Navaho (Utah, USA) *cin bidaγai* ("wood mustache")	An infusion or poultice is used for frozen body parts (Wyman and Harris 1951)
Quichua (Loja, Ecuador) *musgo de arbol*	Used for inflated, sore stomach in children. Boiled in water with honey and drunk. Must not be collected from eucalyptus or pine (Abel 2009, pers. comm.)
Quechua (Ancash, Peru) *barba de piedra, puma champa*	Bath against cold: In the form of a steam bath, mixed up with other plants. Treatment for *nervios* (a psychological condition) and *susto* (when someone's *anima* dissociates from their body): cover a handful of this plant with some water and put to boil, then drink small glasses until feeling better. To treat *susto*, it is necessary to perform a *choq'ma* ritual first (Gonzales de la Cruz et al. 2014)
Santa Catarina and Rio Grande do Sul (Brazil)	Used for genitourinary diseases. Mixed with *yerba mate* (*Camellia sinensis*) and drunk as an infusion. A saxicolous *Usnea* sp. is used; if it is more yellow it is more effective (Londoño-Castañeda et al. 2017)
***Usnea* sect. *Neuropogon* spp. (Nees and Flot.) Mont. [syn. *Neuropogon* spp.]**	
Mapuche-Tehuelche (Argentina/Chile) *barba de piedra*; *flor de piedra*	Used for coughs. Medicine for unspecified gastrointestinal, respiratory, cardiovascular, obstetric-gynecological, and genitourinary afflictions, as well as cultural syndromes (Estomba et al. 2006; Molares and Ladio 2014)
***Usnea aciculifera* Vain.**	
China 尖刺松萝 (*jiān cì sōng luó*)	Used for bladder infection, painful urination, urinary retention, swelling, and edema in heart and kidneys (Wang and Qian 2013)
***Usnea angulata* Ach. [syn. *Usnea sulcata*]**	
Chorote (Dry Chaco, Argentina) *sá'ti-wálisa*	Decoction is drunk for pertussis, common cough, and stomach ache. Lichen put in footwear to prevent sweating and treat foot fungus (Scarpa 2009)
Criollos (Dry Chaco, Argentina) *sajasta*	Anti-inflammatory: poultice prepared of lichen mixed with goat fat and held over inflammation until cured. Treatment for

(continued)

Table 2.7 (continued)

Culture and *folk name*	Traditional use
	infection: decoction of lichen in 2 L water is drunk once a day for 3 consecutive days. Treatment of *pasmo* (possibly tetanus) and *asoleadura* ("exposure to sun"): lichen placed both in shoe underneath foot and between head and hat until symptoms disappear (Scarpa 2004)
***Usnea articulata* L. Hoffm. [syn. *Usnea flavescens*, *Usnea atlantica*]**	
Iraqw (Tanzania) *hewas*	Treatment for stomachache. A handful of *hewas* is chewed fresh and the juice swallowed; it is bitter but relieves the pain after a while. *U. himantodes* is also used (Kokwaro 1976)
Samoa	Used for wounds and shin bruises (Brooker et al. 1987)
Canary Islands *barbas*	Used as a disinfectant, along with other *Usnea* spp. (Darias et al. 1986)
***Usnea baileyi* (Stirt.) Zahlbr.**	
Ayurvedic medicine (India)	Occasional adulterant in *charila* (SEE *Parmotrema nilgherrense*). Mixed with other aromatic herbs, such as *Valeriana jatamansi* for favoring and curing tobacco, along with *U. longissima*, *U. subsordida*, *Hypotrachyna nepalensis*, *H. cirrhata*, and *Ramalina inflata* (Shah 1998)
Himalayas (Sikkim, India)	Medicine for respiratory health (O'Neill et al. 2017)
***Usnea barbata* (L.) Weber ex F.H. Wigg.**	
Xhosa (South Africa)	Used to treat mammary infections in cattle: udder is washed several times with decoction of lichen. Used to treat wounds in humans: decoction. Used for indigestion in humans: tincture or decoction taken orally several times daily (Afolayan et al. 2002; Madamombe and Afolayan 2003)
Nepal	Endangered medicinal lichen banned from raw export (Bhattarai 1999)
Ati (Philippines) *tagahumok puti*	Used for wounds: chopped and mixed with coconut oil, spread over wound. Used for abdominal pain: drink decoction (Madulid et al. 1989)
West Malaysia	Used for colds and strengthening after childbirth (Foxworthy 1922)
Karo (Indonesia) *tai angin*	Used as an aphrodisiac, *oukup* (sauna to restore health of mothers postpartum), rheumatism, and syphilis (Silalahi et al. 2015)

(continued)

Table 2.7 (continued)

Culture and *folk name*	Traditional use
Europe (early modern era)	Probably synonymous with *Usnea* spp. in early modern era pharmacopoeias, which adopted the Ancient Greek uses for insomnia, nausea, and uterine afflictions (see European use of *Usnea* spp.). Used for internal and external bleeding, whooping cough, jaundice, and promoting hair growth (Lightfoot 1777; Willemet 1787; Luyken 1809)
Abejar (Spain)	Used as drying agent and antiseptic for cracks and irritations of the feet (Bustinza and Caballero 1947)
Mbayá (Brazil) *memby rakú í ja* ("master of the energy of creatures")	Liquid made from it is given to women to cure sterility (Cadogan 1949)
***Usnea ceratina* Ach.**	
China 角松萝 (*jiǎo sōng luó*), 红髓松萝 (*hóng suǐ sōng luó*)	Used for coughs, inflamed lungs, pulmonary tuberculosis, hepatitis, headache due to *heat*, infection due to injury, inflamed lymph channels, mastitis, and snakebites (Wang and Qian 2013)
***Usnea dasopoga* (Ach.) Nyl. [syn. *Usnea dasypoga*, *Usnea filipendula*]**	
Java	Unspecified medicinal use (Uphof 1959)
Lhasa (Tibet)	Used in traditional Tibetan medicine (Salick et al. 2006)
Primorsky and Sakhalin (Russian Far East)	Powdered form used to treat wounds and some infections (Moskalenko 1986)
***Usnea densirostra* Taylor**	
Argentina *yerba de la piedra, barba de piedra*	Tea applied externally as astringent, antiseptic, and anti-inflammatory. Also use *U. durietzii* (Bandoni et al. 1972; Garcia et al. 1990; Vitto et al. 1997; Correche et al. 2008)
Uruguay *yerba de la piedra*	Unspecified medicine (Osorio 1982)
***Usnea diffracta* Vain. [syn. *Dolichousnea diffracta*]**	
China 环裂松萝 (*huán liè sōng luó*), 云雾草 (*yúnwù cǎo*, "cloud herb"), 老君须 (*lǎo jūn xū*, "old man beard"), 关公须 (*guāngōng xū*), 龙须草 (*lóng xū cǎo*, "dragon beard herb"), 白龙须 (*bái lóng xū*, "white dragon beard"), 仙人头发 (*xiānrén tóufǎ*, "fairy hair"), 石丝线 (*shí sīxiàn*, "stone thread"), 山挂面 (*shān guàmiàn*, "mountain noodles")	Used for *phlegm-heat*, warm malaria, cough, tuberculosis of the neck or lungs, headache, dim vision, swelling, pus oozing from breasts or sores, burns and scalds, snakebite, *wind-damp* impediment, traumatic injuries, bone fracture, bleeding from external injuries, vomiting blood, blood in feces, bleeding from uterus, menstrual disorders, vaginal discharge, swelling of female genitalia, urinary tract afflictions, and ascarid or schistosoma parasitic infections. Collect throughout the year, then

(continued)

Table 2.7 (continued)

Culture and *folk name*	Traditional use
	dry. Drink decoction; or apply decoction or powdered lichen to the affected area. Other names include 破茎松萝 (*pò jīng sōng luó*), 节松萝 (*jié sōng luó*), 节茎松萝 (*jié jīng sōng luó*), 女萝 (*nǚ luó*), 接筋草 (*jiē jīn cǎo*), 天蓬草 (*tiān péng cǎo*), 天速草 (*tiān sù cǎo*), 金钱草 (*jīnqián cǎo*), 海风藤 (*hǎifēng téng*), 金丝藤 (*jīn sī téng*), 飞仙翅 (*fēi xiān chì*), 松上寄生 (*sōng shàng jìshēng*), 树挂 (*shùguà*), 雪风藤 (*xuě fēng téng*) (Hu et al. 1980; Wang and Qian 2013)
Jianghua (China) 松金藤 (*sōng jīn téng*), 松萝 (*sōng luó*)	Herbal tea, food, medicinal bath for treating rheumatism, from Uygar medicine (Jin et al. 2018; Luo et al. 2018)
Tibet *gser.skud* ("gold thread")	Cures fevers of the lungs, liver, and channels, as well as fever caused by poisoning (Clark 1995)
Korea 송낙 (*song-nag*)	Used to induce menstruation (Pusan) and treat tuberculosis of the neck (Gongju) (Lee 1966; Lee et al. 1977)
***Usnea durietzii* Mot. [syn. *Neuropogon durietzii*]**	
Argentina *barba de la piedra*	In San Luis same as Argentine use of *U. densirostra* (Vitto et al. 1997). In Mendocina recorded as an abortative and toxin (Ruíz Leal 1972).
***Usnea florida* (L.) F. H. Wigg.**	
China 松萝 (*sōng luó*), 花松萝 (*huā sōng luó*)	Used for aching in sinews and bones, stopping bleeding or infection from external injuries, skin diseases, painful urination, coughs, tuberculosis of the lungs or neck, heart palpitations, and edema. Drink decoction; or apply decoction or powdered lichen to the affected area (Wang and Qian 2013)
Europe (early modern)	Decoction used for colds and coughs (Willemet 1787)
Kallaway (Bolivia) *qaqa sunkha*	Unspecified medicinal use (Abdel-Malek et al. 1996)
Mapuche (Chile)	Infusion used for diarrhea (Houghton and Manby 1985)
***Usnea himalayana* C. Bab.**	
Japan *nayonayo saruogase*	Burned as a "lichen cigarette" (Ohmura 2003)
***Usnea himantodes* Stirt. [syn. *Eumitria himantodes*, *Usnea gigas*, *Usnea africana*]**	
Iraqw (Tanzania)	Same as Iraqw use of *U. articulata* (Kokwaro 1976)

(continued)

Table 2.7 (continued)

Culture and *folk name*	Traditional use
Usnea hirta **(L.) F. H. Wigg.**	
Europe (early modern)	Used to heal wounds and to prevent hair loss (Willemet 1787)
Usnea laevis **(Eschw.) Nyl.**	
Sierra Nevada del Cocuy (Colombia) *barba de piedra* or *barbepiedra*	The entire lichen is milled, and the powder is applied in the armpits, feet, etc. (Rodríguez et al. 2018)
Venezuelan Andes *barba de piedra* or *tusinya*	Used for infections, dermatosis, fungal infections, tuberculosis, and pneumonia (Marcano 1991; Marcano et al. 1999)
Usnea longissima **Ach. [syn. *Dolichousnea longissima*]**	
Northern Anatolia (Turkey)	For treating cancer, tuberculosis, and ulcers (Yazici and Aslan 2003; Odabasoglu et al. 2006)
Unani medicine (India) *ushna*	Sometimes used as *ushna* in Unani medicine (Razzack and Fazal 1993; Rauf et al. 2011). See Unani use of *Usnea* spp.
Ayurvedic medicine (India)	Same as Ayurvedic use of *U. baileyi* (mixed in tobacco) and an occasional adulterant in charila (see *Parmotrema nilgherrense*) (Shah 1998)
Madhya and Himachal Pradesh (India) *urmil* (Himachal)	Used to heal bone fractures, mixed with other ingredients (Baiga, Madhya). Washed, air-dried, soaked overnight in salted water, and placed over the affected part (Himachal) (Brij Lal and Upreti 1995; Sharma 1997)
Dolpo (Nepal) *shingbal*, *shing bel*, or *ser kue* (Amchi); *thangbu balto* (Kham); *jhyau* (Nepali)	Medicinal properties are *kha* (bitter) and *ka* (astringent), potency is *sil* (cool). Used to treat wounds, lungs, liver, channel fever, and fever from poisoning. Also used as incense (Lama et al. 2001)
Mongolia 色日古德 (*sè rì gǔ dé*)	Used for injuries of legs and back, bone fractures, diseases of women, clearing *heat*, removing toxins, headache, stomach and intestine issues, diarrhea, sore muscles and tendons, and lung issues (Bai et al. 2014; La et al. 2015)
China 长松萝 (*zhǎng sōng luó*), 天蓬草 (*tiānpéng cǎo*, "canopy herb"), 云雾草 (*yúnwù cǎo*, "cloud herb"), 老君须 (*lǎo jūn xū*, "old man beard"), 胡须草 (*húxū cǎo*, "beard herb"), 龙须草 (*lóng xū cǎo*, "dragon beard herb"), 茶须 (*chá xū*, "tea beard"), 树毛 (*shù máo*, "tree hair")	Same use in China as *U. diffracta* (Hu et al. 1980; Wang and Qian 2013). Other names include 松萝 (*sōng luó*), 女萝 (*nǚ luó*), 金钱草 (*jīnqián cǎo*), 关公须 (*guāngōng xū*), 石须 (*shí xū*, "stone beard"), 海风藤 (*hǎi fēng téng*), 雪风藤 (*xuě fēng téng*), 金丝藤 (*jīn sī téng*), 松毛 (*sōng máo*), 仙人头发 (*xiānrén tóufǎ*, "fairy hair"), 石丝线 (*shí sīxiàn*, "stone thread"), 松上寄生 (*sōng shàng jìshēng*), 树挂 (*shùguà*), 山挂面 (*shān guàmiàn*), 树发七 (*shù fā qī*), 过山龙 (*guò shān lóng*), 飞天蜈蚣 (*fēitiān wúgōng*), 尼山翅 (*ní shān chì*)

(continued)

Table 2.7 (continued)

Culture and *folk name*	Traditional use
Nahki (nw Yunnan, China) 松萝 (*sōng luó*), 长松萝 (*zhăng sōng luó*)	Soaked in boiling water, for moistening the lungs and relieving cough. Whole lichen eaten as healthcare food (Salick et al. 2006; Zhang et al. 2015, 2016)
Shuhi (nw Yunnan) *ʃupu'nu*	Decoction for abdominal pain in women (Weckerle et al. 2006)
Lisu (nw Yunnan) *jianmingyan*	Decoction drunk for inflamed lymph nodes and rheumatism. Crushed and applied as poultice for injuries from falls and infected hair follicles (Ji et al. 2004)
Haida (BC, Canada)	Same as Haida use of *Alectoria sarmentosa* (Turner 1998, 2004a)
Ditidaht (BC, Canada) *p'u7up*; or *Indian bandage*	All *Usnea* spp. and *Alectoria sarmentosa* used for wound dressing, but *U. longissima* is preferred. Wrapped around wound and left for a while (Turner et al. 1983)
***Usnea nidifica* Tayl.**	
China	Unspecified medicine (Wang and Qian 2013)
Rarotonga (Cook Is.) *'uru nū* (Mangiai); *remu nū* (Mauke)	Online reference to medicinal usage on Mangiai: thallus chewed and applied to cuts (to stop bleeding) and stings (McCormack 2007). Whistler (1990) records *remu* as a general term for lichens, mosses, and seaweeds, but records no use
***Usnea orientalis* Motyka.**	
Eastern Nepal	Whole plant ground and applied to cuts and wounds (Shrestha et al. 2016)
***Usnea pectinata* Tayl. [syn. *Eumitria pectinata*]**	
China 拟长松萝 (*nĭ zhăng sōng luó*), 大蓬草 (*dà péng căo*), 篦齿松萝 (*bìchĭ sōng luó*)	Used for stopping bleeding from external injuries, relieving pain, bloody feces, and swelling (Wang and Qian 2013)
Usnea plicata (L.) Weber	
Tripolitania (Libya) *scíba*	Ingredient in medicinal decoction called *scíba*, along with *Pseudevernia furfuracea*, *Ramalina calicaris*, and *R. farinacea* (Natale and Pollio 2012)
Saami (Scandinavia)	Put on sores on feet after walking long distances (Linnaeus 1737)
Europe (early modern)	An astringent for internal and external use (Lightfoot 1777), for whooping cough (Lindley 1838), jaundice, strengthening stomach and abdominal cavity, and restraining abortion (Luyken 1809). Also recommend *U. barbata*. See European use of *Usnea* spp.

(continued)

Table 2.7 (continued)

Culture and *folk name*	Traditional use
Usnea sikkimensis **Biswas sp. nov.**	
Sikkim and Darjeeling (India) *darimataghosa* (Bengali)	Used for lung troubles, hemorrhages, and asthma; powdered and used to strengthen hair (Biswas 1956). (May be a European use)
Himalayas (Sikkim, India)	Used to bandage surface wounds, skin eruptions, and boils; inserted into nostril to stop nose bleeds; put in shoes to prevent or treat blisters (Lepchas). Medicine for respiratory health (Pradhan and Badola 2008; O'Neill et al. 2017)
Usnea strigosa **(Ach.) Eaton**	
Kimi (Amusa, Papua New Guinea) *oleazu*	Concoction taken orally for headaches (Jorim et al. 2012)
Usnea subfloridana **Stirt.**	
Leitrim (Ireland)	Treatment for sore eyes. Mixed with tobacco and butter, boiled, cooled, and applied as lotion to eyes (Allen and Hatfield 2004)
China 亚花松萝 (*yà huā sōng luó*)	Used for painful and reddened eyes, bleeding from external injuries, and swelling (Wang and Qian 2013)
Usnea subsordida Stirt.	
Ayurvedic medicine (India)	Same as Ayurvedic use of *U. baileyi* (mixed in tobacco) and an occasional adulterant in *charila* (see *Parmotrema nilgherrense*) (Shah 1998)
Usnea trichodeoides **Vain. [syn. *Dolichousnea trichodeoides*]**	
China 粗皮松萝 (*cūpí sōng luó*), 天蓬草 (*tiānpéng cǎo*, "canopy herb"), 变红长松萝 (*biàn hóng zhǎng sōng luó*), 树发七 (*shù fā qī*, "tree hair seven")	Used for coughs; pulmonary tuberculosis; headaches; blurred vision; inflamed cornea; swellings, sores, and pus discharge; bleeding from external injuries; bloody feces; uterine bleeding; menstrual disorders; and vaginal discharge. Drink decoction; or apply decoction or powdered lichen to the affected area (Wang and Qian 2013)

Table 2.8 Cetrarioid lichens (Parmeliaceae) used in traditional medicines

Culture and *folk name*	Traditional use
Cetraria islandica (L.) Ach.	
Europe (1500s to present)	Medicinal lichen in European pharmacopoeias from the 1500s to the present. Common throughout Europe and Greenland, mostly for pulmonary and digestive uses. Bitterness may be removed with weak alkaline solution or by heating in water once or twice. Prepared as decoction or infusion in water or milk, ingested as a drink or a jelly. Sometimes mixed with sugar, lemon, or other flavouring. Also taken as a powder, or used for salves. Used as a strengthening tonic and to treat pulmonary tuberculosis, coughing blood, whooping cough, asthma, chronic congestion, indigestion, chronic diarrhea, dysentery, uterine cysts, kidney stones, edema, wounds, and scurvy. Also used as a laxative, a sedative, and as an aphrodisiac (Matthiolus 1586; Ray 1686; Linnaeus 1737; Scopoli 1760; Cramer 1780; Willemet 1787; Withering 1801; Regnault 1802; Lindley 1838; Anonymous 1845; Pereira 1850; Rink and Lindorff 1856)
Pallars (Spain) *liquen de bosc*	Drunk as tea for congestion, tuberculosis, asthma, inflammation, and high blood pressure (Muntané 1991; González-Tejero et al. 1995; Agelet and Vallès 2003)
Ubaye Valley (France)	Decoction used for lung ailments and as an emollient (Novaretti and Lemordant 1990)
Northern Italy *lic* (Kyé, Roccaforte Mondovì); *èrbo d'la vélho, licchia, lichene, pan d'chabbre* (Waldensian); *lica, lichen, erba sechia, erba bronchite* (Sondrio)	Drunk as tea for a cough suppressant (Occitans). Decoction for cough, bronchitis, and intestinal problems; may be served with milk; may weaken body if excessively consumed (Kyé). Decoction or syrup applied externally for cough and bronchitis, decoction for stomach disorders in cattle, and decoction as a digestive and expectorant (Waldensian). Decoction drunk as digestive and for coughs and bronchitis, and decoction fed to livestock for bronchitis and pneumonia (Sondrio). Used for congestion and for recovery after tuberculosis (Venezia Giulia). Similar use in Trentino (Lokar and Poldini 1988; Mattalia et al. 2013; Pieroni and Giusti 2009; Vitalini et al. 2013; Bellia and Pieroni 2015)
Großwalsertal (Austria) *isländisch moos*	A medicinal drink mixed with the syrup of *Plantago lanceolata* (Grasser et al. 2012)
Eastern Tyrol (Austria) *goasstraube* or *isländisch moos*	A medicinal fodder for cattle, sheep, horses, and pigs used as an immune system booster to maintain good health; as a restorer after

(continued)

Table 2.8 (continued)

Culture and *folk name*	Traditional use
	sickness, weight loss, or delivery; as a performance enhancer to promote weight gain and milk production; to improve skin, udder health, hoof quality, and coat gloss; and for pregnant animals to facilitate delivery, prevent complications with birth, and avoid milk fever (Vogl et al. 2016)
Bosnia and Herzegovina *islandski lišaj, islandska mahovina*; *mašina* (Lukomir)	Decoction is drunk for pulmonary ailments and improving blood. Decoction with *Pimpinella anisum* is taken as syrup for cough and croakiness. Powdered lichen soaked in cold water for urinary tract inflammation, cold and cough, and anemia. Taken for heart or chest pain (Lukomir) (Redžić 2007; Šarić-Kundalić et al. 2010; Ferrier et al. 2014, 2015)
Albanians in Upper Reka (Macedonia) *mishk*	Unspecified medicinal use (Pieroni et al. 2013)
Bulgaria	Used for coughs (Nedelcheva and Draganov 2014)
Tripoli (Lebanon)	Lichen is soaked in alcohol and rubbed externally to treat rheumatism and neuralgia (Marc et al. 2008)
Hryniawa (Ukraine) *hrań*	Soaked in milk and taken internally for cough (Kujawska et al. 2015)
Lithuania	Tea, extract with alcohol, or decoction is drunk for respiratory tract disorders. Decoction applied externally for skin diseases and burns (Pranskuniene et al. 2018)
Estonia	Tea taken as remedy for cancer (Sak et al. 2014), cough, colds, bronchitis, lung diseases, respiratory problems, and fever (Sõukand and Kalle 2013).
Sweden *islandslav*	Used for whooping cough, colds, congestion, asthma, other chest ailments, appetite stimulation, diabetes, nephritis, and tuberculosis. Either decoction or infusion made from dried shredded lichen in either water or milk and drunk either warm or cold. Honey or chocolate sometimes added (Ahmadjian and Nilsson 1963)
Iceland	Listed as important medicinal plant (Whitney et al. 2012)
Ket (Siberia)	Decoction for coughs (Eidlitz 1969)
Lhasa (Tibet)	Used in traditional Tibetan medicine (Salick et al. 2006)
China 岛衣 (*dǎo yī*), 太白树 (*tàibái shù*, "white tree")	Decoction drunk to strengthen stomach and improve digestion (Wang and Qian 2013)

(continued)

Table 2.8 (continued)

Culture and *folk name*	Traditional use
Muslims, Yuan Dynasty, China *khazâz* (Arabic)	Listed in *Hui Hui Yao Fang*, the official list of Islamic medicines for the Yuan Dynasty (1250–1350 CE) (Kong and Chen 1996)
Dehcho (NWT, Canada)	Decoction used to treat tuberculosis. Boiled in water 0.5–1 h, until liquid is red, and one-third cup is taken three times daily (Lamont 1977)
***Cetrelia pseudolivetorum* (Asahina) W.L. Culb. and C.F. Culb.**	
China 拟橄揽斑叶梅 (*nǐ gǎn lǎn bān yè méi*), 假漂红梅衣 (*jiǎ piào hóng méi yī*), 裂缘梅衣 (*liè yuán méi yī*)	Used for blurred vision, bleeding from uterus, traumatic bleeding, sores, and chronic psoriasis. Drink decoction; or apply decoction or powdered lichen to the affected area (Wang and Qian 2013)
***Flavocetraria cucullata* (Bellardi) Kärnefelt and A. Thell**	
Pallars (Spain)	Drunk as tea to treat symptoms of asthma (Agelet and Vallès 2003)
***Flavocetraria nivalis* (L.) Kärnefelt and Thell**	
Europe	Although not as commonly used in Europe as *Cetraria islandica*, some practitioners thought it had similar properties (Tychsen 1799; Lindley 1838)
Kallawaya (Qollahuayas, Bolivia)	Prepared in tea for treatment of motion sickness and heart attacks (Bastien 1983)
***Masonhalea richardsonii* (Hook.) Kärnefelt [syn. *Cornicularia richardsonii*]**	
Tlingit (Alaska, BC)	Used as a treatment for inflammation of the lungs (Garibaldi 1999)
***Nephromopsis pallescens* (Schaer.) Park [syn. *Cetrariopsis pallescens*]**	
China 皮革肾岛衣 (*pígé shèn dǎo yī*), 树衣 (*shù yī*, "tree clothes")	Eaten, and has an unspecified medicinal use (Wang and Qian 2013)
***Vulpicida canadensis* (Räsänen) J.E. Mattsson and M.J. Lai**	
Ulkatcho (BC, Canada) *dahgha* ("limb hair")	Medicine for coughs and colds, drink tea made from a couple handfuls of *dahgha* in 1 L water. Also chewed fresh to help the lungs (Hebda et al. 1996)
***Vulpicida juniperinus* (L.) J.E. Mattsson and M.J. Lai**	
Early modern Europe	Used to treat jaundice (Regnault 1802)
Scandinavia	Possibly used to poison wolves along with *Letharia vulpina* (Uphof 1959)
***Vulpicida pinastri* (Scop.) J.E. Mattsson and M.J. Lai**	
Scandinavia	Possibly used to poison wolves along with *Letharia vulpina* (Smith 1921)
China 黄花岛衣 (*huánghuā dǎo yī*)	Used for pulmonary tuberculosis, wounds, oozing pus, skin infections, cancer, and spasms (Wang and Qian 2013)

Table 2.9 Parmelioid lichens (Parmeliaceae) used in traditional medicines

Culture and *folk name*	Traditional use
Unidentified parmelioid lichens	
Rastafari (Cape area, South Africa) *klipbolm*	Used for cancer, women's problems, aiding fertility, and inducing abortion (Aston Philander 2011). See also KhoiSan use of *Xanthoparmelia hottentotta*
Nepal *jhau*	Extract and decoction are applied to treat moles (Gaire and Subedi 2011)
Bhutan *rdo dreg, brag gi me tog, khrag gi ut pal, dbugs med pags pa, pha vang dreg pa, brtan po'i dreg pa*	Unidentified parmelioid lichen on rock used in *Sowa Rigpa* traditional medicine to treat poisoning, fever, diarrhea, vomit reflex, impaired vision, and lymphatic disorders. Also used to restore appetite and promote healthy skin (Yeshi et al. 2018). See also Bhutan use of *Parmelia saxatilis*
Lëpushë (Albania) *urth*	Ground, mixed with sugar, then applied externally to heal eye inflammations, especially in veterinary practice (Pieroni et al. 2005)
Lucca (Italy)	Decoction for coughs, cleansing liver, and anti-inflammatory (Pieroni 2000)
Piaroa (Amaz., Venezuela) *odoche jupacua* (iguana toe)	Used to treat gonorrhea or painful urination. Boiled into a tea and drunk 3–4 times a day for a week (Azenha et al. 1998)
Guahibo (Amaz., Venezuela)	Boiled in water and applied to insect bites or cuts and wounds (Azenha et al. 1998)
Canoparmelia ecaperata **(Mull. Arg.) Elix and Hale**	
India *jhulla* or *charila* (Kumaon)	Used to treat colds and coughs (Singh et al. 2012)
Flavoparmelia caperata **(L.) Hale**	
Rarámuri (Mexico) *řeté cajéra*	Dried, crushed, and dusted on burns (Pennington 1963)
China 皱梅衣 (*zhòu méi yī*), 地花 (*de huā*, "ground flower")	Decoction drunk to clear *heat* (Wang and Qian 2013)
Hypotrachyna cirrhata **(Fr.) Divakar, A. Crespo, Sipman, Elix, and Lumbsch [syn. *Everniastrum cirrhatum*, *Parmelia kamtschadalis*, *Evernia cirrhata*]**	
India *jhulla* or *chadilla* (Kumaon)	Ayurvedic use is the same as *Usnea baileyi* (mixed in tobacco) and an occasional adulterant in *charila* (see *Parmotrema nilgherrense*) (Shah 1998). When burnt the smoke relieves headache and the powder is a good cephalic snuff (Biswas 1947; Nadkarni and Nadkarni 1955). Used to treat colds and fever (Kumaon) (Singh et al. 2012)
Rasuwa (Nepal) *jhyau*	Paste of whole plant applied to cuts and wounds. Unidentified parmelioid lichen also used (Uprety et al. 2010)

(continued)

Table 2.9 (continued)

Culture and *folk name*	Traditional use
Mbowamb (New Guinea) *ugl*	"*Evernia* sp." is chewed and then applied to skin eruptions (Stopp 1963)
Hypotrachyna nepalensis **(Taylor) Divakar, A. Crespo, Sipman, Elix, and Lumbsch** [**syn.** ***Everniastrum nepalense*****,** ***Parmelia nepalensis***]	
Ayurvedic medicine (India)	Same as Ayurvedic use of *Usnea baileyi* (mixed in tobacco) and an occasional adulterant in *charila* (see *Parmotrema nilgherrense*) (Shah 1998)
Himalayas (Sikkim, India)	Medicine for respiratory health (O'Neill et al. 2017)
Taplejung (Nepal) *jhyau*	Used like *Ramalina* spp. for antiseptic, burns, and wounds. Applied as powder in tincture of iodine after applying the leaf juice of *Artemisia dubia* or *Eupatorium adenophorum* (Poudel 2008). Banned from raw export (Bhattarai 1999)
Kathmandu (Nepal) *kalo jhyau*	Used for toothache, sore throat, and pain (Kumar et al. 1996)
Parmelia omphalodes **(L.) Ach.**	
Britain *crottle*, *crotal*, *dark crottle*, or *fiasgag nan creag* (Gaelic: "rock lichen")	In Scotland, they wore socks dyed with *crottle* if walking long distance or sprinkled it on their hose to stop their feet from getting inflamed (Cameron 1900; MacIntyre 1999). Used for a soup to strengthen invalids in Ireland; and for a poultice for cuts, sores, and burns in Ireland and Scotland (McGlinchey 1986; Allen and Hatfield 2004). Probably used interchangeably with *P. saxatilis*.
Europe (early modern era)	Used to stop bleeding and stop hemorrhaging during surgery. Put into nose to stop nosebleeds (Gedner 1756; Willemet 1787)
Parmelia saxatilis **(L.) Ach.**	
Bhutan *sbrul-pag*	Used in *Sowa Rigpa* traditional medicine to heal white leprosy, sore foot, and skin diseases (Wangchuk et al. 2016)
China 石梅表 (*shí méi biǎo*), 石花 (*shí huā*, "stone flower"), 乳花 (*rǔ huā*, "milk flower"), 梅衣 (*méi yī*, "plum clothes"), 石衣 (*shí yī*, "stone clothes"), 石濡 (*shí rú*), 梅藓 (*méi xiǎn*, "plum moss"), 梅花衣 (*méihuā yī*, "plum blossom clothes"), 藻纹梅花衣 (*zǎo wén méihuā yī*), 地衣 (*dì yī*, "ground clothes")	Used for blurred vision, vomiting blood, jaundice, bleeding from uterus, chronic dermatitis, and oral ulcers in children. Drink decoction in wine or apply powder to the affected area (Hu et al. 1980; Wang and Qian 2013)
Dalarna (Sweden) *stenlav*	Used to remove warts (Ahmadjian and Nilsson 1963)
Britain *crottle*; or *light crottle*	Probably used interchangeably with *P. omphalodes* as *crottle*.

(continued)

Table 2.9 (continued)

Culture and *folk name*	Traditional use
Foula (Shetland Is., Britain) *old man*	Mixed with tobacco and smoked in the eighteenth century, a practice still remembered in 1966 (Hawksworth 2003)
Europe *usnea cranii humani*, *muscus cranii humani*, or *muscus ex cranio humano* (Latin); *moss of a dead man's skull* (English); *usnée humaine* (French); *muschio del cranio* (Italian)	An important medicine as early as the late 1500s (Gerarde 1597) and throughout the 1600s (Parkinson 1640; Ray 1686), but various authors think it quackery by the 1700s (Quincy 1724; Diderot et al. 1765). In early drawings it is distinctly *Usnea*-like, but later authors recognize two distinct types: *Usnea*-like or crust-like (James 1748). In modern times, it has been identified as either *Parmelia saxatilis* (Smith 1921) or *Physcia* sp. (Llano 1948), although it is likely any lichen or moss found on a skull (Modenesi 2009). When collected off the skull of criminals (alt. someone who died a violent death), it was very valuable as cure for epilepsy, to stop bleeding, and (if powdered and given in sweet wine) for whooping cough in children. Also mixed with *mumia* (the exudate from a mummy) to make *unguentum armarium*, a salve that was applied to a weapon to heal a wound that it had caused
***Parmelia sulcata* Taylor**	
Metís (Alberta, Canada)	Rubbed on gums of teething babies to relieve discomfort (Marles et al. 2000)
Saanich (BC, Canada) *smexdáles*	Medicinal properties depend on the type of tree it is growing on. Possibly the lichen traditionally used for birth control. Not differentiated from *Lobaria pulmonaria* (Turner and Hebda 2012)
***Parmelinella salacinifera* (Hale) Marcelli and Benatti**	
Pankararu (Brazil) *flor de pedra*	Part of a medicinal mixture of *flor de pedra* (see *Parmotrema wrightii*) (Londoño-Castañeda et al. 2017)
***Parmotrema abessinicum* (Nyl. ex Kremp.) Hale [syn. *Parmelia abessinica*]**	
Bellary District (India) *rathipuvvu* ("rock flower")	Eaten medicinally (Llano 1948)
***Parmotrema andinum* (Müll. Ar.) Hale [syn. *Parmelia hyporysalea*]**	
Ayurvedic medicine (India)	Occasional adulterant in *charila* (see *Parmotrema nilgherrense*) (Chanda and Singh 1971)
***Parmotrema nilgherrense* (Nyl.) Hale [syn. *Parmelia nilgherrensis*]**	
Ayurvedic medicine (India) *charila* (Hindi); *shaileya* or *shilapushpa* ("rock flower," Sanskrit); hazaz-al sakhr ("*rock scab*," Arabic); *shailaja* (Bengali); *chadila* (Urdu); *pathar phool* (Gujarati); *dagad phool* (Gujarati	An important drug in many old Ayurvedic texts that is still used today. The first record is in the Atharvaveda (1500 BCE). Although in some areas of India, high-quality *charila* is mostly *Parmotrema nilgherrense*, the lichen

(continued)

Table 2.9 (continued)

Culture and *folk name*	Traditional use
and Marathi); *kallu hoovu* (Kannada); *rati puvvu* (Telugu); *sheeleyam* (Malayalam); *kapashwe* (Tamil)	mixture can also contain *Parmotrema chinense*, *P. perforatum*, *P. perlatum*, *Hypotrachyna cirrhata*, and *H. nepalensis*, with the occasional adulterants *Ramalina farinacea*, *R. inflata*, *Usnea baileyi*, *U. longissima*, *U. subsordida*, *Parmelia hyporysalea*, *Anaptychia* spp., and *Leptogium* spp. It is used for indigestion, loss of appetite, flatulence, diarrhea, stomach disorders, kidney stones, painful urination, hemorrhoids, involuntary semen emission, lack of menstruation, painful menstruation, enlarged spleen, bronchitis, congestion, shortness of breath, excessive salivation, fevers, headaches, sore throats, toothaches, broken bones, musculoskeletal pain, rheumatism, reducing swelling, leprosy, scabies, soothing irritated skin, and prenatal and postnatal care. Also used as an aphrodisiac, diuretic, sedative, astringent, antiseptic, antibiotic, and a demulcent to reduce inflammation. It is powdered and applied on wounds to promote healing, smoked to relieve headaches, used as incense, used as a cephalic snuff, used in medicated oils, applied as a poultice to renal and lumbar regions to induce urination, and applied as a liniment to the head for headaches (Dutt 1877; Chanda and Singh 1971; Kumar and Upreti 2001; Karadi 2010; Prasad 2013; Rauf et al. 2006) An ingredient in *spemen*, which is used for treating infertility in men (Pardanani et al. 1976).
Ayurvedic medicine (other countries) *jhoola* (Nepal)	Nepal: soup as an aphrodisiac, paste applied externally for kidney stones. China: soup for male infertility, paste applied externally for kidney stones. Malaysia: soup as an aphrodisiac and for seminal weakness. Afghanistan: used for chest disorders, paste applied to wounds for healing. Saudi Arabia: cephalic snuff for headaches and as a painkiller (Kumar et al. 1996; Karadi 2010)
Himalayas (India) *Jhulla* (Kumaon)	Used to treat coughs and throat problems (Kumaon) and unspecified respiratory problems (Sikkim) (Singh et al. 2012; O'Neill et al. 2017)
Kathmandu (Nepal) *kalo jhyau*	Used for toothache, sore throat, and pain (Kumar et al. 1996)
***Parmotrema perforatum* (Jacq.) A. Massal. [syn. *Parmelia perforata*]**	

(continued)

Table 2.9 (continued)

Culture and *folk name*	Traditional use
Ayurvedic medicine (India)	Commonly used as *charila* (see *P. nilgherrense*) (Nadkarni and Nadkarni 1955; Chanda and Singh 1971). Imported for medicine (Younos et al. 1987) and used for diuretic treatments (Biswas 1947).
***Parmotrema perlatum* (Huds.) M. Choisy [syn. *Parmelia perlata*]**	
Ayurvedic medicine (India)	Commonly used as *charila* (see *P. nilgherrense*) (Nadkarni and Nadkarni 1955; Chanda and Singh 1971; Rauf et al. 2006)
***Parmotrema reticulatum* (Taylor) M. Choisy [syn. *Parmelia reticulata*]**	
Himalayas (India) *jhulla* or *chadilla* (Kumaon)	Used to treat coughs and throat problems (Kumaon) and unspecified respiratory problems (Sikkim) (Singh et al. 2012; O'Neill et al. 2017)
Tepehuan and mestizos (Chihuahua, Mexico) *ódai yoošígai* or *flor de piedra* ("rock flower")	Tea drunk to relieve discomfort from kidney disorders or venereal disease. The tea is commonly prepared in the late afternoon and left for one night before being drunk (Pennington 1969)
***Parmotrema saccatilobum* (Taylor) Hale**	
Milne Bay (Papua New Guinea)	Vegetable oil extract of fresh or dried lichen is used for superficial joint and muscle pain or inflammation (Bugni et al. 2009)
***Parmotrema sancti-angelii* (Lynge) Hale [syn. *Parmelia sancti-angelii*]**	
Gond and Oran (Uttar Pradesh, India) *jhavila*	Salve used to treat skin disease called *sem*. Burn 30–50 g of *jhavila* and mix ash with mustard or linseed oil (Brij Lal and Upreti 1995)
***Parmotrema subtinctorium* (Zahlbr.) Hale [syn. *Parmelia subtinctoria*]**	
China 亚大叶梅 (*yà dà yè méi*), 亚燃料梅衣 (*yà ránliào méi yī*)	Used for bleeding from external injury, localized swelling, and pain (Wang and Qian 2013)
***Parmotrema tinctorum* (Nyl.) Hale [syn. *Parmelia tinctoria*]**	
China 大叶梅 (*dà yè méi*, "big plum leaf"), 白石花 (*bái shí huā*, "white stone flower"), 石花 (*shí huā*, "stone flower"), 梅衣 (*méi yī*, "plum clothes"), 石衣 (*shí yī*, "stone clothes"), 虾蟆皮 (*há má pí*, "toad skin")	Used for blurred vision, bleeding from uterus, bleeding from external injuries, sores and swelling, chronic dermatitis, and localized swelling. Drink decoction or apply powdered lichen to the affected area (Wang and Qian 2013)
***Parmotrema wrightii* L. I. Ferraro and Elix**	
Pankararu (Brazil) *flor de pedra*	Part of a medicinal mixture with *Parmelinella salacinifera* and *Heterodermia galactophylla*. Neither *P. salacinifera* nor *H. galactophylla* are added when used for pregnant women, and *H. galactophylla* is not added when used for children. Aqueous extract to treat digestive problems like diarrhea and vomiting. Used in a

(continued)

Table 2.9 (continued)

Culture and *folk name*	Traditional use
	smoker to treat epilepsy and cultural diseases (Londoño-Castañeda et al. 2017)
Parmotrema zollingeri **(Hepp) Hale [syn.** ***Parmelia zollingeri*****]**	
Ati (Philippines) *kalas*	Used as medicine for children with high fever and suffering from convulsions. Burn *kalas* and let the child smell the fumes (Madulid et al. 1989)
Punctelia borreri **(Sm.) Krog**	
China 粉斑星点梅 (*fěn bān xīng diǎn méi*, "pink spotted plum"), 粉斑梅衣 (*fěn bān méi yī*, "spotted plum clothes")	Used for blurred vision, bleeding from uterus, bleeding from external injuries, sores and swelling, and chronic dermatitis. Drink decoction or apply powdered lichen to the affected area (Wang and Qian 2013)
Xanthoparmelia spp. (Vain.) Hale	
Navajo (AZ and NM, USA) New Mexico: *tschétláat* ("rock covering"), *nihaλá·d* ("earth moss") or *céλá·d* ("rock moss"); Arizona: *owa'si* ("rock flower") or *owa'huru'suki* ("rock manure")	New Mexico: Elmore (1943) records *tschétláat* as remedy for impetigo. Wyman and Harris (1941) record widespread use of *nihaλá·d* or *céλá·d* chewed for canker, swollen gums, decayed teeth, etc. (May include *Peltigera* sp.). Arizona: Whiting (1939) records an unidentified rock lichen called *owa'si* or *owa'huru'suki* used for sore mouth, gums, and toothache. See also: Hopi use of yellow rock lichen; Tewa use of rock and ground lichen
Xanthoparmelia conspersa **(Ehrh. ex Ach.) Hale**	
Xhosa (South Africa) *ubulembu belitye*	To treat syphilis eruptions: powder and apply externally to eruptions (perhaps after they are scarified); may be also used internally (sources disagree). To treat both known and suspected snakebites: drink one tablespoon of lichen in cold water; also scarify bite and sprinkle powdered lichen on it to draw out a *humour* (Smith 1888; Watt and Breyer-Brandwijk 1962). See also Xhosa use of "unidentified rock lichen"
Iroquois (Ontario, Canada)	Used for inflamed gums and raw throat caused by fever. Mix in 1 cup cold water with the bark of the tree it was collected off, *Coptis trifolia*, and *Fraxinus nigra*. Take one teaspoon, leave in mouth until water is warm, and then swallow. Repeat for entire cup (Herrick 1995)
O'odham (Arizona, USA) *jievut hiawsik* or *jewed hiósig* ("earth flower")	Traditional use described by Curtin (1949). Lipp (1995) identified the lichen as *X. conspersa*, but Hawksworth (2003) disagrees. Carried as good luck charm, but overuse will make you sick. Mixed with tobacco and smoked to "make young men crazy." Also ground into a powder and

(continued)

Table 2.9 (continued)

Culture and *folk name*	Traditional use
	sprinkled on sores or cuts, but not bound, as it would cause blisters. Applied over several days to heal rattlesnake bite
Xanthoparmelia convoluta **(Kremp.) Hale [*Xanthomaculina convoluta*]**	
Khoikhoi (Namibia)	Infusion taken as remedy for rheumatism and arthritis (Watt and Breyer-Brandwijk 1962). See also Topnaar [a Khoikhoi tribe] use of *X. hottentotta*.
Xanthoparmelia hottentotta **(Ach.) A. Thell et al. [syn. *Xanthomaculina hottentotta*]**	
Unspecified (Namibia)	Used to treat inflammation of udder for goats and sheep. Dried, roasted, and powdered; mixed with aromatic shrubs, fungal spores, and very fine quartz dust; then added to tail-fat to make an ointment for the udder (Epstein 1937)
Topnaar (Kuiseb, Namibia) I *ui* II *khaob*; or *uijkhao.b.*	Decoction drunk to cure coughs and to relieve stomach and chest pains (van Damme et al. 1992)
KhoiSan (Cape area, South Africa) *klipblom*, *klipmos*, or *klipbuchu*	Infusion is drunk for cough, sore throat, colds, fertility, syphilis in men, back pain, abdominal pain, and kidney and bladder diseases. Infusion used as mouthwash for oral thrush and teething in children. Also used as a female medicine for *maak baarmoeder skoon* ("cleaning the womb"), treating general pains, and as an ointment for burns and wounds (Laidler 1928; van Wyk et al. 2008; de Beer and van Wyk 2011; van Wyk and Gorelik 2017).
Xanthoparmelia scabrosa **(Taylor) Hale**	
New Age herbalism	Currently sold as "traditional Chinese medicine," as an aphrodisiac and a cure for male impotence. No record of this use was found prior to 2007 (Tshiteya 2007)
Xanthoparmelia tinctina **(Maheu and Gillet) Hale**	
China 暗腹黄梅 (*àn fù huáng méi*, "dark belly yellow plum"), 暗腹梅衣 (*àn fù méi yī*, "dark belly plum clothes")	Used for blurred vision, bleeding from uterus, bleeding from external injuries, sores and swelling, and chronic dermatitis. Drink decoction; or apply decoction or powdered lichen to the affected area (Wang and Qian 2013)

Table 2.10 Other Parmeliaceae lichens used in traditional medicines

Culture and *folk name*	Traditional use
***Anzia opuntiella* Müll. Arg.**	
China 仙人掌绵腹衣 (*xiānrénzhǎng mián fù yī*, "cactus cotton belly")	Used for blurred vision, bleeding from uterus, traumatic bleeding, sores, and chronic psoriasis. Drink decoction; or apply decoction or powdered lichen to the affected area (Wang and Qian 2013)
***Anzia ornata* (Zahlbr.) Asahina**	
China 瘤绵腹衣 (*liú mián fù yī*, "lumpy cotton belly")	Same as Chinese use of *A. opuntiella* (Wang and Qian 2013)
***Evernia divaricata* (L.) Ach.**	
China 柔扁枝衣 (*róu biǎn zhī yī*, "soft flat twig clothes")	Same use in China as *E. mesomorpha* (Wang and Qian 2013)
***Evernia mesomorpha* Nyl.**	
Chipewyan (SK, Canada) *k'i tsa^{n}jú* ("birch lichen")	Treatment for snow blindness. Harvest from birch, boil, cool, and drop decoction into eyes (Marles 1984)
China 扁枝衣 (*biǎn zhī yī*, "flat twig clothes")	Used for coughs, pneumonia, hot flashes due to pulmonary tuberculosis, hepatitis, headaches, infection due to trauma, inflammation of the breasts, and snakebites (Wang and Qian 2013)
***Evernia prunastri* (L.) Ach.**	
Ancient Greece **σπλάγχνον** (*splanchnon*, "intestines"); or **βρῦον** (*brýon*, "moss")	An arboreal lichen recorded by Dioscorides in Section 1.21 of *De Materia Medica* (50–70 CE) that is best if sweeter-smelling, whiter, and growing on cedar. Probably refers to *E. prunastri* and *Pseudevernia furfuracea*, with *E. prunastri* being preferred (Richardson 1974). Used in ointments for an astringent; decoction used either hot or cold for washing the vulva for diseases of the womb; and used as a remedy against fatigue (López Eire et al. 2006)
Europe (early modern era) *oak moss*; *mousse chêne*; or *eichenmoss*	Ancient Greek uses adopted in Europe starting in 1500s, but originally applied to all fruticose arboreal lichens. By late 1700s, these uses often applied to *Usnea* spp., but sometimes to *Evernia prunastri*. Used for uterine and anal prolapse, and for preventing abortion (Quincy 1724; Willemet 1787; Luyken 1809). Also used for intestinal weakness, fevers, and pulmonary afflictions (Lindley 1838; Uphof 1959). Mixed with *Pseudevernia furfuracea* and *Hypogymnia physodes* as *lichen quercinus virdes*, a popular drug in early modern Europe (Senft 1911; Llano 1948).
Morocco لحية الشيخ (*lahiat alshaykh*, "sheikh's beard": Arabic); *mousse de chêne* (French)	Unspecified medicinal use (Doukkali et al. 2015)

(continued)

Table 2.10 (continued)

Culture and *folk name*	Traditional use
Guezoul (Tiaret, Algeria) *lahiate chibani*; *mousse du chêne* (French)	Used to treat coughs and to stimulate the appetite (Miara et al. 2013)
***Hypogymnia hypotrypa* (Nyl.) Rass.**	
China 黄袋衣 (*huáng dài yī*), 粉黄袋梅衣 (*fěn huáng dài méi yī*)	Used for dim vision, bleeding from uterus, bleeding from external injury, chronic dermatitis, and sores. Drink decoction with 3–9 g lichen one time; or apply decoction or powdered lichen to the affected area (Wang and Qian 2013)
***Hypogymnia physodes* (L.) Nyl. [syn. *Parmelia physodes*]**	
Europe (early modern era)	Mixed with *Evernia prunastri* and *Pseudevernia furfuracea* as *lichen quercinus virdes*, a popular drug in early modern Europe (Senft 1911; Llano 1948).
Neshnabé (WI, USA) *wa'kwûnûk* ("egg bush")	Eaten raw as a cure for constipation (Smith 1933)
***Letharia vulpina* (L.) Hue [syn. *Evernia vulpina*]**	
Sweden *ulf-mossa*	Used for wolf poison in Sweden. Pulverized, mixed with fat and flesh, warmed in pan over fire, and fresh blood and cheese added to create an odor. Sometimes mixed with powdered glass or strychnine. Put under skin of carcass, wolf dies within 24 h of ingestion. Older, drier lichen is more potent (Withering 1801; Schade 1954)
Niitsitapii (Alberta, Canada) *e-simatch-sis* ("yellow dye")	Infusion of the lichen and bone marrow for stomach disorders like ulcers. Lichen was blackened in a fire and rubbed on a rash, eczema, and wart sores (McClintock 1910; Hellson and Gadd 1974)
Syilx (BC, Canada) *kwarē'uk*; or *kwerníkw*	Weak decoction drunk for internal problems and strong decoction used to wash external sores and wounds (Teit and Boas 1928; Turner et al. 1980)
Umatilla and Cayuse (Oregon, USA) *laxpt*; or *maqa'hl*	Boiled and then applied as a poultice for open sores, boils, bruises, swellings, arthritis, and eye problems. Also used for saddle sores on horses. Liquid also drunk for hemorrhaging (Hunn 1990, 2005)
Achomawi (California, USA)	Used for poison arrows. Tips imbedded in masses of wet lichen and left for up to a year, rattlesnake venom sometimes added (Merriam 1966)
Yuki and Wailaki (CA, USA) *ōl-gät'-i*	Medicine for inflammation and to dry up running sores (Chesnut 1902; Mead 1972)
***Pseudevernia furfuracea* (L.) Zopf [syn. *Evernia furfuracea*, *Borrera furfuracea*]**	
Ancient Egypt	Found in a vase in a tomb dated to the fourteenth to sixteenth century BCE, along with other medicinal plants (Müller 1881). Used, along with other botanicals, to stuff mummies (Baumann 1960). Does not currently grow in the area, potentially imported from elsewhere.

(continued)

Table 2.10 (continued)

Culture and *folk name*	Traditional use
Ancient Greece	See Ancient Greek use of *Evernia prunastri*.
Tripolitania (Libya) *scíba*	Ingredient in medicinal decoction called *scíba* from the early 1900s, along with *Usnea plicata*, *Ramalina calicaris*, and *R. farinacea* (Natale and Pollio 2012)
Europe (early modern era) *treemoss*; *mousse d'arbre* (French)	A substitute for quinine (Willemet 1787). Used for fevers and as an astringent (Lindley 1838). Mixed with *Evernia prunastri* and *Hypogymnia physodes* as *lichen quercinus virdes*, a popular drug in early modern Europe (Senft 1911; Llano 1948).
Kutahya (Turkey)	Used for wounds, eczema, and hemorrhoids. Put in healing cream with clay (Güvenç et al. 2012)
Pallars (Catalonia, Spain) *liquen*	Drunk as tea for asthma, congestion, and high blood pressure (Agelet and Vallès 2003)
Alfacar, Víznar (AN, Spain) *musgo*	Used for respiratory ailments. Washed, boiled for a considerable time, then drunk (González-Tejero et al. 1995)
Jaén (AN, Spain) *líquen de pino*, or *muedos*	In Villanueva del Arzobispo, it is collected and sold for very good medicines (Fernández Ocaña 2000)

Table 2.11 Physciaceae used in traditional medicines around the world

Culture and *folk name*	Traditional use
***Anaptychia* spp. Körber**	
Ayurvedic medicine (India)	Occasional adulterant in *charila* (see *Parmotrema nilgherrense*) (Chanda and Singh 1971)
***Anaptychia ciliaris* (L.) Körb. ex A. Massal.**	
Erkowit (Sudan) *bakour*	Mixed with other plants and burned to clean air and repel insects (Khalid et al. 2012; Adam et al. 2018)
***Heterodermia diademata* (Taylor) D. D. Awasthi**	
Nepali (Sikkim, India) *dhungo ku seto jhua*	Used for cuts and injuries. Leaves of *Ageratina adenophora* are made into paste and put on cuts and then plastered with paste of lichen thalli to protect it from water and any other infection (Saklani and Jain 1994).
Taplejung, Solukhumbu, and Gorkha (Nepal) *jhyauu* or *jhulo* (Nepali); *yangben* (Limbu, Rai); *lunhokva* (Bantawa Rai); *myann* (Sherpa); *shingdrak* (Himali Bhotia)	To treat fresh wounds and to stop bleeding: the lichen is mixed with extracts or juices of *Artemisia vulgaris* or *Eupatorium odoratum*. As an antiseptic and healing agent for cattle: the lichen is powdered and applied externally on fresh cuts and wounds (only in Gorkha) (Devkota et al. 2017)

(continued)

Table 2.11 (continued)

Culture and *folk name*	Traditional use
***Heterodermia galactophylla* (Tuck.) W.L. Culb.**	
Pankararu (Brazil) *flor de pedra*	Part of a medicinal mixture of *flor de pedra* (see *Parmotrema wrightii*) (Londoño-Castañeda et al. 2017)
***Physcia* spp. (Schreber) Michaux**	
Europe (early modern era) *usnea cranii humani*	See European use of *usnea cranii humani* under *Parmelia saxatilis* (Llano 1948)

Table 2.12 Ramalinaceae used in traditional medicines around the world

Culture and *folk name*	Traditional use
***Ramalina* spp. Ach.**	
Ancient Greece Λειχήν (*leikhēn*) or βρύον (*brýon*, "moss")	A cryptogam growing on wet rocks is recorded by Dioscorides in section 4.53 of *De Materia Medica* (50–70 CE). Could be any saxicolous lichen or bryophyte. Early European herbals interpret it as a thalloid liverwort (e.g., de L'Obel 1576). Recent interpretation is as *Ramalina* sp., as this matches with the original drawing (López Eire et al. 2006; Yavuz 2012). Applied as a poultice, it stops bleeding, relieves inflammation, and cures *lichen* (the skin disease). Mixed with honey it cures jaundice, and smeared on the mouth and tongue, it relieves colds and congestion. Pliny records a similar cryptogam in his *Naturalis Historia* (77 CE) that is dry, is white, and grows on rocks near streams. It is put on wounds to stop bleeding and used to cure jaundice and impetigo. It has been interpreted as a thalloid liverwort, *Peltigera canina*, or *Ochrolechia parella* (Bostock and Riley 1855; Yavuz 2013).
Taplejung (Nepal)	Same as Taplejung use of *Hypotrachyna nepalensis* (topical antiseptic) (Poudel 2008; Devkota et al. 2017)
Kanikkars (Tamil Nadu, India) *kalchadai*	Used in combination with dried elephant milk, silt stone, and *Cuminum cyminum* extract to cure small pox (Nagendra Prasada et al. 1996)
***Ramalina bourgeana* Mont. ex Nyl. [syn. *Niebla bourgeana*]**	
Almeria (Spain) *flor de piedra* ("stone flower")	Decoction used as diuretic to treat renal lithiasis. One cup taken daily until patient is better (González-Tejero et al. 1995; Martínez-Lirola et al. 1996)
***Ramalina calicaris* (L.) Fr.**	

(continued)

Table 2.12 (continued)

Culture and *folk name*	Traditional use
Tripolitania (Libya) *scíba*	Ingredient in medicinal decoction called *scíba*, along with *R. farinacea*, *Usnea plicata*, and *Pseudevernia furfuracea* (Natale and Pollio 2012)
***Ramalina capitata* (Ach.) Nyl.**	
Pallars (Spain) *liquen*	Drunk as tea to relieve symptoms of asthma (Agelet and Vallès 2003)
***Ramalina conduplicans* Vain.**	
Yi, Dai, and Han (s. Yunnan, China) 硬枝树花 (*yìng zhī shù huā*), 树花 (*shù huā*, "tree flower"), 树须 (*shù xū*, "tree beard"), 石花菜 (*shí huācài*, "stone cauliflower")	Cold dish served at marriage banquets, couples who eat it will love each other more and never separate. Boiled in water with soda for 10–20 min, soaked in new water for 1–2 days, and served with chili powder, salt, and other seasonings (Wang et al. 2001). Medicine to reduce inflammation (Wang and Qian 2013).
***Ramalina farinacea* (L.) Ach.**	
Ayurvedic medicine (India)	Occasional adulterant in *charila* (see *Parmotrema nilgherrense*) (Shah 1998)
Tripolitania (Libya) *scíba*	Ingredient in medicinal decoction called *scíba*, along with *R. calicaris*, *Usnea plicata*, and *Pseudevernia furfuracea* (Natale and Pollio 2012)
Nigeria	Aqueous extract for treating mental disorders. Tinctures for treatment of ringworm tinea (Esimone and Adikwu 1999)
***Ramalina inflata* Hooker f. and Taylor**	
Ayurvedic medicine (India)	Same as Ayurvedic use of *Usnea baileyi* (mixed in tobacco) and an occasional adulterant in *charila* (see *Parmotrema nilgherrense*) (Shah 1998)
***Ramalina menziesii* Taylor**	
Pomo (California, USA)	Used as baby diapers (Goodrich et al. 1980)
***Ramalina roesleri* (Hochst.) Hue**	
China 肉刺树花 (*ròucì shù huā*)	Used for traumatic injuries, bleeding, and swelling (Wang and Qian 2013)
***Ramalina sinensis* Jatta**	
Kumaon (India) *jhulla*	Used to treat colds and coughs (Singh et al. 2012)
Yunnan (China) 中国树花 (*Zhōngguó shù huā*)	Same as Chinese use of *R. conduplicans* (Wang et al. 2001)
***Vermilacinia flaccescens* (Nyl.) Spjut and Hale [syn. *Niebla flaccescens*]**	
Quechua (Pampallacta, Peru) *papel-papel*	Drink infusion for coughs. Thallus also chewed with coca leaves for magic rituals (Velasco-Negueruela et al. 1995)

Table 2.13 Stereocaulaceae used in traditional medicines around the world

Culture and *folk name*	Traditional use
Stereocaulon exutum **Nyl.**	
China 裸珊瑚枝 (*luǒ shānhú zhī*, “bare coral branch”), 石寄生 (*shí jìshēng*, “stone parasite”)	Same as Chinese use of *S. paschale* (Wang and Qian 2013)
Stereocaulon himalayense **Asahina and I.M. Lamb**	
Lepchas (Darjeeling, India) *dhungo-ku-jhau*	Thalli pounded and boiled in water; take 100 mL twice daily after meals for burning sensation when urinating or other urinary trouble; decoction also used for tongue blisters (Saklani and Jain 1994)
Himachal (India) *chanchal*	Decoction used to treat urinary infections. Entire lichen boiled in water or goat’s milk (Sharma 1997)
Stereocaulon paschale **(L.) Hoffm.**	
Mistissini Cree (Quebec) *wapskirnok*	Used to treat rheumatism/arthritis associated with diabetes (Fraser 2006; Leduc et al. 2006)
China 东方珊瑚枝 (*Dōngfāng shānhú zhī*, “Oriental coral branch”), 石寄生 (*shí jìshēng*, “stone parasite”)	Used for spontaneous external bleeding, other bleeding, and dizziness. Drink decoction. Other names include 指状珊瑚衣 (*zhǐ zhuàng shānhú yī*), 东方珊瑚衣 (*Dongfang shanhú yi*), 东方衣 (*Dōngfāng yī*) (Hu et al. 1980; Wang and Qian 2013)
Stereocaulon vulcani **(Bory) Ach.**	
Réunion *fleur de roche*; or *fleur galet*	Boiled to treat ulcers. Roasted and used to treat cankers. Mixed with sulfur, *Hubertia ambavilla* [endemic shrub], and coconut oil to make an ointment for wounds. Used in a drink to stop vaginal discharges. Boiled in water with a handful of *Hylocereus undatus* roots [cactus], *Tribulus cistoides*, and a piece of *Argemone mexicana* root [poppy] and drunk to treat syphilis (Lavergne 1989)

Table 2.14 Collemataceae used in traditional medicines around the world

Culture	Traditional use
Leptogium **spp. (Ach.) Gray**	
Ayurveda (India)	Occasional adulterant in *charila* (see *Parmotrema nilgherrense*) (Chanda and Singh 1971)

Table 2.15 Lobariaceae used in traditional medicines around the world

Culture and *folk name*	Traditional use
***Crocodia aurata* Link (Galloway and Elix) [syn. *Pseudocyphellaria aurata*]**	
Ambavaniasy (Madagascar)	Used as tea to treat indigestion (Sharnoff 1997)
***Dendriscosticta wrightii* (Tuck.) B. Moncada and Lücking [syn. *Sticta wrightii*]**	
China 深杯牛皮叶 (*shēn bēi niúpí yè*), 老龙皮 (*lǎo lóng pí*, "old dragon skin"), 宽叶牛皮叶 (*kuān yè niúpí yè*)	Used for indigestion, and edema from kidney inflammation (Wang and Qian 2013)
***Lobaria* spp. (Schreber) Hoffm. [partial syn. *Sticta* spp.]**	
Bhutan	Pulverized and made into a paste to cure skin diseases (Søchting 1999)
Northwest Yunnan (China) *qingwapi*	Whole plant used to treat indigestion (Ju et al. 2013)
Gitksan (BC, Canada) *gwilalh ganaaw* ("frog blankets")	Used as arthritis medicine, a tonic, and a spiritual health-promoting and purification treatment. Aqueous infusion used as tea or a bath (Johnson 1997)
Haida (BC, Canada) *k̲ayd gyaa'ad* ("tree blanket")	Ingredient in several different medicinal mixtures. Also called *hlk'inx̲a k̲wii'awaay* ("forest cloud"); or *xil k̲wii.awaa* ("cloud leaves") (Turner 2004a)
Nuxalk *sts'wakt-aak*	Used for stomach pains, but not diarrhea, constipation, or vomiting. Only collected from *Cornus stolonifera* [dogwood] or *Pyrus diversifolia* [crabapple], boiled, and five cups of hot decoction are drunk daily. Decoction also used as an eyewash. Also, plant pulverized and applied to skin (Smith 1929; Turner 1973)
Makah (Washington, USA) *didi'dichia* ("growing on rocks")	When found on rocks it is used for running sores that are hard to heal, especially sores on the leg caused by bruises from walking among rocks (Densmore 1939). The identity of this lichen is uncertain: most *Lobaria* and *Sticta* species grow on trees
***Lobaria isidiosa* (Müll. Arg.) Vain.**	
China 裂芽肺衣 (*liè yá fèi yī*), 老龙皮 (*lǎo lóng pí*, "old dragon skin"), 珊瑚芽肺衣 (*shānhú yá fèi yī*)	Used for indigestion, reducing inflammation, relieving pain, burns and scalds, edema due to kidney inflammation, and malnutrition in children (Hu et al. 1980; Wang and Qian 2013)
***Lobaria kurokawae* Yoshim.**	
China 光肺衣 (*guāng fèi yī*), 老龙皮 (*lǎo lóng pí*, "old dragon skin"), 石龙衣 (*shí lóng yī*, "stone dragon clothes"), 老龙七 (*lǎo lóng qī*, "old dragon seven")	Same as Chinese use of *L. pulmonaria*, but not used for severe itching of skin (Wang and Qian 2013)
***Lobaria orientalis* (Asahina) Yoshim.**	
China 东方肺衣 (*Dōngfāng fèi yī*), 青蛙皮 (*qīngwā pí*, "frog skin")	Same as Chinese use of *L. pulmonaria* (Wang and Qian 2013)

(continued)

Table 2.15 (continued)

Culture and *folk name*	Traditional use
Lobaria oregana **(Tuck.) Müll. Arg.**	
Gitga'at (BC, Canada) *nagaganaw* ("frog dress")	Boiled with juniper and used as medicine for sore throats. Best for medicine if collected off *Abies lasiocarpa* [fir] (Turner and Thompson 2006)
Lobaria pulmonaria **(L.) Hoffm.**	
Europe (early modern era) *muscus pulmonarius* (Latin); *lungwort, lungs of oak, oak lung* (English); *hazelraw* (Scotland); *crotal coille* (Ireland); *lungenkraut* (German)	Its use for lung ailments goes back at least as far as the late 1400s (von Kaub 1485) and was widespread throughout Europe through to the 1600s (de L'Obel 1576; Parkinson 1640; Ray 1686). Its popularity then waned, only being used in certain areas like the Scottish Highlands and New Forest (England), but many authors remained convinced of its efficacy (Watson 1756; Withering 1801; Pereira 1850; Wise 1863; Cameron 1900; de Crespigny and Hutchinson 1903). It was mainly used in lung ailments (e.g., tuberculosis, asthma, coughs, spitting blood), but also for liver diseases, as an appetite stimulant, for diarrhea, for heavy menstrual flow, to stop bleeding, and for jaundice. It was usually boiled with water or milk and drunk, or made into an ointment for external use. It was also used for lung ailments in livestock in England, Germany, and Sweden (De Grey 1639; Willemet 1787; Drummond 1861).
Molise (Italy)	Applied to cuts as an antiseptic and healing agent (Guarrera et al. 2008)
Turkey	Used as a laxative cream and for treatment of respiratory disorders (Atalay et al. 2014)
India *golmataghosa* (Bengal)	Used for hemorrhages, lung troubles, asthma, and strengthening hair. The hill men use it for curing eczema on the head and cleaning hair (Biswas 1956)
Afghanistan *gul-i-sang* ("stone flowers")	Applied to newborn child's navel to dry and heal wound. Used as contraceptive, four different methods: (1) consume the lichen with water during menstrual period (Kabul); (2) dry, grind, and pop the resulting powder into the mouth like snuff for 3 days during menstrual period (Kunduz); (3) grind and consume 24 h after giving birth (Kabul); (4) men consume the lichen (Kabul) (Hunte et al. 1975)
China 肺衣 (*fèi yī*), 兜衣 (*dōu yī*), 蛤蚂七 (*há mă qī*), 石龙皮 (*shí lóng pí*, "stone dragon skin"), 石龙衣 (*shí lóng yī*, "stone dragon clothes"), 牛皮叶 (*niúpí yè*, "leather leaf")	Used for indigestion, malnutrition in children, abdominal distention, ascarid infestation, burns and scalds, edema due to kidney inflammation, local swelling, reducing inflammation, relieving pain, and severe

(continued)

Table 2.15 (continued)

Culture and *folk name*	Traditional use
	itching of skin. Drink decoction or apply powder to the affected area (Hu et al. 1980; Wang and Qian 2013)
Manila (Philippines)	Added to jelly made from a red algae (possibly *Hydropuntia edulis* or *Betaphycus gelatinus*) and sugar and used to treat chest ailments (von Martens 1866)
Nlaka'pamux (BC, Canada) *ʔes-tə/kʷl'-ít tək p'ə·/p'éy'łe tək /qʷzém* ("yellowish frog moss")	Previously used medicinally, details forgotten (Turner et al. 1990)
Coast Tsimshian (BC, Canada)	Used medicinally (Johnson 2006)
Hesquiat (BC, Canada) *ƛacƛastuphc̓um*	Applied to the faces of children when their skin is peeling. Also used as medicine for coughing up blood (Turner and Efrat 1982)
Saanich (BC, Canada)	Same as Saanich use of *Parmelia sulcata* (possible birth control) (Turner and Hebda 2012)
Cuijingo (Puebla, Mexico)	Medicine for lung ailments (Hersch-Martínez 1997)
Lobaria quercizans **Michaux [syn.** ***Sticta glomulifera*** **in N.A.]**	
Menominee (Wisconsin, USA) *wakûn*	Eaten as a tonic and as medicine for run-down systems. Only picked off hard maple or hemlock trees. Cooked in soups (Smith 1923)
Lobaria retigera **(Bory) Trevis.**	
China 网肺衣 (*wǎng fèi yī*), 老龙皮 (*lǎo lóng pí*, "old dragon skin"), 蛇皮苔 (*shé pí tái*, "snake skin moss")	Same as Chinese use of *L. pulmonaria*. Other names include 石龙皮 (*shí lóng pí*, "stone dragon skin"), 石龙衣 (*shí lóng yī*, "stone dragon clothes"), 癞肚皮 (*lài dùpí*), 石花 (*shí huā*, "stone flower") (Hu et al. 1980; Wang and Qian 2013)
Lobaria sublaevis **(Nyl.) Yoshim.**	
China 亚平肺衣 (*yà píng fèi yī*), 老龙皮 (*lǎo lóng pí*, "old dragon skin")	Used for indigestion, edema, inflammation, and pain relief (Wang and Qian 2013)
Lobaria yunnanensis **Yoshim.**	
China 云南肺衣 (*Yúnnán fèi yī*); 青蛙皮 (*qīngwā pí*, "frog skin": nw Yunnan)	Same as Chinese use of *L. pulmonaria* (Wang and Qian 2013)
Lobaria virens **(With.) J.R. Laundon [syn.** ***Lobaria laetevirens*****]**	
Europe (early modern era)	Occasionally listed in old European pharmacopoeias (Gioanetto 1993)
Sticta **spp. (Schreber) Ach.**	
Makah (Washington, USA)	See Makah use of *Lobaria* spp. (Densmore 1939)
Nuxalk	See Nuxalk use of *Lobaria* spp. (Smith 1929)

Table 2.16 Nephromataceae used in traditional medicines around the world

Culture and *folk name*	Traditional use
Nephroma arcticum **(L.) Torss.**	
Yup'ik (Alaska) *kusskoak*	Infusion with hot water is fed to a person in weak condition to make him strong, a very effective medicine (Oswalt 1957)

Table 2.17 Peltigeraceae used in traditional medicines around the world

Culture and *folk name*	Traditional use
Peltigera **spp. Willd.**	
Dena'ina (Alaska, USA) *k'udyika'a*	Decoction drunk for tuberculosis and prolonged bleeding. *Umbiliaria* spp. are also used (Kari 1987)
Haida (BC, Canada) *hlk'inx̲a̲ k̲wii'awaay* ("forest cloud") or *xil k̲wii.awaa* ("cloud leaves")	Ingredient in several different medicinal mixtures (Turner 2004a)
Oweekeno (BC, Canada) *x̌xwpı̓gà*	Thallus pounded, mixed with spruce pitch, and used to dress wounds (Compton 1993)
Ditidaht (BC, Canada) *ƛaƛx̣ʔa·ʔaq* ("flat against the rock") or *ƛ̓i·ƛ̓i·dqʷaqsibak̓kʷ* ("resembles whale baleen")	A grey *Peltigera* growing on rocks that was used to induce urination. Picked, washed, squashed, and eaten (Turner et al. 1983)
Navajo (NM, USA) *nihaλá·d* ("earth moss")	May be chewed like *Xanthoparmelia* sp. for cankers, swollen gums, and decayed teeth (Wyman and Harris 1941)
Peltigera aphthosa **(L.) Willd.**	
China 绿皮地卷 (*lǜ pí de juǎn*), 老龙皮 (*lǎo lóng pí*, "old dragon skin")	Used to improve digestion (Wang and Qian 2013)
Europe (early modern era)	As early as the 1700s, it was used as medicine to treat thrush (mouth ulcers) in children. Make an infusion in milk and drink. It induces vomiting in large doses. Also used to expel worms (Gedner 1756; Strandman 1769; Willemet 1787; Withering 1801; Luyken 1809; Lindley 1838)
Tlingit (Alaska, USA)	Dried, powdered, and used to treat burns and scalds (Emmons 1991)
Ditidaht (BC, Canada) *tı̓tı̓diččʔa·* ("rocks growing on rocks")	Chewed and eaten for tuberculosis. Also used as poultice for sores on the legs (Turner et al. 1983)
Nlaka'pamux (BC, Canada) *p'ə·/p'éy'łe tək /qʷzém* ("frog moss") or *p'ə·/p'éy'łeh=éy'st* ("frog's rocks")	Used to rub on beestings (Turner et al. 1990)
Peltigera britannica **(Gyelnik) Holt.-Hartw. and Tønsberg**	
Ditidaht (BC, Canada)	Possibly same as Ditidaht use of *P. aphthosa* (Turner et al. 1983)

(continued)

Table 2.17 (continued)

Culture and *folk name*	Traditional use
***Peltigera canina* (L.) Willd.**	
Britain *lichen cinereus terrestris* (Latin); *dog lichen* or *ash-colored ground liverwort* (English); *lus ghoinnich* or *gearan* (Gaelic, from Cameron 1900)	A plant called *the star of the earth* was used as a cure for rabies in Britain as early as the 1600s. In the oldest record, this is definitely a vascular plant (De Grey 1639), but later authors decided that it was *P. canina* (Gourdon 1687; Dampier and Sloane 1698). The remedy was popularized in 1720 by Dr. Mead and in the London Pharmacopoeia for 1721. It enjoyed a short period of renown (Mortimer 1735; Hartley et al. 1737), before people began to become suspicious of its efficacy (Steward 1738; Ranby and Peters 1744; Layard 1757; Lightfoot 1777). It was still being used in some areas in Wales in the early 1800s (Trevelyan 1909; Allen and Hatfield 2004). Dried lichen and black pepper were pulverized and mixed into warm milk. This remedy was called *pulvis antilyssus*.
Lithuania *meškapėdė šunvaistė*	Name means "dog's medicine bear paw" (Kolosova et al. 2017)
China 犬地卷 (*quǎn de juǎn*)	Used for rabies and jaundice (Wang and Qian 2013)
India *ghosa* (Bengali)	Used for rabies and jaundice in India (Biswas 1956); Tonic and medicine for liver complaints in the Himalayas (Subramanian and Ramakrishnan 1964)
Hesquiaht (BC, Canada)	Unspecified medicine (Turner and Efrat 1982)
***Peltigera membranacea* (Ach.) Nyl. [syn. *Peltigera canina* var. *membranacea*]**	
Kwakwaka'wakw (BC, Canada) *tl'extl'ekw'és* ("seaweed of the ground")	Used as a love charm (Boas 1921)
***Peltigera polydactylon* (Neck.) Hoffm. [syn. *Peltigera polydactyla*]**	
Himalayas (Sikkim, India) *Jhau* (Lepchas)	Used as antiseptic and to stop bleeding, thalli made into paste and put on cuts (Lepchas). Used to treat unspecified respiratory problems (Saklani and Jain 1994; O'Neill et al. 2017)
Himachal Pradesh (India) *sharda*	Lichen is washed, pounded, and boiled in goat's milk; the resulting mash is soaked in cow's urine to be used as an antiseptic over cuts and bruises (Sharma 1997)
China 多指地卷 (*duō zhǐ de juǎn*), 多指地卷 (*duō zhǐ de juǎn*), 牛皮叶 (*niúpí yè*, "leather leaf"), 地卷草 (*de juǎn cǎo*)	Used for traumatic injuries and to strengthen the constitution (Wang and Qian 2013)
Iroquois (Ontario, Canada)	Tea used to induce vomiting and as an anti-love medicine. Either makes loved one return or unbewitches you (Herrick 1995)

Table 2.18 Teloschistaceae used in traditional medicines around the world

Culture and *folk name*	Traditional use
***Polycauliona candelaria* (L.) Frödén, Arup, and Søchting [syn. *Xanthoria candelaria*]**	
Europe (early modern era)	Boiled with milk to treat jaundice, along with *Xanthoria parietina* (Tonning 1769).
***Rusavskia elegans* (Link) S. Y. Kondr. and Kärnefelt [syn. *Xanthoria elegans*]**	
Afghanistan *rabooch* (ɾa'bɔtʃ, "stone dirt") (Wakhi); *engelchek* (энгелчек) (Kyrgyz)	Ground and applied to infected wounds for humans and livestock (Wakhi). Ground and applied to wounds on livestock; or ground, mixed in butter, and fed to livestock (especially yak calves) with diarrhea (Kyrgyz). Other lichen species may also be used (Soelberg and Jäger 2016)
***Teloschistes flavicans* (Sw.) Norm.**	
China 黄枝衣 (*huáng zhī yī*, "yellow branch clothes"), 赛桂 (*sài guì*, in Tibet)	Used to clear *heat* in the lung and liver, and to remove toxins (Wang and Qian 2013)
***Xanthomendoza fallax* (Hepp ex Arnold) Søchting, Kärnefelt and S. Y. Kondr [syn. *Xanthoria fallax*]**	
China Lhasa (Tibet)	Used in traditional Tibetan medicine (Salick et al. 2006)
***Xanthoria parietina* (L.) Th. Fr.**	
Andalusia (Spain) *flor de piedra* ("stone flower") or *rompepiedra* ("stone breaker")	Decoction in wine for menstrual complaints (Campohermoso). Decoction in water for kidney disorders (Barranquete, Cueva de los Medinas, Joya, Pozo de los Frailes, and Puebloblanco). Decoction in water for toothaches (Fernan Pérez and Joya). An analgesic for several pains (Fuente del Escribano). Ingredient in a cough syrup with *Ceratonia siliqua, Ficus carica*, and *Prunus amygdalus* fruits; *Olea europaea* and *Origanum vulgare* leaves and flowers; and lots of sugar or honey (San Isidro Jiménez) (González-Tejero et al. 1995)
Europe (early modern era) *common yellow wall moss*	Boiled with milk to treat jaundice, along with *Polycauliona candelaria* (Tonning 1769). Used for diarrhea and dysentery, to stop bleeding, and as a quinine replacement for malaria (Willemet 1787; Luyken 1809; Lindley 1838; Pereira 1850; Lebail 1853). Also used for hepatitis (Gioanetto 1993).
China 石黄衣 (*shí huáng yī*, "stone yellow clothes")	Used medicinally as an antibacterial (Wang and Qian 2013)

Table 2.19 Roccellaceae used in traditional medicines around the world

Culture and *folk name*	Traditional use
***Roccella* sp.**	
Ancient Greece *φῦκος θαλάσσιον (phŷkos thalássion*, "marine phycos"), *ballaris*, *irane*, or *gnomeusilum*	A cryptogam growing on seashore rocks is recorded by Dioscorides in section 4.99 of *De Materia Medica* (50–70 CE). Recommended for inflammations and gout in the feet that needs to be reduced (López Eire et al. 2006). Possibly a marine algae, but identified by Richardson (1974) as *Roccella* sp.
Sicuani (Peru)	Two *Roccella* sp. sold in indigenous market: one for coughs and one for fever (Sharnoff 1997)
***Roccella decipiens* Darb. [syn. *Roccella babingtonii*]**	
Seri (Sonora, Mexico) *heecoj* or *?èekkiix*	Tea: shortness of breath and fever. Ground, moistened, and strained: filtrate put on a burn or sore. Ground with clay and water: fever and diarrhea. Ground and mixed with water: bathe child with fever (Felger and Moser 1985)
***Roccella fuciformis* (L.) DC.**	
Pondicherry (India) *mathaghasa* ("to rub on skull")	Used to clean hair and cure eczema on the skull and back or the ear (Biswas 1947)
***Roccella phycopsis* Ach. [*Roccella tinctoria*]**	
France *orseille* or *orchal*	Remedy for tickling in the throat (France). Used in Mauritius for a medicinal broth (may refer to all fruticose lichens) (de Candolle 1816; Lebail 1853)
Madras (India)	Unspecified drug (Biswas 1947)

Table 2.20 Ophioparmaceae used in traditional medicines around the world

Culture and *folk name*	Traditional use
***Ophioparma lapponica* (Räs.) R. W. Rogers and Hafellner**	
China 拉普兰红盘衣 (*Lāpŭlán hóng pán yī*), 岩石赤星衣 (*yánshí chì xīng yī*)	Same Chinese use as *O. ventosa* (Wang and Qian 2013)
***Ophioparma ventosa* (L.) Norman**	
China 红盘衣 (*hóng pán yī*), 石霜 (*shí shuāng*, "stone frost"), 赤星衣 (*chì xīng yī*)	Used externally to stop bleeding from external injury, relieve pain (Wang and Qian 2013)

Table 2.21 Umbilicariaceae used in traditional medicines around the world

Culture and *folk name*	Traditional use
***Lasallia papulosa* (Ach.) Llano [syn. *Umbilicaria papulosa*]**	
Ekuanitshit (Quebec, Canada) *uâkuanâpishku*	Tea used for urinary problems (Clément 1990; Uprety et al. 2012)
***Umbilicaria* spp. Hoffm.**	
Dena'ina (Alaska, USA) *qalnigi jegha* ("rock ear")	Decoction drunk for tuberculosis and prolonged bleeding. Also used *Peltigera* spp. (Kari 1987)
Inuit (Quebec, Canada)	Used as a tea to treat tuberculosis (Stevens et al. 1984; Sharnoff 1997)
***Umbilicaria esculenta* (Miyoshi) Minks [syn. *Gyrophora esculenta*]**	
Japan 石耳 (*iwatake*)	An esteemed food that promotes longevity when eaten (Kawagoe 1925; Sato 1968)
Korea 석버섯(*seogi-beoseot*, "stone ear mushroom"), 석(*seogi*, "stone ear")	Used to treat dysentery, anal fistula, stomachache, fever, night blindness, and cancer (Lee 1966; Kim et al. 2006)
China 美味石耳 (*měiwèi shí ěr*, "delicious stone ear"), 石耳 (*shí ěr*, "stone ear"), 石壁花 (*shíbì huā*, "stone wall flower"), 地耳 (*de ěr*, "earth ear"), 石木耳 (*shí mù'ěr*, "stone black fungus"), 岩萌 (*yán méng*, "cliff bud"), 灵芝 (*língzhī*), 胳衣 (*gē yī*)	Used for tuberculosis, spontaneous external bleeding, intestinal bleeding, rectal hernia into the vagina, bloody and cloudy urination, vaginal discharge, snakebites, and cuts. Drink decoction or apply externally to the affected area (Hu et al. 1980; Wang and Qian 2013)
***Umbilicaria hypococcinae* (Jatta) Llano**	
China 红腹石耳 (*hóng fù shí ěr*, "red-bellied stone ear"), 红石耳 (*hóng shí ěr*, "red stone ear"), 石耳子 (*shí ěr zi*, "stone ear's child"), 茁底石耳 (*zhuó dǐ shí ěr*), 红腹脐衣 (*hóng fù qí yī*)	Used for indigestion, distention, and pain in stomach duct and abdomen, dysentery, and malnutrition in children. Drink decoction (Wang and Qian 2013)
***Umbilicaria mammulata* (Ach.) Tuck.**	
Attikamekw (Quebec, Canada) *asine-wakunik*	During difficult childbirth the lichen is boiled and placed on a woman's stomach (Raymond 1945)
Nihitahawak Cree (Saskatchewan, Canada) *asinīwākon*	Made into soup as nourishment for sick person, as it will not upset the stomach. Lichen cleaned, broken into small pieces, and very hot water poured over it and water discarded. Lichen then added to fish broth and cooked 5–10 min, soup thickened as it cooled (Leighton 1985)
***Umbilicaria muhlenbergii* (Ach.) Tuck. [syn. *Actinogyra muhlenbergii*]**	
Chipewyan (Saskatchewan, Canada) *thetsíⁿ*	Used to expel tapeworms. Lichen is burned slightly in a frying pan, mashed well, and then boiled to make a syrup which is drunk. It can be chewed for the same purpose (Marles 1984; Marles et al. 2000)
Cree (Manitoba, Canada) *asinīwāhkona*, *wakoonak*, or *asinīwākon*	Decoction given to someone with a stomachache to "clean out the stomach" (Marles et al. 2000)

(continued)

Table 2.21 (continued)

Culture and *folk name*	Traditional use
Tłı̨chǫ (NWT, Canada) *kwechi*	Soup eaten as a tonic and for breathing problems (Rebesca et al. 1994; Uprety et al. 2012)
***Umbilicaria nanella* Frey et Poelt**	
China 小黑腹石耳 (*xiǎo hēi fù shí ěr*), 黑石耳 (*hēi shí ěr*, "black stone ear"), 黑脐衣 (*hēi qí yī*), 黑石耳子 (*hēi shí ěr zi*), 裂叶石耳 (*liè yè shí ěr*), 盘形石耳 (*pán xíng shí ěr*)	Used for indigestion, stomachache, dysentery, malnutrition in children, expelling ascarid parasites, vaginal discharges, glomus tumors, and reducing swelling. Drink decoction (Wang and Qian 2013)
***Umbilicaria vellea* (L.) Ach.**	
China 绒毛石耳 (*róngmáo shí ěr*, "fluffy stone ear")	Used for eye infections, bloody feces, and rectal hernia into the vagina (Wang and Qian 2013)

Table 2.22 Icmadophilaceae used in traditional medicines around the world

Culture and *folk name*	Traditional use
***Siphula* sp. Fr.**	
Northern Peru *pelo de piedra*	Unspecified medicine. Oral aqueous application (Bussmann and Sharon 2006a)
***Thamnolia subuliformis* (Ehrh.) W. Culb.**	
Nakhi (nw Yunnan, China) 雪地茶 (*xuě dì chá*, "snow ground tea"), 太白茶 (*tàibái chá*, "white tea"), 石白茶 (*shí bái chá*, "stone white tea"), 太白针 (*tàibái zhēn*, "white needle"),雪茶 (*xuě chá*, "snow tea"), 白雪茶 (*bái xuě chá*, "white snow tea")	Used for inflammation. Boiling water added to dry thalli in cup and infusion is drunk after 3–5 min. May be same as Nakhi use of *T. vermicularis* (Wang et al. 2001; Fu et al. 2005)
***Thamnolia vermicularis* (Sw.) Ach. ex Schaerer [syn. *Cladonia vermicularis*]**	
Nakhi (nw Yunnan, China) *xiare*, 地茶 (*de chá*, "earth tea"), 雪茶 (*xuě chá*, "snow tea"), 太白茶 (*tàibái chá*, "white tea"), 石白茶 (*shí bái chá*, "stone white tea"), 太白针 (*tàibái zhēn*, "white needle"), 太白菜 (*tàibái cài*, "white vegetable"), 蛔样地衣 (*huí yàng dìyī*), (*bái xuě chá*, "white snow tea")	A widely recognized medicinal plant, tea used to tranquilize the mind and clear *heat*. Also used for sunstroke, eye irritation, coughs, sore throat, inflammation, high blood pressure, fevers, epilepsy, and a decrease in vital energy. Boiling water added to dry thalli in cup and infusion is drunk after 3–5 min (Wang et al. 2001; Jiang et al. 2002; Fu et al. 2005; Salick et al. 2006; Byg et al. 2010; Ju et al. 2013; Wang and Qian 2013)
Ayurvedic (Uttarakhand and Himachal Pradesh, India) *swarn*	Germicide to preserve milk and other dairy products. Lichen is dried and burned, and milk is exposed to the smoke (Sharma 1997)
Bhotia (Uttarakhand, India) *chhai dhoop*	Used to preserve buttermilk. A handful of lichen is put in a wide cup containing burning coal and the smoke directed into the milk. It kills the 1–2 mm long white worms that grow in milk (Upreti and Negi 1996)

(continued)

Table 2.22 (continued)

Culture and *folk name*	Traditional use
South America ***contrayerba blanca***	Used to stimulate the stomach (Lindley 1838)
Sierra Nevada del Cocuy (Colombia) *neblilla* or *neblina*	Taken as infusion or chewed for toothache and asthma. Consumed fresh when walking in mountains to obtain energy for physically demanding activities, alleviate thirst, and cleanse the chest (Rodríguez et al. 2018)
Aymara (Bolivia) *wari kunka*	Infusion drunk for cough and hoarseness (Macía et al. 2005)

Table 2.23 Megasporaceae used in traditional medicines around the world

Culture and *folk name*	Traditional use
***Aspicilia* spp.**	
Tehran (Iran) شیر زاد (*sher ziad*), *chir zadi*, or *agalactie*	Ingredient in wine and medicinal compounds in ninth- to thirteenth-century Arabic writings (Crum 1993). Used to increase the flow of human milk (Hooper 1937). Likely includes numerous vagrant species, including *Aspicilia esculenta*, *A. jussuffii*, *A. fruticulosa*, and *A. hispida* (Sohrabi and Ahti 2010).

Table 2.24 Pertusariaceae used in traditional medicines around the world

Culture and *folk name*	Traditional use
***Lepra albescens* (Hudson) Hafellner [syn. *Pertusaria albescens*, *Variolaria discoidea*]**	
Europe (early modern era)	Used to treat malaria, along with *P. amara* (Lindley 1838)
***Lepra amara* (Ach.) Hafellner [syn. *Pertusaria amara*, *Variolaria faginea*]**	
Europe (early modern era)	Used to treat malaria, along with *P. albescens* (Lindley 1838)
***Pertusaria pertusa* (Weigel) Tuck. [syn. *Pertusaria communis*]**	
Europe (early modern era)	Cure for malaria, more effective for men. Also used for intermittent toothache, and powdered and used to kill worms (Lebail 1853)
***Varicellaria velata* (Turner) Schmitt and Lumbsch [syn. *Pertusaria velata*]**	
China 包被鸡皮衣 (*bāo bèi jī píyī*), 石箱 (*shí xiāng*, “stone box”)	Used to stop bleeding and relieve pain. External use only (Wang and Qian 2013)

Table 2.25 Verrucariaceae used in traditional medicines around the world

Culture and *folk name*	Traditional use
Dermatocarpon miniatum **(L.) W. Mann**	
China 皮果衣 (*pí guǒ yī*), 白石耳 (*bái shí ěr*, "white stone ear"), 黑石耳 (*hēi shí ěr*, "black stone ear"), 石耳子 (*shí ěr zi*, "young stone ear"), 岩菇 (*yán gū*, "rock mushroom")	Used for high blood pressure, as a diuretic, for expelling parasites, for malnutrition in children, for dysentery, for improving digestion, and for abdominal distention. Drink decoction or eat as soup (Wang and Qian 2013)

Table 2.26 Hygrophoraceae used in traditional medicines around the world

Culture and *folk name*	Traditional use
Dictyonema huaorani **Dal-Forno, Schmull, Lücking, and Lawrey**	
Huaorani (Amazon, Ecuador) *nęnęndapę*	Mixed with several species of bryophyte, made into an infusion, and drunk by shaman to cause hallucinations and call on *wenae* (malevolent spirits) to curse people. Also causes sterility (Davis and Yost 1983; Schmull et al. 2014)

Table 2.27 Unidentified lichens used in traditional medicines around the world

Culture and *folk name*	Traditional use
Xhosa (South Africa) *mthafathafa*	An unidentified rock lichen is used to treat gonorrhea. Fresh lichen is crushed and mixed with water, and infusion is drunk. Lichen also dried over fire and crushed, and powder is applied to wound's infected area (Matsiliza and Barker 2001). See also KhoiSan use of *Xanthoparmelia hottentotta*
Vhasenzi and Vhalemba (Venda, South Africa)	Unidentified lichen growing on *Englerophytum magalismontanum* tree is an ingredient in medicine. Also prepared and burnt to invoke ancestral spirits during *malombo* (Vhasenzi) or *mbila* (Vhalemba) dances (Mabogo 1990)
Tommo So (Dogonland, Mali) *nyengegi*	Unidentified epiphytic lichen/moss growing on *Adansonia digitata* trees is powdered, mixed with *Lannea microcarpa* seed oil, and then applied to deep boils (Inngjerdingen et al. 2004)
Trentepohlia jolithus [*Lepraria iolithus*]	A non-lichenized algae considered a lichen in early literature. Used for small pox and measles (Luyken 1809)
New Forest (England) *brighten*	An unidentified lichen is recommended for weak eyes (Wise 1863)
Slieve Aughty (Ireland) *dub-cosac*	An unidentified lichen is good for heart trouble (Allen and Hatfield 2004)
Saami (Sweden)	Unidentified ground lichens. A piece of sod probably containing lichens is rubbed on the affected area with an accompanying incantation to treat psoriasis, small pox, chicken pox, and eczema (DuBois and Lang 2013)
Ogliastra (Sardina, Italy) *lana d'arbore*	Unidentified arboreal lichen is applied externally as an antiseptic and to relieve foot perspiration and irritation (Loi et al. 2004)

(continued)

Table 2.27 (continued)

Culture and *folk name*	Traditional use
Brahuis (Balochistan, Pakistan)	An unidentified rock lichen that is extremely bitter is used medicinally in diseases of languor and oppression of the life force. The lichen is dried and crushed. They swallow the powder, and then drink water (Masson 1842; Hooper 1937)
Rotuma (Fiji) *rimi*	A grey lichen found on coconut tree trunks is used to make medicine used in treating high fevers and/or convulsions (McClatchey 1993)
Dena'ina (Alaska, USA) *sheh tsadn nde*	A large foliose lichen is used for coughs, tuberculosis, and general sickness. Boil and drink decoction. Also used for bleeding that won't stop (Garibaldi 1999)
Tlingit (Alaska, USA)	Lichens from the ground in the woods are used for sores. Crushed and then heated on rocks with seal oil and mountain goat tallow (de Laguna 1972)
Chipewyan (Alberta, Canada)	White crustose lichens on aspen bark, along with the dead tree periderm, are scraped off and put on cuts and deep wounds to stop bleeding (Marles et al. 2000)
Niitsitapii (Alberta, Canada)	Mixed with kinnikinnick leaves and shredded willow bark to make a smoking mixture (Russell 1973). Cited by Siegel (1989) who added the claim that it was narcotic (Siegel 2013, pers. comm.) and was then cited by Pollan (2001) who added the claim that it was hallucinogenic
Nihitahawak Cree (Saskatchewan, Canada)	White crustose lichens on aspen bark, along with the dead tree periderm, are scraped off and used to stop bleeding and to treat venereal disease (Leighton 1985)
Algonquin (Quebec, Canada)	White crustose lichens on birch bark used for diaper rash and other skin rashes (Black 1980)
Tewa (California, USA) *k̂uk'owà* ("rock skin"); *năŋ'a* ("earth clothing")	*k̂uk'owà* is pulverized and applied to lips for cold sores, rubbed on sores about a child's mouth, and put into the cavity of a decayed tooth to stop pain. *năŋ'a* is applied to teeth and gums to cure toothache (Robbins et al. 1916). See also Hopi use of *Xanthoparmelia* sp.
Northern Paiute (Nevada, USA) *tuh-botza-yo-caw-son*; or *lizard semen*	Black, orange, and yellow lichens on rocks are used as important antibiotics and fungicides. Powdered material is applied as a healing agent to sores, especially mouth sores of children (Train et al. 1941; Sharnoff 1997)
Western Shoshone (Nevada, USA) *timbe-boon-goo*	Black, orange, and green lichens on rocks. Diarrhea medicine: soak overnight in water and drink the solution. Smallpox medicine: powder and boil with *Purshia* leaves and dried mountain rat urine; drink half cup of solution morning and night (Train et al. 1941)
Southern Paiute (Nevada, USA) *timpapsuchicu*	Unidentified lichen is ground up, mixed with water, and used as poultice for sores on mouth and elsewhere (Stoffle et al. 1989, 1994)
Northern Paiute and Western Shoshone	Same use for unidentified lichen as Southern Paiute (Stoffle et al. 1989)
Hopi (Arizona, USA)	Yellow lichens on rocks are applied to cheeks to reduce swelling and relieve toothache (Beaglehole and Beaglehole 1935). See also Hopi use of *Xanthoparmelia* sp.

(continued)

Table 2.27 (continued)

Culture and *folk name*	Traditional use
Kewa Pueblo and Hispanics (New Mexico, USA) *yerba de la piedra* (Spanish)	Grey lichens are boiled until green and given to one who talks and laughs to himself. Also good for headaches (Kewa). Also rubbed on gums as cure for inflamed gums or powdered and applied on any kind of sore or injury (Hispanics) (Curtin 1965)
Ka'igwu (Oklahoma, USA)	Lichens on the north side of tree trunk are dried, powdered, and applied to sore gums for abscesses and teething infants. Also mixed with smoking tobacco for a mildly soporific effect (Vestal and Schultes 1939)
Seri (Sonora, Mexico) *hast yamása* ("rock lichen")	Grey foliose and orange crustose lichens on rocks are taken as a tea to induce vomiting (Felger and Moser 1985)
Huastec (Mexico) *tsakam k'uthay* ("little *Tillandsia usneoides*")	An unidentified arboreal lichen is used as an unspecified obstetrical-gynecological medicine and for bleeding (Alcorn 1984)
Lacadone (Chiapas, Mexico)	Unidentified lichens are invoked in magical healing of skin eruptions (Sharnoff 1997)
Quichua (Loja, Ecuador) *musgo de piedra*	There are seven different colours of lichens on rocks. If all seven colours are boiled in a drink, it will cure a person with a chronic illness who is about to die (Abel 2009, pers. comm.)
Loja (Ecuador)	An unidentified lichen is used for an unspecified medicine (Bussmann and Sharon 2006b)
Denís and Kinja (Amazonas, Brazil) *baduhu-tsinã* ("deer snuff")	An unidentified pyrenocarpous lichen on trees is used as a snuff. Yellow powder is collected off the surface of lichen for snuff. Used frequently and induces sneezing (Prance 1972; Milliken et al. 1992)
White crustose lichen Witoto/Bora (Loreto, Peru)	An unidentified white crustose lichen growing on *Rinorea racemosa* is sometimes used (along with other botanicals and ash) to add to the resin of *Virola sebifera* or *V. elongata* to make *oo'-koey*, a hallucinogenic orally ingested paste (Mckenna et al. 1984; UBC 2014)
Chácobo (Beni, Bolivia)	Five unidentified lichens are used to treat chest pain, appendix pain, headache, liver problems, and rheumatism (Boom 1987)
Aymara (Titicaca, Bolivia) *pampa untu* ("wild llama fat")	An unidentified lichen is given to babies as an infusion if they are constipated (La Barre 1948)

References

Abdel-Malek S, Bastien JW, Mahler WF et al (1996) Drug leads from the Kallawaya herbalists of Bolivia. 1. Background, rationale, protocol and anti-HIV activity. J Ethnopharmacol 50:157–166

Abel (2009) Interview with Abel, a jambiyachack (Quichua healer) of Ñamarín (Saraguro, Loja, Ecuador), conducted by S. Crawford on August 2019

Acharius E (1810) Lichenographia universalis. Apud I.F. Danckwerts, Gottingae

Adam M, Elhassan GOM, Yagi S et al (2018) In vitro antioxidant and cytotoxic activities of 18 plants from the Erkowit Region, Eastern Sudan. Nat Prod Bioprospect 8:97–105

Adams F (1847) The seven books of Paulus Ægineta. The Sydenham Society, London

Afolayan AJ, Grierson DS, Kambizi L et al (2002) In vitro antifungal activity of some South African medicinal plants. S Afr J Bot 68:72–76

Agelet A, Vallès J (2003) Studies on pharmaceutical ethnobotany in the region of Pallars (Pyrenees, Catalonia, Iberian Peninsula). Part III. Medicinal uses of non-vascular plants. J Ethnopharmacol 84:229–234

Ahmadjian V, Nilsson S (1963) Swedish lichens. American Swedish Historical Foundation, pp 38–47

Alcorn JB (1984) Huastec Mayan ethnobotany. University of Texas Press, Austin

Allen DE, Hatfield G (2004) Medicinal plants in folk tradition: an ethnobotany of Britain and Ireland. Timber Press, Portland

Amoreux PJ (1787) Recherches et Expérences sur les Diverses Especies de Lichens, Dont on peut faire usage en Médecine et dans les Arts. In: Hoffmann GF et al (eds) Mémoires sur l'utilité des lichens dans la médecine et dans les arts. Chez Piestre et Delamollière, Lyon, pp 1–103

Anonymous (1845) Pharmacopoea universalis. Landes Industrie Comptoir, Weimar

Articus K (2004) Phylogenetic studies in *Usnea* (Parmeliaceae) and allied genera. Comprehensive summaries of Uppsala Dissertations from the Faculty of Science and Technology 931. Acta Universitatis Upsaliensis, Upsala

Aston Philander L (2011) An ethnobotany of Western Cape Rasta bush medicine. J Ethnopharmacol 138:578–594

Atalay F, Halici MB, Mavi A et al (2014) Antioxidant phenolics from *Lobaria pulmonaria* (L.) Hoffm. and *Usnea longissima* Ach. lichen species. Turk J Chem 35(4):647–661

Azenha G, Iturriaga T, Michelangeli FI, Rodriguez E (1998) Ethnolichenology, biochemical activity, and biochemistry of Amazonian lichen species. Cornell University Undergraduate Research Program on Biodiversity 1:8–14

Bai L, Bao HY, Bau T (2014) Isolation and identification of a new benzofuranone derivative from *Usnea longissima*. Nat Prod Res 28:534–538

Bandoni AL, Mendiondo ME, Rondina RV, Coussio JD (1972) Survey of Argentine medicinal plants. I. Folklore and phytochemical screening. Lloydia 35:69–80

Bank TH II (1953) Botanical and ethnobotanical studies in the Aleutian Islands: II. Health and medical lore of the Aleuts. Pap Mich Acad Sci Arts Lett 38:415–431

Bastien JW (1983) Pharmacopeia of Qollahuaya Andeans. J Ethnopharmacol 8:97–111

Bauhin J, Cherler JH (1650) Historiae plantarum universalis. Tomus I [section 2]. Liber VII. Typographia Caldoriana, Ebroduni

Baumann BB (1960) The botanical aspects of Ancient Egyptian embalming and burial. Econ Bot 14:84–104

Beaglehole E, Beaglehole P (1935) A note on Hopi sorcery. Mem Am Anthropol Assoc 44:5–10

Bellia G, Pieroni A (2015) Isolated, but transnational: the glocal nature of Waldensian ethnobotany, Western Alps, NW Italy. J Ethnobiol Ethnomed 11:37

Berihuete-Azorín M (2013) First archaeobotanical approach to plant use among Selknam hunter-gatherers (Tierra del Fuego, Argentina). Archaeol Anthropol Sci 5:255–266

Best E (1905) Maori medical lore. J Polynesian Soc 14:1–23

Bhattarai NK (1999) Medicinal plants and the Plant Research Division of Nepal. Med Plant Conserv 5:7–8
Biswas K (1947) The lichen flora of India. J Roy Asiatic Soc Bengal Sci 13:75–113
Biswas K (1956) Common medicinal plants of Darjeeling and the Sikkim Himalayas. West Bengal Government Press, Alipore, India
Black MJ (1980) Algonquin ethnobotany: an interpretation of aboriginal adaptation in Southwestern Quebec. National Museum of Man Mercury Series 65. Ottawa, Canada
Black PL, Arnason JT, Cuerrier A (2008) Medininal plants used by the Inuit of Qikiqtaaluk (Baffin Island, Nunavut). Botany 86:157–163
Boas F (1921) Ethnology of the Kwakiutl. Annual Report of the Bureau of American Ethnology to the Secretary of the Smithsonian Institution 35
Boom BM (1987) Ethnobotany of the Chácobo Indians, Beni, Bolivia. New York Botanical Garden, Bronx
Bostock J, Riley HT (1855) The natural history of Pliny the Elder. Taylor and Francis, London
Brodo IM, Hawksworth DL (1977) *Alectoria* and allied genera in North America. Opera Botanica 42:1–164
Brooker SG, Cooper RC (1962) New Zealand medicinal plants. Unity Press, Aukland
Brooker SG, Cambie RC, Cooper RC (1987) New Zealand medicinal plants. Heinemann, Aukland
Bugni TS, Andjelic CD, Pole AR et al (2009) Biologically active components of a Papua New Guinea analgesic and anti-inflammatory lichen preparation. Fitoterapia 80:270–273
Bussmann RW, Sharon D (2006a) Traditional medicinal plant use in Northern Peru: tracking two thousand years of healing culture. J Ethnobiol Ethnomed 2:47
Bussmann RW, Sharon D (2006b) Traditional medicinal plant use in Loja province, Southern Ecuador. J Ethnobiol Ethnomed 2:44
Bustinza F, Caballero L (1947) Contribución al estudio de los antibióticos precedentes de líquenes. Anales del Jardín Botánico de Madrid 7:511–548
Byg A, Salick J, Law W (2010) Medicinal plant knowledge among lay people in five Eastern Tibet villages. Hum Ecol 38:177–191
Cadogan L (1949) Sintesis de la medicina racional y mistica Mbyá-Guaraní. América Indígena 9:21–35
Cameron J (1900) The Gaelic names of plants, 2nd edn. John MacKay, Glasgow
Chanda S, Singh A (1971) A crude lichen drug (charila) from India. J Res Indian Med 6:209–215
Chesnut VK (1902) Plants used by the Indians of Mendocino County, California. Systematic and geographic botany, and aboriginal uses of plants. Government Printing Office, Washington, DC, pp 295–408
Christanell A, Vogl-Lukasser B, Vogl CR, Gütler M (2010) The cultural significance of wild-gathered plant species in Kartitsch (Eastern Tyrol, Austria) and the influence of socioeconomic changes on local gathering practices. In: Santayana MP, Pieroni A, Puri RK (eds) Ethnobotany in the new Europe: people, health, and wild plant resources. Berghahn Books, New York
Clark B (1995) The quintessence tantras of Tibetan medicine. Snow Lion Publication, Ithaca, NY
Clément D (1990) L'ethnobotanique montagnaise de Mingan. Collection Nordicana 53. Université Laval, Canada
Compton BD (1993) Upper North Wakashan and Southern Tsimshian ethnobotany: the knowledge and usage of plants and fungi among the Oweekeno, Hanaksiala (Kitlope and Kemano), Haisla (Kitamaat) and Kitasoo peoples of the Central and North Coasts of British Columbia. Ph.D. thesis, University of British Columbia, Canada
Correche E, Kurdelas R, Gomez Lechon MJ et al (2008) Antibacterial, cytotoxic and antioxidant activities of extracts and secondary metabolites obtained from shrubs and lichens of Argentine. In: Vinardell MP, Mitjans M (eds) Pharmacological applications of South America plants. Research Signpost, Kerala, pp 112–123
Cramer GCP (1780) Dissertatio inauguralis medica de lichene islandico. Typis Kunstamannianis, Erlangae

Crawford S (2007) Ethnolichenology of *Bryoria fremontii*: wisdom of elders, population ecology, and nutritional chemistry. M.Sc. thesis, University of Victoria, Canada
Crawford SD (2015) Lichens used in traditional medicine. In: Ranković B (ed) Lichen secondary metabolites. Springer, Cham, pp 27–80
Crum H (1993) A lichenologist's view of lichen manna. Contrib Univ Mich Herb 19:293–306
Culpeper N (1788) The English physician. Printed for P. M'Queen et al., London
Curtin LSM (1949) By the prophet of the earth: ethnobotany of the Pima. University of Arizona Press, Tucson
Curtin LSM (1965) Healing herbs of the upper Río Grande. Southwest Museum, Los Angeles
Dampier G, Sloane H (1698) Part of a Letter from Mr. George Dampier. Philos Trans R Soc London 20:49–52
Darias V, Bravo L, Barquín E et al (1986) Contribution to the ethnopharmacological study of the Canary Islands. J Ethnopharmacol 15:169–193
Davis EW, Yost JA (1983) Novel hallucinogens from eastern Ecuador. Bot Mus Leafl Harv Univ 29:291–295
de Beer JJJ, van Wyk B-E (2011) An ethnobotanical survey of the Agter–Hantam, Northern Cape Province, South Africa. S Afr J Bot 77:741–754
De Candolle MAP (1816) Essai sur les proprieties médicales des plantes, compares avec leurs forms extérieures et leur classification naturelle. Crochard, Paris
De Crespigny RC, Hutchinson H (1903) The new forest: its traditions, inhabitants and customs. John Murray, London
De Grey T (1639) The compleat horse-man, and expert ferrier. Thomas Harper, London
de L'Obel M (1576) Plantarum, seu, Stirpium historia. Ex officina C. Plantini, Antverpiæ
De Laguna F (1972) Under Mount Saint Elias: the history and culture of the Yakutat Tlingit, Part 1. Smithsonian Contributions to Anthropology 7. Washington, DC
de Tournefort JP (1694) Elémens de botanique ou méthode pour connaitre les plantes. L'Imprimerie Royale, Paris
Densmore F (1939) Nootka and Quileute Music. Bureau of American Ethnology Bulletin 124. Government Printing Office, Washington, DC
Devkota S, Chaudhary RP, Werth S, Scheidegger C (2017) Indigenous knowledge and use of lichens by the lichenophilic communities of the Nepal Himalaya. J Ethnobiol Ethnomed 13 (15):1–10
Diderot D, d'Alembert JLR, Mouchon P (1765) Encyclopédie. Briasson, Paris
Dillenius JJ (1742) Historia muscorum. E theatro Sheldoniano, Oxonii
Dorstenius T (1540) Botanicon. Egenolph, Francofurti
Doukkali Z, Bouidida H, Srifi A et al (2015) Les plantes anxiolytiques au Maroc. Études ethnobotanique et ethno-pharmacologique. Phytothérapie 13:306–313
Drummond AT (1861) On the economical uses of *Sticta pulmonaria* Hoffm. Ann Botan Soc Canada 1:81–84
DuBois TA, Lang JF (2013) Johan Turi's animal, mineral, vegetable cures and healing practices: an in-depth analysis of Sami (Saami) folk healing one hundred years ago. J Ethnobiol Ethnomed 9:57
Dutt UC (1877) The materia medica of the Hindus, compiled from Sanskrit medical works. Thacker, Spink & Co., Calcutta
Eidlitz K (1969) Food and emergency food in the circumpolar area. Studia Ethnographica Upsaliensia 32. Almqvist & Wiksells, Uppsala
Elmore FH (1943) Ethnobotany of the Navajo. Monographs of the School of American Research 8. University of New Mexico Press, Albuquerque
Emmons GT (1991) The Tlingit Indians. Anthropological Papers of the American Museum of Natural History 70. University of Washington Press, Seattle.
Epstein H (1937) Animal husbandry of the Hottentots. Onderstepoort J Vet Sci Anim Ind 9:631–666

Esimone CO, Adikwu MU (1999) Antimicrobial activity and cytotoxicity of *Ramalina farinacea*. Fitoterapia 70:428–431

Esslinger TL (2018) A cumulative checklist for the lichen-forming, lichenicolous and allied fungi of the Continental United States and Canada, version 22. Opuscula Philolichenum 17:6–268

Estomba D, Ladia A, Lozada M (2006) Medicinal wild plant knowledge and gathering patterns in a Mapuche community from North-western Patagonia. J Ethnopharmacol 103:109–119

Felger RS, Moser MB (1985) People of the desert and sea: ethnobotany of the Seri Indians. University of Arizona Press, Tucson

Fernández Ocaña AM (2000) Estudio etnobotánico en el Parque Natural de las Sierras de Cazorla, Segura y Las Villas. Investigación química de un grupo de especies interesantes. Ph.D. thesis, University of Jaén, Spain

Ferrier J, Šačiragić L, Chen ECH et al (2014) Ways the Lukomir Highlanders of Bosnia and Herzegovina treat diabetes. In: Pieroni A, Quave CL (eds) Ethnobotany and biocultural diversities in the Balkans. Springer, New York, pp 13–27

Ferrier J, Saciragic L, Trakić S et al (2015) An ethnobotany of the Lukomir Highlanders of Bosnia & Herzegovina. J Ethnobiol Ethnomed 11:81

Foxworthy FW (1922) Minor forest products of the Malay Peninsula. Malayan For Rec 2:1–217

Fraser MH (2006) Ethnobotanical investigation of plants used for the treatment of type 2 diabetes by two Cree communities in Quebec: quantitative comparisons and antioxidant evaluation. M. Sc. thesis, McGill University, Canada

Fu H, Wang L, Chen Y, Liao R (2005) A study on nutritional components of two different lichen teas from Yunnan. Nat Prod Res Dev 17:340–343

Gabriel L, White HE (1954) Food and medicines of the Okanakanes. Report of the Okanagan Historical Society of Vernon, British Columbia 18:24–29

Gaire BP, Subedi L (2011) Medicinal plant diversity and their pharmacological aspects of Nepal Himalayas. Pharm J 3:6–17

Garcia GH, Campos R, de Torres RA (1990) Antiherpetic activity of some Argentine medicinal plants. Fitoterapia 61:542–546

Garibaldi A (1999) Medicinal flora of the Alaska natives. University of Alaska Anchorage, USA

Garrett JT (2003) The Cherokee herbal: native plant medicine from the four directions. Bear & Co., Vermont

Garth TR (1953) Atsugewi Ethnobotany. Anthropological Records [University of California] 14:129–212

Gedner C (1756) Cui bono? In: Linnaeus C (ed) Amoenitates Academicæ. Sumtu & literis Laurentii Salvii, Holmiæ, pp 231–256

Gerarde J (1597) The herball or general historie of plantes. John Norton, London

Gill SJ (1983) Ethnobotany of the Makah and Ozette people, Olympic Peninsula, Washington (USA). Washington State University, Pullman

Gioanetto F (1993) Aspetti etnofarmacologici e nutrizionali dei licheni. Notiziario [Società Lichenologica Italiana] 5:29–37

Goldie WH (1904) Maori Medical Lore. Trans Proc Roy Soc N Z 37:1–120

Gonzales de la Cruz M, Baldeón Malpartida S, Beltrán Santiago H et al (2014) Hot and cold: medicinal plant uses in Quechua speaking communities in the high Andes (Callejón de Huaylas, Ancash, Perú). J Ethnopharmacol 155:1093–1117

González-Tejero MR, Martínez-Lirola MJ, Casares-Porcel M, Molero-Mesa J (1995) Three lichens used in popular medicine in eastern Andalucia (Spain). Econ Bot 49:96–98

Goodrich J, Lawson C, Lawson VP (1980) Kashaya Pomo plants. American Indian Monograph Series 2. University of California, USA

Gourdon R (1687) A receipt to cure mad dogs, or men or beasts bitten by mad dogs. Phil Trans R Soc London 16:298

Grasser S, Schunko C, Vogl CR (2012) Gathering "tea" – from necessity to connectedness with nature. Local knowledge about wild plant gathering in the Biosphere Reserve Grosses Walsertal (Austria). J Ethnobiol Ethnomed 8:31

Guarrera PM, Lucchese F, Medori S (2008) Ethnophytotherapeutical research in the high Molise region (Central-Southern Italy). J Ethnobiol Ethnomed 4:7
Güvenç A, Küpeli Akkol E, Süntar İ et al (2012) Biological activities of *Pseudevernia furfuracea* (L.) Zopf extracts and isolation of the active compounds. J Ethnopharmacol 144:726–734
Hale BW, DePriest PT (1999) Mason E. Hale's list of epithets in the parmelioid genera. Bryologist 102:462–544
Hart J (1974) Plant taxonomy of the Salish and Kootenai Indians of Western Montana. M.A. thesis, University of Montana, USA
Hart J (1976) Montana – native plants and early peoples. The Montana Historical Society, USA
Hartley D, Sandys F, Wollaston F (1737) Another case of a person bit by a mad-dog. Phil Trans R Soc London 40:274–276
Hawksworth DL (2003) Hallucinogenic and toxic lichens. Int Lichenological Newsl 36:33–35
Hebda RJ, Turner NJ, Birchwater S et al (1996) Ulkatcho food and medicine plants. Ulkatcho Indian Band, Anahim Lake, Canada
Hellson JC, Gadd M (1974) Ethnobotany of the Blackfoot Indians. National Museum of Man Mercury Series 19. Ottawa, Canada
Herrick JW (1995) Iroquois medical botany. Syracuse University Press, New York
Hersch-Martínez P (1997) Medicinal plants and regional traders in Mexico: physiographic differences and conservational challenge. Econ Bot 51:107–120
Hooper D (1937) Useful plants and drugs of Iran and Iraq. Botanical Series 9(3). Field Museum of Natural History, Chicago
Houghton PJ, Manby J (1985) Medicinal plants of the Mapuche. J Ethnopharmacol 13:89–103
Hu S, Kong YC, But PPH (1980) An enumeration of the Chinese Materia Medica. The Chinese University Press, Hong Kong
Hunn ES (1990) Nch'i-Wána: "The Big River": Mid-Columbia Indians and their land. University of Washington Press, Seattle
Hunn G (2005) Unpublished 1976–1980 ethnobotany field notes.
Hunte P, Safi M, Macey A, Kerr GB (1975) Folk methods of fertility regulation; and the traditional birth attendant (the dai). US Agency for International Development, Buffalo
Inngjerdingen K, Nergård CS, Diallo D et al (2004) An ethnopharmacological survey of plants used for wound healing in Dogonland, Mali, West Africa. J Ethnopharmacol 92:233–244
James R (1748) Dictionnaire universel de medecine. Braisson, David l'aîné, Durand, Paris
Ji H, Shengji P, Chunlin L (2004) An ethnobotanical study of medicinal plants used by the Lisu people in Nujiang, Northwest Yunnan. China Econ Bot 58(Suppl 1):S253
Jiang B, Zhao Q-S, Peng L-Y et al (2002) Constituents from *Thamnolia vermicularis*. Acta Bot Yunnanica 24:525–530
Jin B, Liu Y, Xie J et al (2018) Ethnobotanical survey of plant species for herbal tea in a Yao autonomous county (Jianghua, China): results of a 2-year study of traditional medicinal markets on the Dragon Boat Festival. J Ethnobiol Ethnomed 14:58
Johnson LM (1997) Health, wholeness, and the land: Gitksan traditional plant use and healing. Ph. D. thesis, University of Alberta, Canada
Johnson LM (2006) Gitksan medicinal plants-cultural choice and efficacy. J Ethnobiol Ethnomed 2:29
Jorim RY, Korape S, Legu W et al (2012) An ethnobotanical survey of medicinal plants used in the eastern highlands of Papua New Guinea. J Ethnobiol Ethnomed 8:47
Ju Y, Zhuo J, Liu B, Long C (2013) Eating from the wild: diversity of wild edible plants used by Tibetans in Shangri-la region, Yunnan, China. J Ethnobiol Ethnomed 9:28
Karadi K (2010) A pharmacognostical and analytical study of Shaileya (*Parmelia* species) W.S.R. to different market samples. M.D. thesis, Rajiv Gandi University of Health Sciences, India
Kari PR (1987) Tanaina Plantlore. US National Park Service, Anchorage, AK
Kawagoe S (1925) The market fungi of Japan. Trans Br Mycol Soc 10:201–206

Kay MS (1995) Environmental, cultural, and linguistic factors affecting Ulkatcho (Carrier) botanical knowledge. M.Sc. thesis, University of Victoria, Canada.
Kerry-Nicholls JH (1886) The origin, physical characteristics, and manners and customs of the Maori race. J Anthropol Inst G B Irel 15:187–209
Khalid H, Abdalla WE, Abdelgadir H et al (2012) Gems from traditional north-African medicine: medicinal and aromatic plants from Sudan. Nat Prod Bioprospect 2(3):92–103
Kim H, Song M-J, Potter D (2006) Medicinal efficacy of plants utilized as temple food in traditional Korean Buddhism. J Ethnopharmacol 104:32–46
Kiringe JW (2008) A survey of traditional health remedies used by the Maasai of Southern Kaijiado District, Kenya. Ethnobot Res Appl 4:61–74
Kokwaro JO (1976) Medicinal plants of East Africa. East African Lit. Bur., Nairobi
Kolosova V, Svanberg I, Kalle R et al (2017) The bear in Eurasian plant names: motivations and models. J Ethnobiol Ethnomed 13:14
Kong YC, Chen DS (1996) Elucidation of Islamic drugs in Hui Hui Yao Fang: a linguistic and pharmaceutical approach. J Ethnopharmacol 54:85–102
Kujawska M, Łuczaj Ł, Typek J (2015) Fischer's Lexicon of Slavic beliefs and customs: a previously unknown contribution to the ethnobotany of Ukraine and Poland. J Ethnobiol Ethnomed 11:85
Kumar K, Upreti DK (2001) *Parmelia* spp. (lichens) in ancient medicinal plant lore of India. Econ Bot 55:458–459
Kumar S, Banskota AH, Manandhar MD (1996) Isolation and identification of some chemical constituents of *Parmelia nepalensis*. Planta Med 62:93–94
La Barre W (1948) The Aymara Indians of the Lake Titicaca Plateau, Bolivia. Mem Am Anthropol Assoc 68:1–250
La XN, Liang H, Ba GN, Tai BD (2015) Chemical constituents from *Usnea longgisima*, a traditional Mongolian medicine (II). Zhong Yao Cai 38(12):2541–2542
Laidler PW (1928) The magic medicine of the Hottentots. S Afr J Sci 25:433–447
Lal B, Upreti DK (1995) Ethnobotanical notes on three Indian lichens. Lichenologist 27:77–79
Lama YC, Ghimire SK, Aumeeruddy-Thomas Y (2001) Medicinal plants of Dolpo: Amchis' knowledge and conservation. WWF Nepal Program, Kathmandu
Lamont SM (1977) The Fisherman Lake Slave and their environment: a story of floral and faunal resources. M.Sc. thesis, University of Saskatchewan, Canada
Lavergne R (1989) Plantes medicinales indigenes tisanerie et tisaneurs de la Reunion. Ph.D. thesis, Université des Sciences et Techniques du Languedoc, France
Layard DP (1757) An essay on the nature, causes, and cure of the contagious distemper among the horned cattle in these kingdoms. John Rivington et al., London
Lebail JBEF (1853) Des lichens, considérés sous le point de vue économique, médical, et physiologique (nutrition). M.D. thesis, Faculté de Médecine de Paris, France
Leduc C, Coonishish J, Haddad P, Cuerrier A (2006) Plants used by the Cree Nation of Eeyou Istchee (Quebec, Canada) for the treatment of diabetes: a novel approach in quantitative ethnobotany. J Ethnopharmacol 105:55–63
Lee SJ (1966) Korean folk medicine. Seoul National University, Seoul
Lee EB, Yun HS, Woo WS (1977) Plants and animals used for fertility regulation in Korea. Kor J Pharmacognosy 8:81–88
Leighton AL (1985) Wild plant use by the Woods Cree (Nihithawak) of east-central Saskatchewan. National Museum of Man Mercury Series 101. Ottawa, Canada
Lev E (2007) Drugs held and sold by pharmacists of the Jewish community of medieval (11–14th centuries) Cairo according to lists of materia medica found at the Taylor–Schechter Genizah collection, Cambridge. J Ethnopharmacol 110:275–293
Li F, Zhuo J, Liu B et al (2015) Ethnobotanical study on wild plants used by Lhoba people in Milin County, Tibet. J Ethnobiol Ethnomed 11:23
Lightfoot J (1777) Flora Scotica. B. White, London
Lindley J (1838) Flora Medica. Longman et al., London

Linnaeus C (1737) Flora Lapponica. Apud Salomonem Schouten, Amstelaedami

Linnaeus C (1753) Species plantarum. Impensis Laurentii Salvii, Holmiæ

Lipp FJ (1995) Ethnobotanical method and fact: a case study. In: Schultes RE, Reis S (eds) Ethnobotany: evolution of a discipline. Dioscorides Press, Oregon, pp 52–59

Llano GA (1948) Economic uses of lichens. Econ Bot 2:15–45

Loi MC, Poli F, Sacchetti G et al (2004) Ethnopharmacology of Ogliastra (Villagrande Strisaili, Sardinia, Italy). Fitoterapia 75:277–295

Lokar LC, Poldini L (1988) Herbal remedies in the traditional medicine of the Venezia Giulia Region (North East Italy). J Ethnopharmacol 22:231–279

Londoño-Castañeda PA, Buril MLL, Rego-Cunha IP et al (2017) Lichens used in the traditional medicine by the Pankararu Indigenous Community, Pernambuco-Brazil. Global J Sci Front Res C Biol Sci 17(4)

López Eire A, Cortés Gabaudan F, Gutiérrez Rodilla BM, Vázquez de Benito MC (2006) Estudios y Traducción. Dioscórides. Sobre los remedios medicinales. Manuscrito de Salamanca. Ediciones Universidad, Spain

Luo B, Liu Y, Liu B et al (2018) Yao herbal medicinal market during the Dragon Boat Festival in Jianghua County, China. J Ethnobiol Ethnomed 14:61

Luyken JA (1809) Tentamen historiae lichenum in genere cui accedunt primae lineae distributionis novae. Henry Dieterich, Gottingen

Mabogo DEN (1990) The ethnobotany of the Vhavenda. M.Sc. thesis, Department of Botany, University of Pretoria, South Africa

Macdonald C (1974) Medicines of the Maori. Collins, Auckland

Macía MJ, García E, Vidaurre PJ (2005) An ethnobotanical survey of medicinal plants commercialized in the markets of La Paz and El Alto, Bolivia. J Ethnopharmacol 97:337–350

MacIntyre D (1999) The role of Scottish native plants in natural dyeing and textiles. University of Edinburgh, Scotland

Madamombe IT, Afolayan AJ (2003) Evaluation of antimicrobial activity of extracts from South African *Usnea barbata*. Pharm Biol 41:199–202

Madulid DA, Gaerlan FJM, Romero EM, Agoo EMG (1989) Ethnopharmacological study of the Ati tribe in Nagpana, Barotac Viejo, Iloilo. Acta Manilana 38:25–40

Marc EB, Nelly A, Annick D-D, Frederic D (2008) Plants used as remedies antirheumatic and antineuralgic in the traditional medicine of Lebanon. J Ethnopharmacol 120:315–334

Marcano V (1991) Propriedades biomedicas de las embriofitas asifonogamas andinas: II liquenes. Rev Fac Farm 28:1–13

Marcano V, Rodriguez-Alcocer V, Morales Méndez A (1999) Occurrence of usnic acid in *Usnea laevis* Nylander (lichenized ascomycetes) from the Venezuelan Andes. J Ethnopharmacol 66:343–346

Marles RJ (1984) The ethnobotany of the Chipewyan of Northern Saskatchewan. M.Sc. thesis, University of Saskatchewan, Canada

Marles RJ, Canada, Natural Resources Canada, Canadian Forest Service (2000) Aboriginal plant use in Canada's northwest boreal forest. UBC Press, Vancouver

Marshall AG (1977) Nez Perce social groups: an ecological interpretation. Ph.D. thesis, Washington State University, USA

Martínez-Lirola MJ, González-Tejero MR, Molero-Mesa J (1996) Ethnobotanical resources in the province of Almería, Spain: Campo de Nijar. Econ Bot 50:40–56

Masson C (1842) Narrative of various journeys in Balochistan, Afghanistan and the Panjab. Richard Bentley, London

Matsiliza B, Barker NP (2001) A preliminary survey of plants used in traditional medicine in the Grahamstown area. S Afr J Bot 67:177–182

Mattalia G, Quave CL, Pieroni A (2013) Traditional uses of wild food and medicinal plants among Brigasc, Kyé, and Provençal communities on the Western Italian Alps. Genet Resour Crop Evol 60:587–603

Matthiolus PA (1586) De plantis epitome utilissima. Johann Feyerabend, Frankfurt

McClatchey WC (1993) The traditional Rotuman medicinal system and ethnopharmacopoea. M.Sc. thesis, Brigham Young University, USA
McClintock W (1910) The old north trail. MacMillan, London
McCormack G (2007) Cook Islands biodiversity database, version 2007.2. Cook Islands Natural Heritage Trust, Rarotonga. cookislands.bishopmuseum.org
McGlinchey C (1986) The last of the name. Edited by B Friel, Blackstaff, Belfast
Mckenna DJ, Towers GHN, Abbott FS (1984) Monoamine oxidase inhibitors in South American hallucinogenic plants part 2: constituents of orally-active Myristicaceous hallucinogens. J Ethnopharmacol 12:179–211
McKennan RA (1959) The upper Tanana Indians. Yale University Publ in Anthrop 55:1–226
Mead GR (1972) The ethnobotany of the California Indians: a compendium of the plants, their users, and their uses. Museum of Anthropology, University of Northern Colorado, USA
Merriam CH (1966) Ethnographic notes on California Indian tribes. University of California Archaeological Research Facility, USA
Miara MD, Hammou MA, Aoul SH (2013) Phytothérapie et taxonomie des plantes médicinales spontanées dans la région de Tiaret (Algérie). Phytothérapie 11:206–218
Milliken W, Miller RP, Pollard SR, Wandelli EV (1992) Ethnobotany of the Waimiri Atroari Indians of Brazil. Royal Botanical Gardens, Kew, England
Mitchell ME (2015) Contentious cures: the rise and decline of lichens as European materia medica. Pharm Hist 57:55–66
Modenesi P (2009) Skull lichens: a curious chapter in the history of phytotherapy. Fitoterapia 80:145–148
Molares S, Ladio A (2014) Medicinal plants in the cultural landscape of a Mapuche-Tehuelche community in arid Argentine Patagonia: an eco-sensorial approach. J Ethnobiol Ethnomed 10:61
Mortimer C (1735) A narration of the experiments made June 1, 1734. Phil Trans R Soc London 39:313–360
Moskalenko SA (1986) Preliminary screening of far-eastern ethnomedicinal plants for antibacterial activity. J Ethnopharmacol 15:231–259
Moxham TH (1986) The commercial exploitation of lichens for the perfume industry. In: Brunke EJ (ed) Progress in essential oil research. Walter de Gruyter, Berlin, pp 491–503
Müller J (1881) Lichenologische Beiträge XIV. Flora Odor Allgemeine Botanische Zeitung 64:513–527
Muntané J (1991) Aportació al coneixement de l'Ethnobotánic de Cerdanya. Ph.D. thesis, University of Barcelona, Spain
Muthee JK, Gakuya DW, Mbaria JM et al (2011) Ethnobotanical study of anthelmintic and other medicinal plants traditionally used in Loitoktok district of Kenya. J Ethnopharmacol 135:15–21
Nadkarni KM, Nadkarni AK (1955) Indian materia medica. Popular Book Depot, Bombay
Nagendra Prasada P, Ranjit Singh AJA, Narayanan LM, Natarajan CR (1996) Ethnobotany of the Kanikkars of South Tamilnadu – I. Ethnobotany in South Asia. Scientific Publishers, Jodhpur, India, pp 292–298
Natale AD, Pollio A (2012) A forgotten collection: the Libyan ethnobotanical exhibits (1912–14) by A. Trotter at the Museum O. Comes at the University Federico II in Naples, Italy. J Ethnobiol Ethnomed 8:4
Nedelcheva A, Draganov S (2014) Bulgarian medical ethnobotany: the power of plants in pragmatic and poetic frames. In: Pieroni A, Quave CL (eds) Ethnobotany and biocultural diversityies in the Balkans. Springer, New York, pp 45–65
Nissen K (1921) Lapponian lichen names. In: Lynge B (ed) Studies on the lichen flora of Norway. Jacob Dybwad, Oslo, pp 238–247
Novaretti R, Lemordant D (1990) Plants in the traditional medicine of the Ubaye Valley. J Ethnopharmacol 30:1–34
O'Neill AR, Badola HK, Dhyani PP, Rana SK (2017) Integrating ethnobiological knowledge into biodiversity conservation in the Eastern Himalayas. J Ethnobiol Ethnomed 13:21

Odabasoglu F, Cakir A, Suleyman H et al (2006) Gastroprotective and antioxidant effects of usnic acid on indomethacin-induced gastric ulcer in rats. J Ethnopharmacol 103:59–65
Ohmura Y (2003) What species of Japanese lichens are edible? Lichen: News Bulletin of the Lichenological Society of Japan 13:6–9
Osorio HS (1982) Contribution to the lichen flora of Uruguay XVII. The scientific name of the "Yerba de la Piedra". Phytologia 52:217–220
Oswalt WH (1957) A western Eskimo ethnobotany. Anthropol Pap Univ Alaska 6:16–36
Pardanani DS, DeLima RJ, Rao RV et al (1976) Study of the effects of speman on semen quality in oligospermic men. Indian J Surg 38:34–39
Parkinson J (1640) Theatrum botanicum. Tho. Cotes, London
Pennington CW (1963) The Tarahumar of Mexico: their environment and material culture. University of Utah Press, USA
Pennington CW (1969) The Tepehuan of Chihuaha: their material culture. University of Utah Press, USA
Pereira J (1850) The elements of materia medica and therapeutics, vol II, part II. Longman, Brown, Green, and Longmans, London
Pérez-Nicolás M, Vibrans H, Romero-Manzanares A et al (2017) Patterns of knowledge and use of medicinal plants in Santiago Camotlán, Oaxaca, Mexico. Econ Bot 71:209–223
Pieroni A (2000) Medicinal plants and food medicines in the folk traditions of the upper Lucca Province, Italy. J Ethnopharmacol 70:235–273
Pieroni A, Giusti ME (2009) Alpine ethnobotany in Italy: traditional knowledge of gastronomic and medicinal plants among the Occitans of the upper Varaita valley, Piedmont. J Ethnobiol Ethnomed 5:32
Pieroni A, Dibra B, Grishaj G et al (2005) Traditional phytotherapy of the Albanians of Lepushe, Northern Albanian Alps. Fitoterapia 76:379–399
Pieroni A, Rexhepi B, Nedelcheva A et al (2013) One century later: the folk botanical knowledge of the last remaining Albanians of the upper Reka Valley, Mount Korab, Western Macedonia. J Ethnobiol Ethnomed 9:22
Pollan M (2001) The botany of desire: a plant's eye view of the world. Random House, New York
Poudel P (2008) Medicinal plants of change VDC of Taplejung. Eastern Nepal, Kathmandu
Powers S (1877) Aboriginal botany. In: Tribes of California. Government Printing House, Washington, pp 419–431
Pradhan BK, Badola HK (2008) Ethnomedicinal plant use by Lepcha tribe of Dzongu valley, bordering Khangchendzonga Biosphere Reserve, in North Sikkim, India. J Ethnobiol Ethnomed 4:22
Prance GT (1972) Ethnobotanical notes from Amazonian Brazil. Econ Bot 26:221–237
Pranskuniene Z, Dauliute R, Pranskunas A, Bernatoniene J (2018) Ethnopharmaceutical knowledge in Samogitia region of Lithuania: where old traditions overlap with modern medicine. J Ethnobiol Ethnomed 14:70
Prasad R (2013) Ayurveda therapeutics workshop – explore traditional recipes from kottakkal arya vaidya sala. Handout 2. AVS Kottakkal Australia
Quincy J (1724) Pharmcopoeia Officianalis & Extemporanea. E. Bell in Cornhill, W. Taylor in Paternoster-Row, and J. Osborn in London
Ranby, Peters C (1744) The case of a person bit by a mad dog. Phil Trans R Soc London 43:257–262
Randlane T, Saag A, Thell A, Ahti T (2013) Third world list of cetrarioid lichens – in a new databased form, with amended phylogenetic and type information. Cryptogam Mycol 34:79–84
Rauf A, Afaq SH, Latif A (2006) Pharmacognostical standardization of "Ushna" a cardiotonic drug mentioned in Adviya Qalbiya. In: Abdin MZ, Abrol YP (eds) Traditional systems of medicine. Narrosa House, New Delhi, pp 226–235
Rauf A, Latif A, Rehman S, Afaq SH (2011) In-vitro antibacterial screening of extracts of *Usnea longissima* lichen. Int J Appl Biol Pharm Technol 2:14–18

Ray J (1686) Historia plantarum. Mariæ Clark and Henricum Faithorne Regiæ Societatis Typographum, London

Raymond M (1945) III. Notes ethnobotaniques sur les Tête-de-Boule de manouan. Études Ethnobotaniques Québécoises 55:113–154

Razzack HMA, Fazal HMU (1993) The concept of birth control in Unani medicine. Central Council for Research in Unani Medicine, New Dehli

Rebesca MA, Romie D, Johnson M, Ryan J (1994) Traditional Dene medicine part I: Report. Dene Cultural Institute, Lac La Martre, NWT, Canada

Redžić SS (2007) The ecological aspect of ethnobotany and ethnopharmacology of population in Bosnia and Herzegovina. Collegium Antropologicum 31(3):869–890

Regnault JB (1802) Observations on pulmonary consumption, or an essay on the Lichen islandicus, considered both as an aliment and a medicine in that disorder. J. Smeeton, London

Richardson DHS (1974) The vanishing lichens: their history and importance. Hafner Press, New York

Rink HJ, Lindorff JFT (1856) Help to the patients. Originally published in Greenland. Translated and reprinted by U. Søchting in Graphis Scripta 3:24 (1990)

Robbins WW, Harrington JP, Freire-Marreco BW (1916) Ethnobotany of the Tewa Indians. Smithson Inst Bur Am Ethnol Bull 55:1–124

Rodríguez MA, Angueyra A, Cleef AM, van Andel T (2018) Ethnobotany of the Sierra Nevada del Cocuy-Güicán: climate change and conservation strategies in the Colombian Andes. J Ethnobiol Ethnomed 14:34

Rout J, Kar A, Upreti DK (2005) Traditional remedy for kidney stones from a high altitude lichen: *Cladonia rangiferina* L. Wigg (reindeer moss) of Eastern Himalaya. Ethnobotany 17:164–166

Ruíz Leal A (1972) Flora Popular Mendocina. Deserta 3:36–38

Russell A (1973) Horns in the high country. Knopf, New York

Sak K, Jürisoo K, Raal A (2014) Estonian folk traditional experiences on natural anticancer remedies: from past to the future. J Pharm Biol 52:855–866

Saklani A, Jain SK (1994) Cross-cultural ethnobotany of Northeast India. Deep Publishers, New Dehli

Salick J, Byg A, Amend A et al (2006) Tibetan medicine plurality. Econ Bot 60(3):227–253

Šarić-Kundalić B, Dobeš C, Klatte-Asselmeyer V, Saukel J (2010) Ethnobotanical study on medicinal use of wild and cultivated plants in middle, south and west Bosnia and Herzegovina. J Ethnopharmacol 131:33–55

Sato M (1968) An edible lichen of Japan, *Gyrophora esculenta* Miyoshi. Nova Hedwigia 16:505–509

Scarpa GF (2004) Medicinal plants used by the Criollos of Northwestern Argentine Chaco. J Ethnopharmacol 91:115–135

Scarpa GF (2009) Etnobotánica médica de los indígenas chorote y su comparación con la de los criollos del Chaco semiárido (Argentina). Darwiniana 47(1):92–107

Schade A (1954) Über Letharia vulpina (L.) Vain. und ihre Vorkommen in der Alten Welt. Berichte der Bayerischen Botanischen Gesellschaft 30:108–126

Schmull M, Dal-Forno M, Lücking R et al (2014) *Dictyonema huaorani* (Agaricales: Hygrophoraceae), a new lichenized basidiomycete from Amazonian Ecuador with presumed hallucinogenic properties. Bryologist 117(4):386–394

Scopoli GA (1760) Flora Carniolica. Sumptibus J.T. Trattner, Viennae

Senft E (1911) The so-called "Lichen Quercinus virides". Pharmazeutische Post 43:1017–1019

Shah NC (1998) Lichens of economic importance from the hills of Uttar Pradesh, India. J Herbs Spices Med Plants 5:69–76

Sharma GK (1997) Ethnomedicinal flora: ayurvedic system of medicine in a remote part of the Indo-Tibetan Himalayas. J Tenn Acad Sci 72:53–54

Sharnoff SD (1997) Lichens and people. www.lichen.com/

Shou-Zhong Y (1998) The divine farmer's materia medica: a translation of the Shen Nong Ben Cao Jing. Blue Poppy Press, Boulder, CO

Shrestha N, Shrestha S, Koju L et al (2016) Medicinal plant diversity and traditional healing practices in eastern Nepal. J Ethnopharmacol 192:292–301

Siegel RK (1989) Intoxication: life in pursuit of artificial paradise. Pocket Books, New York

Siegel RK (2013) Inquiry about use of lichens as narcotics, corresp with S Crawford, June 29, 2013

Siegfried EV (1994) Ethnobotany of the northern Cree of Wabasca/Desmarais. M.A. thesis, University of Calgary, Canada

Silalahi M, Nisyawati WEB et al (2015) The local knowledge of medicinal plants trader and diversity of medicinal plants in the Kabanjahe traditional market, North Sumatra, Indonesia. J Ethnopharmacol 175:432–443

Singh H, Husain T, Agnihotri P et al (2012) Biodiversity conservation through traditional beliefs system: a case study from Kumaon Himalayas, India. Int J Conserv Sci 3(1):33–40

Smith A (1888) A contribution to South African Materia Medica. J. C. Jula & Co., Cape Town

Smith AL (1921) Lichens. Cambridge University Press, London

Smith HH (1923) Ethnobotany of the Menomini Indian. Bull Public Mus Milwaukee 4:1–174

Smith HI (1929) Materia medica of the Bella Coola and neighbouring tribes of British Columbia. Bull – Natl Mus Can 56:47–68

Smith HH (1932) Ethnobotany of the Ojibwe Indians. Bull Public Mus Milwaukee 4:327–525

Smith HH (1933) Ethnobotany of the Forest Potawatomi Indians. Bull Public Mus Milwaukee 7:1–230

Smith GW (1973) Arctic Pharmacognosia. Arctic 26:324–333

Søchting U (1999) Lichens of Bhutan: biodiversity and use. Department of Mycology, University of Copenhagen, Copenhagen

Soelberg J, Jäger AK (2016) Comparative ethnobotany of the Wakhi agropastoralist and the Kyrgyz nomads of Afghanistan. J Ethnobiol Ethnomed 12:2

Sohrabi M, Ahti T (2010) Nomenclatural synopsis of the Old World's "manna" lichens (*Aspicilia*, Megasporaceae). Taxon 59:628–636

Sõukand R, Kalle R (2013) Where does the border lie: locally grown plants used for making tea for recreation and/or healing, 1970s–1990s Estonia. J Ethnopharmacol 150:162–174

Stevens J, Palliser J, Avataq Cultural Institute (1984) Traditional medicine project = Project sur la médecine traditionelle. Avataq Cultural Institute, Inukjuak, Quebec

Steward T (1738) Concerning the virtues of the star of the earth. Phil Trans R Soc London 40:449–462

Stoffle RW, Evans MJ, Halmo DB et al (1989) Native American plant resources in the Yucca Mountain Area, Nevada. Science Applications International Corp., Las Vegas, NV

Stoffle RW, Evans MJ, Halmo DB et al (1994) Native American cultural resources on Pahute and Rainier Mesas, Nevada Test Site. Desert Research Institute Technical Report No. 84

Stopp K (1963) Medicinal plants of the Mt. Hagen people (Mbowamb) in New Guinea. Econ Bot 17:16–22

Strandman P (1769) Purgantia indigena. In: Linnaeus C (ed) Amoenitates Academicæ. Sumtu & literis Laurentii Salvii, Holmiæ, pp 293–310

Stubbs RD (1966) An investigation of the edible and medicinal plants used by the Flathead Indians. M.A. thesis, University of Montana, USA

Subramanian SS, Ramakrishnan S (1964) Amino-acids of *Peltigera canina.* Curr Sci 33:522

Teit JA, Boas F (1900) The Thompson Indians of British Columbia. Am Mus Nat Hist Mem 2:163–392

Teit JA, Boas F (1928) The Salishan tribes of the western plateaus. Rep Bur Am Ethnol Secr Smithson Inst 45:23–296

Thell A, Feuerer T, Kärnefelt I et al (2004) Monophyletic groups within the Parmeliaceae identified by ITS rDNA, β-tubulin and GADPH sequences. Mycol Prog 3:297–314

Thell A, Crespo A, Divakar PK et al (2012) A review of the lichen family Parmeliaceae – history, phylogeny and current taxonomy. Nord J Bot 30:641–664

Tonning H (1769) Rariora Norvegiæ. In: Linnaeus C (ed) Amoenitates Academicæ. Sumtu & literis Laurentii Salvii, Holmiæ, pp 466–496

Train P, Archer WA, Henrichs JR (1941) Medicinal uses of plants by Indian tribes of Nevada. Contributions Toward a Flora of Nevada 33. U.S. Dept. of Agriculture, Washington, DC

Trevelyan M (1909) Folk-lore and Folk-stories of Wales. Elliot Stock, London

Truong C, Divakar PK, Yahr R et al (2013) Testing the use of ITS rDNA and protein-coding genes in the generic and species delimitation of the lichen genus *Usnea* (Parmeliaceae, Ascomycota). Mol Phylogenet Evol 68:357–372

Tshiteya RM (2007) Herbal medicines for common ailments: a quick reference guide. Natural Remedies Inc., Alexandria, VA

Turner NJ (1973) The ethnobotany of the Bella Coola Indians of British Columbia. Syesis 6:193–220

Turner NJ (1998) Plant technology of first peoples in British Columbia. RBCM handbook. UBC Press, Vancouver

Turner NJ (2004a) Plants of Haida Gwaii. Sono Nis Press, Winlaw, BC

Turner NJ (2004b) Expert report: Tsilhqot'in and Xeni Gwet'in plant use and occupancy. Presented in the William vs. Her Magesty the Queen, Xeni Gwet'in/Tsilhqot'in Land Rights Trial. B.C. Supreme Court, Canada

Turner NJ, Efrat BS (1982) Ethnobotany of the Hesquiat Indians of Vancouver Island. Cultural Recover Paper 2. B.C. Prov. Mus., Canada

Turner NJ, Hebda RJ (2012) Saanich ethnobotany: culturally important plants of the WSANEC people. RBCM, Canada

Turner NJ, Thompson JC (2006) "Nwana"a lax Yuup: plants of the Gitga'at people. Cortex Consulting, School of Environmental Studies, and Coasts Under Stress, Victoria, Canada

Turner NJ, Bouchard R, Kennedy DID (1980) Ethnobotany of the Okanagan-Colville Indians of British Columbia and Washington. Occas Pap BC Prov Mus 21

Turner NJ, Thomas J, Carlson BF, Ogilvie RT (1983) Ethnobotany of the Nitinaht Indians of Vancouver Island. Occas Pap BC Prov Mus 24:1–165

Turner NJ, Thompson LC, Thompson MT, York AZ (1990) Thompson ethnobotany: knowledge and usage of plants by the Thompson Indians of British Columbia. RBCM, Canada

Turney-High HH (1937) The flathead Indians of Montana. Mem Am Anthropol Assoc 48

Tychsen N (1799) Sammenlignende Forsøg med Lichen nivalis og Lichen islandicus. Nye samling af det Kongelige Danske Videnskabernes Selskabs Skrifter 5:372–390

UBC (University of British Columbia) (2014) Accession No. V172209. Herbarium Database. www.biodiversity.ubc.ca/museum/herbarium/vascular/

Uphof JCT (1959) Dictionary of economic plants. Hafner Publishing, New York

Upreti DK, Chatterjee S (2007) Significance of lichens and their secondary metabolites: a review. In: Ganguli BN, Deshmukh SK (eds) Fungi: multifaceted microbes. Anamaya Publishers, New Dehli, pp 169–188

Upreti DK, Negi HR (1996) Folk use of *Thamnolia vermicularis* (Swartz) Ach. in Lata Village of Nanda Devi Biosphere Reserve. Ethnobotany [India] 8:92–95

Upreti DK, Bajpai R, Nayaka S, Singh BN (2016) Ethnolichenological studies in India: future prospects. In: Jain AK (ed) Indian ethnobotany: emerging trends. Scientific Publisher, Jodhpur, pp 195–233

Uprety Y, Asselin H, Boon EK et al (2010) Indigenous use and bio-efficacy of medicinal plants in the Rasuwa District, Central Nepal. J Ethnobiol Ethnomed 6:3

Uprety Y, Asselin H, Dhakal A, Julien N (2012) Traditional use of medicinal plants in the boreal forest of Canada: review and perspectives. J Ethnobiol Ethnomed 8:7

van Damme P, van den Eynden V, Vernemmen P (1992) Plant uses by the Topnaar of the Kuiseb Valley Namib Desert. Afrika Focus 8:223–252

van Wyk B-E, Gorelik B (2017) The history and ethnobotany of Cape herbal teas. S Afr J Bot 110:18–38

van Wyk B-E, de Wet H, Van Heerden FR (2008) An ethnobotanical survey of medicinal plants in the southeastern Karoo, South Africa. S Afr J Bot 74:696–704

Vartia KO (1973) Antibiotics in lichens. In: Ahmadjian V, Hale ME (eds) The lichens. Academic Press, New York

Velasco-Negueruela A, Pérez-Alonso MJ, Abaraca GE (1995) Medicinal plants from Pampallakta: an Andean community in Cuzco (Peru). Fitoterapia 66:447–461

Vestal PA, Schultes RE (1939) The economic botany of the Kiowa Indians, as it relates to the history of the tribe. Botanical Museum, Cambridge, MA

Vitalini S, Iriti M, Puricelli C et al (2013) Traditional knowledge on medicinal and food plants used in Val San Giacomo (Sondrio, Italy)—An alpine ethnobotanical study. J Ethnopharmacol 145:517–529

Vitto LAD, Petenatti EM, Petenatti ME (1997) Recursos herbolarios de San Luis (República Argentina) primera parte: plantas nativas. Multequina:49–66

Vogl CR, Vogl-Lukasser B, Walkenhorst M (2016) Local knowledge held by farmers in Eastern Tyrol (Austria) about the use of plants to maintain and improve animal health and welfare. J Ethnobiol Ethnomed 12:40

von Kaub JW (1485) Gart der Gesundheit [The Garden of Health]. Peter Schoeffer, Mainz

von Martens G (1866) Die Preussische Expedition nach Ost-Asien. Botanischer Theil, Berlin

Wang LS, Qian ZG (2013) 中国药用地衣图鉴 [Zhong guo yao yong di yi tu jian = Illustrated Medicinal Lichens of China]. Yunnan ke ji chu ban she, China

Wang L-S, Narui T, Harada H et al (2001) Ethnic uses of lichens in Yunnan, China. Bryologist 104:345–349

Wangchuk P, Namgay K, Gayleg K, Dorji Y (2016) Medicinal plants of Dagala region in Bhutan: their diversity, distribution, uses and economic potential. J Ethnobiol Ethnomed 12:28

Watson W (1756) An account of some of the more rare English plants observed in Leicestershire. Phil Trans R Soc London 49:803–806

Watt JM, Breyer-Brandwijk MG (1962) The medicinal and poisonous plants of Southern and Eastern Africa, 2nd edn. E. & S. Livingstone, Edinburgh

Weber GH, Wiggers FH (1780) Primitiae Florae Holsaticae. Litteris Mich. Frider. Bartschii Acad. Typogr., Kiel.

Weckerle CS, Huber FK, Yongping Y, Weibang S (2006) Plant Knowledge of the Shuhi in the Hengduan Mountains, Southwest China. Econ Bot 60(1):3–23

Wennekens AJ (1985) Traditional plant usage by Chugach Natives around Prince William Sound and on the Lower Kenai Peninsula, Alaska. M.A. thesis, University of Alaska, USA

Whistler WA (1990) Ethnobotany of the Cook Islands: the plants, their Maori names, and their uses. Allertonia 5:347–424

Whiting AF (1939) Ethnobotany of the Hopi. Bull – Mus North Ariz 15

Whitney CW, Gebauer J, Anderson M (2012) A survey of wild collection and cultivation of indigenous species in Iceland. Hum Ecol 40(5):781–787

Willemet R (1787) Lichénographie Économique, ou Histoire des Lichens Utiles dans la Médecine et dans les Arts. In: Hoffmann GF et al (eds) Mémoires sur l'utilité des lichens dans la médecine et dans les arts. Chez Piestre et Delamollière, A Lyon, pp 1–48

Wise JRC (1863) The New Forest: its history and its scenery. Smith, Elder, and Co., London

Withering W (1801) A systematic arrangement of British plants. T. Cadell et al., London

Wyman LC, Harris SK (1941) Navajo Indian medical ethnobotany. Univ N M Bull Anthropol Ser 3:1–76

Wyman LC, Harris SK (1951) The ethnobotany of the Kayenta Navaho: an analysis of the John and Louisa Wetherill ethnobotanical collection. Univ. N. M. Publ. Biol. 5. University of New Mexico Press, USA

Yavuz M (2012) Lichens mentioned by Pedanios Dioscorides. Ethno Medicine 6:103–109

Yavuz M (2013) Lichens in the prescriptions of Pliny the Elder. Oltenia – Studii şi Comunicări Ştiinţele Naturii 29:115–119

Yavuz M, Çobanoğlu G (2010) Ethnological uses and etymology of the word Usnea in Ebubekir Razi's "Liber Almansoris". Br Lichen Soc Bull 106:3–12

Yazici K, Aslan A (2003) Lichens from the regions of Guemueshane, Erzincan and Bayburt (Turkey). Cryptogam Mycol 24:287–300

Yeshi K, Wangdi T, Qusar N et al (2018) Geopharmaceuticals of Himalayan Sowa Rigpa medicine: ethnopharmacological uses, mineral diversity, chemical identification and current utilization in Bhutan. J Ethnopharmacol 223:99–112

Younos C, Fleurentin J, Notter D et al (1987) Repertory of drugs and medicinal plants used in traditional medicine of Afghanistan. J Ethnopharmacol 20:245–290

Zhang C, Hu J (1981) Studies on chemical components of the medicinal lichen "Jin shua ba", *Lethariella cladoniodes* (Nyl) Krog. 西北植物研究 (Xi Bei Zhi Wu Yan Jiu) 1:74–76

Zhang L, Zhang Y, Pei S et al (2015) Ethnobotanical survey of medicinal dietary plants used by the Naxi People in Lijiang Area, Northwest Yunnan, China. J Ethnobiol Ethnomed 11:40

Zhang L, Chai Z, Zhang Y et al (2016) Ethnobotanical study of traditional edible plants used by the Naxi people during droughts. J Ethnobiol Ethnomed 12:39

Zwelfer J (1672) Pharmacopoeia augustana. Apud Vincentium Caimax, Dordrechti

Lichen Secondary Metabolites as Potential Antibiotic Agents

3

Marijana Kosanić and Branislav Ranković

Abstract

It is well known that pathogenic microbes pose serious threats to human health and are increasing in prevalence in institutional health care settings due to the growing resistance that infectious agents have developed against antibiotics. Therefore, new alternatives for combating the spread of infection through antibiotic-resistant microbes are necessary for keeping pace with the evolution of 'super' pathogens. Natural products are proposed as a therapeutic alternative to conventional antimicrobial treatment. Among them, lichen-derived products and their antibiotic properties are of special interest to scientists as up to 50% of all lichens have been reported to possess antibiotic activities. A great number of reports concerning the antimicrobial screening of lichens have appeared in the literature. According to published data, the lichens and their secondary metabolites exhibited the activity against a great number of microorganisms. Therefore, this study represents lichens as very interesting source of bioactive compounds which provide unlimited opportunities for new antimicrobial agents.

3.1 Needing for New Antibiotics from Nature

Before the introduction of antibiotics in the 1940s, infections were rare, but rapidly increased in frequency as the use of antibiotics increased. In fact, most antibiotics that were first used in the 1940s and 1950s are no longer used clinically because nowadays the resistance of infectious beings to these antibiotics is very common. Over time, they have been developing new antibiotics and with the introduction of each, new drug-resistant bacteria appeared rapidly. The process of resistance is augmented by short generation times of bacteria enabling rapid mutation and

M. Kosanić (✉) · B. Ranković
Department of Biology, Faculty of Science, University of Kragujevac, Kragujevac, Serbia

B. Ranković (ed.), *Lichen Secondary Metabolites*,
https://doi.org/10.1007/978-3-030-16814-8_3

selection of resistant strains, and a horizontal transfer of resistance genes. Bacterial pathogens resistant to more than one, or even most clinically used antibiotics, have become common. Today, we moved the mode of use and prescription of antibiotics in order to try to slow the relentless pace of bacterial evolution, but not yet found a solution to this problem. Microbiologists continue to study how bacteria evolve so that we can predict how they will respond to medical treatment and so we can better manage the evolution of infectious diseases.

Bacteria are able to resist the effects of antimicrobials through preventing intracellular access, immediately removing antimicrobial substances through efflux pumps; modifying the antimicrobial agent through enzymatic breakdown or modifying the antimicrobial targets within the bacterial cell to render the substance ineffective. Successful development of resistance often results from a combination of two or more of these strategies (Sheldon 2005).

The first antibiotic resistance mechanism described was that of penicillinase. Its presence and activity was first reported by Abraham and Chain in 1940 shortly after its discovery (Abraham and Chain 1940).

Antimicrobial resistance traits are genetically coded and can either be intrinsic or acquired. Intrinsic resistance is due to innately coded genes which create natural 'insensitivity' to a particular antibiotic. Innate resistance is normally expressed by virtually all strains of that particular bacterial species. Acquired resistance is gained by previously susceptible bacteria either through mutation or horizontally obtained from other bacteria possessing such resistance via transformation, transduction or conjugation. Acquired resistance is limited to subpopulations of a particular bacterial species and may result from selective pressure exerted by antibiotic usage.

The drug resistance of human and animal pathogens is one of the best documented in biological evolution and a serious problem in both developed and developing countries. The consumption of more than 1 ton daily antibiotics in some European countries has resulted in resistance to bacterial populations, thus causing a serious public health problem. In view of this scenario, the search for new antimicrobial substances from natural sources, including lichens, has gained importance in pharmaceutical companies.

Since lichens produce a variety of substances with antimicrobial properties, it is expected that screening programs discover candidate compounds for the development of new antibiotics. However, scientific research to determine the therapeutic potential of lichens is limited; there is a lack of scientific studies that confirm the possible experimental antibiotic properties of a large number of lichens. It is expected that compounds that reach targets different from those used by known antibiotics may be active against resistant pathogens.

3.2 Antimicrobial Activity and Probable Mechanisms of Action of Lichens

Antimicrobial activity of lichens is understood as their ability to eliminate microorganisms or to inhibit their growth.

The antimicrobial properties of lichen extracts and their secondary metabolites are known for long (Burkholder and Evans 1945; Stoll et al. 1950; Vartia 1973; Piovano et al. 2002; Ranković et al. 2007a; Paudel et al. 2008; Schmeda-Hirschmann et al. 2008; Micheletti et al. 2009) and still assess the mechanisms of their effects.

The probable mechanisms of antimicrobial action of lichens are:

Inhibition of Cell Wall Synthesis The peptidoglycan layer is important for cell wall structural integrity, being the outermost and primary component of the wall. Inhibitors of cell wall synthesis act by inhibiting the synthesis of the peptidoglycan layer of bacterial cell walls and thus come to degradation of the cell wall.

Inhibition of Protein Synthesis (Translation) Protein synthesis inhibitors act at the ribosome inhibiting the synthesis of proteins of the pathogen to occur misreading the sequence of amino acids, and thus inhibit the functioning of the pathogenic cells.

Alteration of Cell Membranes Injury bacterial plasma membranes lead to cell death through leakage of cell contents and associated disruption of the cross-membrane potential (which essentially are ion concentration gradients).

Inhibition of Nucleic Acid Synthesis Nucleic acid inhibitors act by inhibiting the production of nucleic acids (DNA and RNA).

Antimetabolite Activity Antimetabolites prevents a cell from carrying out a metabolic reaction. Antimetabolites function by competitive inhibition of enzymes and by erroneous incorporation into nucleic acids. In both cases, the cells become unable to normal functioning.

The secondary metabolites of the lichen are active substances against pathogenic microorganisms. Most known lichen substances with antimicrobial activity are usnic acid, phenolic compounds, triterpenes, steroids, anthraquinones, depsides, depsidones, depsones and for most of them are known mechanisms of their antibiotic action.

Usnic Acid Usnic acid, a compound produced by various lichen species, has been demonstrated previously to inhibit growth of different bacteria and fungi; however, mechanism of its antimicrobial activity not been fully explained.

The mechanisms of the antibiotic activity of usnic acid against Gram-positive bacteria were attributed to its protonophoric properties as an uncoupler of oxidative phosphorylation (Abo-Khatwa et al. 1996). This effect was remarkably more pronounced than that produced by dinitrophenol, a well-known uncoupler agent that increases the permeability of mitochondria to protons, reducing the electrochemical potential and thus inhibiting adenosine triphosphate synthesis. Two research groups found that usnic acid played an active role in the transport of protons through the membranes of isolated mouse liver mitochondria, uncoupling the electron flow through respiratory electron transport chains from the generation of an acidic gradient and thereby inhibiting the synthesis of adenosine triphosphate (Abo-Khatwa et al. 1996; Bouaid and Vicente 1998). The ability of usnic acid to shuttle protons through membranes was confirmed by studies with artificial phospholipid membranes (Bačkor et al. 1998). A model for the function of usnic acid in the control of the intracellular pH of lichens was recently proposed (Hauck and Jurgens 2008) that involves two hypotheses: (1) at the optimum pH near the pK_{a1} value of usnic acid (4,4), buffering is assumed to be compensated by the usnic acid-mediated proton transport into the cell; (2) at low pH (<3.5), the equilibrium between usnic acid and usneate shifts towards the usnic acid and protons is increasingly shuttled into the cells, considering that usnic acid dissociates to usneate at cytosolic pH 7.4. This implies that more protonated molecules would be able to cross the membrane and release protons into the cell. As a result, the intracellular pH would decrease and lead to the death of the cells (Hauck and Jurgens 2008).

In addition, Maciąg-Dorszyńska et al. (2014) assume that inhibition of RNA synthesis may be a general mechanism of antibacterial action of usnic acid, with additional direct mechanisms, such as impairment of DNA replication in *B. subtilis* and *S. aureus*.

Usnic acid has been used as antibiotic (e.g. Binan, Usno) and is still available as a topical antiseptic in some products (e.g. Gessato shaving treatment from Italy, Camillen 60 Fudes spray and nail oil from Germany). It is suggested for application in medical devices since usnic acid inhibits bacterial biofilm formation on polymer surfaces (Francolini et al. 2004). This compound or derivatives are valuable active compounds against serious pathogens such as vancomycin-resistant enterococci, methicillin-resistant *Staphylococcus aureus* (Elo et al. 2007), mycobacteria (Ingolfsdottir et al. 1998) or *Listeria monocytogenes* (Tomasi et al. 2006).

Phenols Most lichen substances with antibiotic activity are phenolic metabolites. Phenols are one of the largest classes of secondary biomolecules, which are characterized by the presence of aromatic rings with hydroxyl group bonded directly to an aromatic hydrocarbon group. Although they are firstly identified in plants (Cowan 1999), their presence was also observed in lichens (Odabasoglu et al. 2004; Kekuda et al. 2011; Ranković et al. 2010, 2014a, b; Mitrović et al. 2011; Kosanić and Ranković 2011; Manojlović et al. 2012). In recent years, there was a causal relationship between the total content of these compounds with biological activities recorded in a large number of lichens, which include anti-inflammatory, antiallergic, anticancer, antihypertensive, antirheumatic and antibacterial activity. Antimicrobial properties of phenolics are explained by the presence of phenol hydroxyl groups, which number is in correlation with their toxicity towards microorganisms (Cowan 1999). The possible mechanisms of their action include inhibition of extracellular microbial enzymes, deprivation of the substrates required for microbial growth or direct action on microbial metabolism through inhibition of oxidative phosphorylation, by sulfhydryl groups and some nonspecific interactions (Cowan 1999).

A correlation between phenolic constituents and antimicrobial activity has been established. Gulluce et al. (2006) found that content of total phenolic of the extracts of *Parmelia saxatilis, Platismatia glauca, Ramalina pollinaria, Ramalina polymorpha* and *Umbilicaria nylanderiana* was strongly related with their antimicrobial activity. A positive correlation was seen between the phenolic content and antibiotic activity of *Lasallia pustulata*, *Parmelia sulcata*, *Umbilicaria crustulosa* and *Umbilicaria cylindrica* (Ranković et al. 2007a). Ranković et al. (2007b) found a high correlation between antimicrobial efficacy of a macrolichens *Cladonia furcata, Parmelia caperata, Parmelia pertusa, Hypogymnia physodes* and *Umbilicaria polyphylla* and their phenolic content. Many other studies also have shown a direct correlation between the phenolic content and the antimicrobial activity (Vartia 1973; Ingolfsdottir et al. 1998; Ranković et al. 2009, 2011).

Terpenes In recent years, more data indicate the presence of terpenes in numerous representatives of lichens (Culberson 1970; Rundel 1978; Abdullah et al. 2007). One of the many functions of these compounds is their antimicrobial activity, but the mechanism of action of terpenes on microorganisms is not fully understood (Cowan 1999). According to their lipophilic nature, it is assumed to act by disrupting membrane functions of microbial cells (Cowan 1999), and some authors believe that they may cause increasing of nonspecific cell membrane permeability for the antibiotic molecule (Byron et al. 2003).

Steroids These compounds are highly often present in lichens. Steroids have been reported to have antibacterial properties, the correlation between membrane lipids and sensitivity for steroidal compound indicates the mechanism in which steroids specifically associate with membrane lipid and exerts its action by causing leakages from liposomes (Epand et al. 2007; Mohammed 2013).

Depsides, tridepsides and **tetradepsides** consist of two, three and four hydroxybenzoic acid residues linked by ester groups. These are the most numerous classes of secondary metabolites in lichens. More than 100 lichen compounds are depsidones, which have an additional ether bond between aromatic rings. **Depsidones** in lichen are believed to arise by oxidative cyclisation of depsides. It has been found that depsidone and depside compounds such as atranorin, divaricatic acid, lecanoric acid, evernic acid, salazinic acid, physodic acid and stictic acid possess important antimicrobial activity (Manojlović et al. 2012; Kosanić et al. 2013, 2014; Ranković et al. 2014a, b).

Anthraquinones and **xanthones** are also important constituents of many lichens. Anthraquinones such as parietin, parietinic acid, emodin, fallacinol and fallacinal were shown to have a high antimicrobial effect (Manojlović et al. 2002). Manojlović et al. (2010) found that the antimicrobial activity of lichen *Laurera benguelensis* is mainly related to the presence of lichexanthone.

No clear mechanisms have been identified that specifically indicate how depsides, tridepsides, tetradepsides, depsidones, anthraquinones and xanthones target microbial invasion.

3.3 Methods to Detect Antimicrobial Activity of Lichens

Methods to detect antimicrobial activity are in vitro procedures used to detect antimicrobial resistance in individual microbial isolates.

For lichens, antimicrobial activity can be examined for their extracts, as well as for their secondary metabolites.

Extraction is a process of separation of active compounds from plant material using different solvents. Extract can be prepared using various methods, such as sonification, heating under reflux, soxhlet extraction, maceration and others. Different solvent systems are available to extract the bioactive compound from natural products. The solvent systems used in extraction are selected on the basis of their capacity to dissolve the maximum amount of desired active constituents and the minimum amount of undesired constituents. The extraction of hydrophilic compounds uses polar solvents such as methanol, ethanol or ethyl acetate. For extraction of more lipophilic compounds, dichloromethane or a mixture of dichloromethane/methanol are used (Sasidharan et al. 2011). Due to the fact that extracts usually occur as a combination of various types of bioactive compounds or phytochemicals with different polarities, it is still required their separation to obtain pure compounds using different separation techniques such as TLC, column chromatography, flash chromatography, Sephadex chromatography and HPLC. The pure compounds are then used for the determination of structure and antimicrobial activity.

Currently, several methods have been applied to measure the in vitro antimicrobial activity of lichens, such as disk diffusion, agar dilution and broth microdilution (Fig. 3.1).

The **disk diffusion** method allows for the simultaneous testing of a large number of antimicrobials in a relatively easy and flexible manner. In this method, the

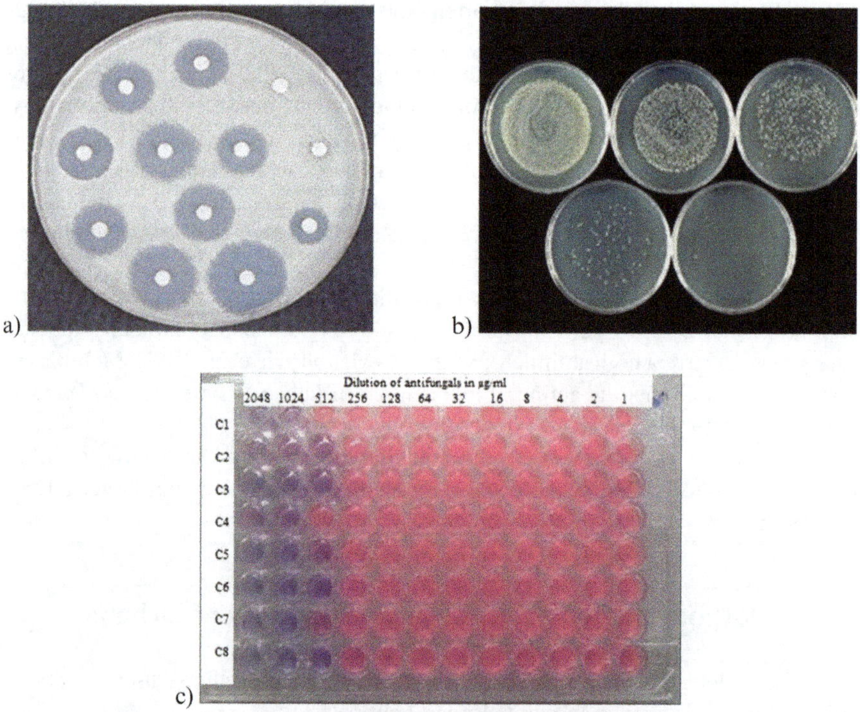

Fig. 3.1 Disk diffusion (**a**), agar dilution (**b**) and broth microdilution (**c**) methods to measure the in vitro antimicrobial activity of test samples

bacterial inoculum is adjusted to certain concentration, inoculated onto the entire surface of a Mueller–Hinton agar (MHA) plate with a sterile cotton-tipped swab to form an even lawn. The paper disks impregnated with diluted antibiotic solution was placed on the surface of each MHA plate using a sterile pair of forceps. Then the plates were incubated aerobically and the diameter of zone inhibition was measured by a ruler or caliper. Based on the diameter of the inhibition zone, the results are then assigned to three categories, susceptible, intermediate or resistant. The bigger the diameter of the inhibition zone, the more susceptible is the microorganism to the antimicrobial.

Agar dilution is a quantitative susceptibility testing method because MIC values can be obtained using the method. In this method, twofold serial dilutions of an antibiotic made in MHA medium and then bacterial suspensions were inoculated on the MHA using a Cathra replicator with 1 mm pins. It has been studied extensively as a method for the bacteria growing aerobically. The advantages of agar dilution include the ability to simultaneously test the susceptibility of a number of bacteria in one plate and the ability to test susceptibility of fastidious organisms since the agar with supplements is able to adequately support the bacteria growth. However, agar dilution is not commonly used in most microbiology laboratories due to the time-consuming and labor-intensive.

Broth microdilution is another quantitative reference method routinely used in clinical laboratories. In this method, susceptibility panel in 96-well microtiter plates were containing various concentration of antimicrobial agents. Then, standardized numbers of bacteria was inoculated into the wells of 96-well microtiter and incubate overnight at 35 °C. This method is often used to obtain minimal inhibitory concentration (MIC) values. The MIC value was observed as the lowest concentration where no viability was observed in the wells of 96-microwell plates after incubation. It is a widely utilized method, allowing for the simultaneous testing of multiple antimicrobials with ease particularly when commercially prepared microtiter trays are used. Compared with agar-based method, broth microdilution can decrease much labor and time. However, limitations of the method primarily are associated with the lack of or poor growth of many anaerobic microorganisms. Testing some fastidious anaerobes gives inconsistent and unreliable results because of poor growth of strains due to excessive exposure to oxygen during the set-up procedure (CLSI 2009).

Among the three methods, disk diffusion seems to be the most popular method used to examine the antimicrobial activity of natural antimicrobials including lichens. Although the method is relatively inexpensive and easy to perform, there are several disadvantages. Since disk diffusion measures the inhibition zone size which is then converted to categories of susceptible/intermediate/resistant, this method is unable to obtain MIC values (Dickert et al. 1981). Also, it has been reported (Klancnik et al. 2010) that this method is not always reliable for determining the antimicrobial activity of natural antimicrobials, i.e. lichen extract, because the polarity of the natural compounds can affect the diffusion of compounds onto the culture medium. Compounds with less polarity diffused slower than more polar ones (Moreno et al. 2006). Due to these concerns, disk diffusion may not be a suitable one to determine the antimicrobial activity of natural compounds. Besides, similar to other agar-based methods, disk diffusion is labor-intensive and time-consuming (Klancnik et al. 2010).

In contrast, agar dilution and broth microdilution methods are able to overcome some of the limitations of the disk diffusion method. Not only are they more convenient for routine antimicrobial susceptibility testing of bacteria in clinical laboratories, they are capable of drawing quantitative conclusions by determining the MIC values for antimicrobials, as opposed to qualitative data generated by the disk diffusion method (Kim and Kim 2007).

Apart from these most commonly used methods, some other methods are also used to determine the antimicrobial activity. One of these is the antimicrobial gradient method (Etest) which combines the principle of dilution methods with that of diffusion methods in order to determine the MIC value. It is based on the possibility of creating a concentration gradient of the antimicrobial agent tested in the agar medium. The Etest is a commercial version of this technique. In the procedure, a strip impregnated with an increasing concentration gradient of the antimicrobial agent from one end to the other is deposited on the agar surface, previously inoculated with the microorganism tested (Balouiri et al. 2016).

Time-kill test is the most appropriate method for determining the bactericidal or fungicidal effect. It is a strong tool for obtaining information about the dynamic interaction between the antimicrobial agent and the microbial strain. The time-kill test reveals a time-dependent or a concentration-dependent antimicrobial effect (Balouiri et al. 2016).

ATP bioluminescence assay is based on the capacity to measure adenosine triphosphate (ATP) produced by bacteria or fungi. As ATP is the chemical form of energy of all living cells, it is present in more or less a constant amount in a cell. Therefore, its quantification is used to estimate the microbial population in a sample. D-luciferin in the presence of the ATP undergoes conversion by luciferase to oxyluciferin that generates light. The quantity of the emitted light is measured by a luminometer and expressed as relative light unit (RLU) which can be converted into RLU/mole of ATP. Thus, there is a linear relationship between cell viability and luminescence measured (Balouiri et al. 2016).

Addition to those listed, some bioautographic techniques, i.e. agar diffusion, direct bioautography and agar overlay assay, also have been described for the investigation of antimicrobial compounds.

3.4 Lichen Extracts as Potential Antibacterial and Antifungal Agents

The screening of lichen extracts has been of great interest to scientists for the discovery of new compound effective in the treatment of microbial infection. There are various reports on the antimicrobial activity of crude lichen extracts.

The first study on the antibiotic properties of lichens was carried out by Burkholder et al. (1944). He tested 42 lichens for antibiotic property and 27 were reported to inhibit growth of bacteria. A number of lichen extracts were screened for antibacterial activity in the 1950s and in many cases activity was confirmed against mycobacteria and Gram-positive organisms (Stoll et al. 1950). A review of the work performed during this period is presented in Vartia (1973). More recent reports include a study describing the antimicrobial screening of lichen extracts and subsequent isolation of compounds with a broad spectrum of activity against filamentous fungi, yeast as well as Gram-positive and -negative bacteria.

Extracts of Andean lichens *Protousnea poeppigii* and *Usnea florida* demonstrated antimicrobial activity against the pathogenic fungi *Microsporum gypseum, Trichophyton mentagrophytes* and *T. rubrum.* Gram-positive *Staphylococcus aureus* and *Bacillus subtilis* were sensitive to methanolic extracts of four different Antarctic lichen species (Paudel et al. 2008). Ranković et al. (2007a, b) tested aqueous, acetone and methanol extracts of *Cladonia furcata, Parmelia caperata, Parmelia pertusa, Hypogymnia physodes, Umbilicaria polyphylla, Lasallia pustulata, Parmelia sulcata, Umbilicaria crustulosa* and *Umbilicaria cylindrica* from Serbia on six species of bacteria and ten species of fungi. The strongest activity was observed with methanol extracts of *Parmelia pertusa* and *Parmelia sulcata* and the weakest activity was manifested by *Parmelia caperata* and *Umbilicaria cylindrica.* Aqueous extracts of all tested lichen species were inactive. *Bacillus mycoides* was the most sensitive bacterial species tested, whereas *Candida albicans* was the most sensitive fungal species examined. Other studies monitored *Ramalina farinacea* and 69 species of lichens from New Zealand and showed their inhibitory effect against a lot of bacteria such as *Bacillus, Pseudomonas, E.coli, Streptococcus, Staphylococcus, Enterococcus* and *Mycobacterium* (Esimone and Adikwn 1999; Perry et al. 1999). Behera et al. (2005) reported that acetone, methanol and light petroleum extracts of lichen *Usnea ghattensis* were effective against *Bacillus licheniformis, B. megaterium, B. subtilis* and *S. aureus*. Also, Karagoz et al. (2009) evaluated aqueous and ethanol extracts of 11 different species from Turkey and determined potent antibacterial activity of aqueous extract of *Peltigera polydactyla* and ethanol extract of *Ramalina farinacea*. Recently, Mitrović et al. (2011) studied antibacterial and antifungal activity of methanol extracts of five lichen species (*Flavoparmelia caperata, Evernia prunastri, Hypogymnia physodes* and *Cladonia foliacea*). The analysis of their antibacterial potential were performed on 15 strains of bacteria and revealed the strongest inhibitory effect, especially on Gram (+) bacteria, of *Hypogymnia physodes* and *Cladonia foliacea*. In case of fungi, *Evernia prunastri* exerted the best effect on yeasts, while *Hypogymnia physodes* were better on filamentous fungi. Similarly, acetone extracts of the lichens *Cladonia furcata, Lecanora atra* and *Lecanora muralis* were studied for their antimicrobial potential (Ranković et al. 2011). The antimicrobial activity was estimated by determination of the minimal inhibitory concentration by the broth microdilution method against six species of bacteria and ten species of fungi. Extract of *Cladonia furcata* was the most active antimicrobial agent with minimum inhibitory concentration values ranging from 0.78 to 25 mg/mL, while the lowest activity showed *Lecanora muralis.* In similar research, antifungal activity of hexane, ethyl acetate and methanol extracts of *Parmelia reticulata* was evaluated against soil-borne pathogenic fungi, namely, *Sclerotium rolfsii, Rhizoctonia solani, R. bataticola,*

Fusarium udum, Pythium aphanidermatum and *P. debaryanum* by Goel et al. (2011). Maximum antifungal activity was exhibited by hexane and ethyl acetate extracts against most of the test pathogens.

Acetone, diethyl ether and ethanol extracts of the lichen *Cetraria aculeata* for their antimicrobial activity have been evaluated. The extracts were found active against *Escherichia coli, Staphylococcus aureus, Aeromonas hydrophila, Proteus vulgaris, Streptococcus faecalis, Bacillus cereus, Bacillus subtilis, Pseudomonas aeruginosa* and *Listeria monocytogenes*. However, no antimicrobial activity against the fungi was detected (Türk et al. 2003). The lichen extract almost increased by twofold in the presence of the stock solution of the colloidal silver concentrate. The ointment containing extract of lichen *Ramalina farinacea* exhibited antimicrobial activities against *Escherichia coli, Salmonella typhi, Aspergillus niger and Candida albicans* (Ofokansi and Esimone 2005).

The aqueous and ethanol extracts prepared from some lichen species were evaluated for antibacterial activity against six standard strains (*Escherichia coli, Pseudomonas aeruginosa, Bacillus subtilis, Klebsiella pneumoniae, Staphylococcus aureus* and *Staphylococcus epidermidis*) and two environmental strains (*Aeromonas*). The aqueous and ethanol extracts showed a variable range of antibacterial activity to both standard strains and environmental strains. Similarly, the aqueous extract of *Peltigera polydactyla* and the ethanol extract of the *Ramalina farinacea* exhibited potent antibacterial activities (Karagoz et al. 2009).

The antimicrobial activity of the acetone, diethyl ether and ethanol extracts of the lichen *Cetraria aculeata* tested against different pathogenic bacteria and fungi showed antimicrobial effect only against bacteria but not against fungi. In related study, *Roccella belangeriana* were extracted from different solvents like acetone, methanol, diethyl ether, ethanol, ethyl acetate, petroleum ether, chloroform and aqueous extracts and tested against 12 bacterial strains. A maximum antibacterial activity was observed from chloroform extracts against *Enterococci* sp. and minimum activity was observed from ethyl acetate extract against *Klebsiella pneumoniae, Enterococci* sp., *Salmonella* sp. and *Shewanella* sp. (Karthikaidevi et al. 2009). In addition, antibacterial and antifungal activity of the acetone, methanol and aqueous extracts of the lichen *Lecanora frustulosa* and *Parmeliopsis hyperopta* has been screened in vitro against the *Bacillus mycoides, Bacillus subtilis, Staphylococcus aureus, Enterobacter cloacae, Escherichia coli, Klebsiella pneumoniae, Aspergillus flavus, Aspergillus fumigatus, Botrytis cinerea, Candida albicans, Fusarium oxysporum, Mucor mucedo, Paecilomyces variotii, Penicillium purpurescens, Penicillium verrucosum* and *Trichoderma harsianum*. Tested lichen species also showed strong activity against both bacteria and fungi (Kosanić et al. 2010). This would perhaps indicate that the lichens would be used in the treatment of various diseases.

According to Schmeda-Hirschmann et al. (2008), dichloromethane and methanol extracts of *Protousnea poeppigii* had strong antifungal effects against the fungal pathogens *Microsporum gypseum, Trichophyton mentagrophytes* and *T. rubrum*. The extracts were also active against the yeasts *Candida albicans*, *C. tropicalis*, *Saccharomyces cerevisiae* and the filamentous fungi *Aspergillus niger*, *A. flavus* and *A. fumigatus*, but with much higher strength. In the same assay, extracts of *Usnea florida* also showed strong antifungal properties. Methanol extracts of five lichens from Antarctica (*Caloplaca regalis*, *Caloplaca* sp., *Lecanora* sp., *Ramalina terebrata*, *Stereocaulon alpinum*) exhibited target-specific antibacterial activity, especially strong against Gram-positive bacteria, compared to previously described lichens (Paudel et al. 2008). Whiton and Lawrey (1982) reported that ascospore germination of *Sordaria fimicola* was significantly inhibited by evernic and vulpinic acids. Aqueous, ethanol and ethyl acetate extracts of *Alectoria sarmentosa* and *Cladonia rangiferina* were found to have moderate antifungal action against different species of fungi, including human pathogens (Ranković and Mišić 2007), ethanol extracts showing the highest activity. Halama and Van Haluwin (2004) reported that acetone extracts of *Evernia prunastri* and *Hypogymnia physodes* showed a strong inhibitory effect on the growth of some plant pathogenic fungi, i.e. *Phytophthora infestans*, *Pythium ultimum* and *Ustilago maydis*.

Antimicrobial features of acetone, methanol and aqueous extracts of lichens of *Cladonia furcata*, *Parmelia caperata*, *Parmelia pertusa*, *Hypogymnia physodes* and *Umbilicaria polyphylla* were investigated by Ranković et al. (2009) by two different methods at the same time. Testing of antimicrobial activities of extracts from five species of lichens was performed by disk diffusion test in relation to Gram-positive and Gram-negative bacteria and fungal organisms, and through determination of minimal inhibitory concentration (MIC) by Broth Tube Dilution method. They found that acetone and methanol extracts of all investigated lichens in different concentrations manifested selective antibacterial and antifungal activity. That activity was more evident in relation to Gram-positive, than Gram-negative bacteria and fungal organisms. Acetone and methanol extracts of lichens *Parmelia pertusa*, *Hypogymnia physodes* and *Umbilicaria polyphylla* inhibited the growth of all tested microorganisms, most of all of lichens *Cladonia furcata* and *Parmelia caperata*. Although the methanol extracts were generally the most active against the test organisms, the lowest MIC value was measured for acetone extract of species *Cladonia furcata* 0.39 mg/mL in relation to bacterium *Bacillus subtilis*. Aqueous extracts of investigated lichens were inactive against all tested organisms.

Santiago et al. (2010) examined antibacterial activity of 63 lichens collected from different sites in Luzon Island, Philippines. Lichen crude extracts were then tested against Gram-positive bacteria (*Bacillus subtilis* and *Staphylococcus aureus*) and Gram-negative bacteria (*Escherichia coli* and *Pseudomonas aeruginosa*) using the paper disk diffusion assay. Their results showed that all 45 tested extracts inhibited at least one of the test bacteria. However, only 38 extracts were found to be very active (>19 mm zone of inhibition) against Gram-positive bacteria. Similarly, Santos et al. (1964) observed that 30 of the 38 lichen extracts tested inhibited at least 1 of the 12 test microorganisms, particularly the Gram-positive test bacteria. Other studies also confirmed several lichens to be active against Gram-positive microorganisms. For example, Saenz et al. (2006) found that four species of lichens, namely, *Ramalina canariensis*, *R. subfarinacea*, *Cladonia firma* and *Lecanora muralis*, were most active against Gram-positive bacteria.

In vitro antifungal activity of acetone, methanol and chloroform extracts of *Parmotrema tinctorum* was investigated against ten plant pathogenic fungi viz.

Aspergillus niger, A. flavus, A. fumigatus, Alternaria alternata, Fusarium oxysporum, F. solani, F. roseum, Ustilago sp., *Albugo candida* and *Penicillium citrinum*, with reference to commercially available synthetic antifungal drug ketoconazole (positive control) using disk diffusion assay (Tiwari et al. 2011). Methanol extract was most effective against all investigated fungi followed by acetone and chloroform extract. Principal component analysis (PCA) concluded that though ketoconazole was effective against five of the investigated fungi, the extracts of *Parmotrema tinctorum* were more effective against rest of the five broad-spectrum plant pathogenic fungi (*A. fumigatus, F. solani, F. roseum, P. citrinum* and *Ustilago* spp.).

In study described by Karthikaidevi et al. (2009), antimicrobial activity of the mangrove lichen *Roccella belangeriana* collected from the Gulf of Mannar Biosphere Reserve area was tested. The lichen was extracted in different solvents: acetone, methanol, diethyl ether, ethanol, ethyl acetate, petroleum ether, chloroform and water and tested against 14 bacterial strains and tri-fungal strains by Agar well diffusion assay. Regarding antibacterial activity, the maximum zone of inhibition was recorded in methanol extracts against *Vibrio cholerae* and the minimum zone of inhibition was in ethyl acetate extract against *Klebsiella pneumoniae, Enterococci* sp., *Salmonella* sp. and *Shewanella* sp. Regarding the antifungal activity, the maximum zone of inhibition was recorded against *Aspergillus niger* and the minimum was noted against *Rhizopus* sp.

Five common lichens (*Cladonia* sp., *Everniastrum* sp., *Parmelia* sp., *Stereocaulon* sp. and *Usnea* sp.) of Darjeeling hills were extracted from different solvents like ethanol, methanol, petroleum ether, chloroform, and aqueous extracts and tested against four Gram-positive and four Gram-negative bacterial strains. Ethanol extracts exerted stronger inhibitory action followed by methanol extracts. Aqueous extracts manifested less activity to the tested microorganisms (Sharma et al. 2012).

Srivastava et al. (2013) investigated antimicrobial activity of the acetone, methanol and ethanol extracts of some common lichen species such as *Usnea longissima, Everniastrum cirrhatum, Peltigera polydactylon and Sulcaria sulcata* were screened in vitro against six clinically important pathogenic bacteria, *Staphylococcus aureus, Streptococcus faecalis, Bacillus cereus, Pseudomonas aeruginosa, Salmonella typhimurium* and *Escherichia coli* by Kirby-Bauer technique of disk diffusion method. Minimum inhibitory concentration was taken out by broth microdilution method according to the NCCLS guidelines. It was found that acetone, methanol and ethanol extracts of the investigated lichens showed relatively strong antimicrobial activity against all the Gram-positive bacteria and two Gram-negative bacteria. The lowest MIC value was observed to be as low as 6.25 μg/mL against *B. cereus* of *U. longissima*.

Antibiotic properties of acetone and methanol extracts from 34 North American lichens were screened against four pathogenic bacteria by Shrestha et al. (2014). The micro-well dilution method was used to determine the minimum inhibitory concentration. Most of the lichen extracts demonstrated inhibitory effects against *Staphylococcus aureus, Pseudomonas aeruginosa* and methicillin-resistant *S. aureus* with MIC values ranging from 3.9 to 500 μg/mL. In addition, extracts from three species,

Letharia Columbiana, Letharia vulpina and *Vulpicida canadensis* (MIC = 125–500-μg/mL) were also effective against *Escherichia coli*. Generally, acetone extractions were found to be more effective than methanol extractions.

In the study described by Vivek et al. (2014), antibacterial potential of three *Parmotrema* species viz., *P. tinctorum, P. grayanum* and *P. praesorediosum* from India was tested against three Gram-positive and five Gram-negative bacteria by Agar well diffusion assay. The lichen extracts showed dose-dependent antibacterial activity. Overall, the lichen extracts were more inhibitory to Gram-positive bacteria than Gram-negative bacteria. *P. grayanum* displayed high inhibitory activity against test bacteria. Balaji and Hariharan (2007) reported marked antimicrobial efficacy of dichloromethane extract of *P. praesorediosum*. Kumar et al. (2010) showed the antibacterial activity of methanol extract of *P. pseudotinctorum*. Sinha and Biswas (2011) reported the antibacterial efficacy of solvent extracts of *P. reticulatum* from Sikkim, India. Verma et al. (2011) found antibacterial efficacy of solvent extracts of *P. nilgherrensis* and *P. sanctiangelii* collected from Karnataka, India. Chauhan and Abraham (2013) showed the inhibitory effect of methanol extract of *Parmotrema* sp. against clinical isolates of bacteria. In addition, Javeria et al. (2013) showed the inhibitory efficacy of solvent extracts of *P. nilgherrense* against drug-resistant bacteria.

In the study described by Ranković et al. (2012), acetone lichen extracts obtained from *Usnea barbata* showed a moderate antibacterial and antifungal activity. It inhibited the microorganisms tested at concentrations from 0.125 to 12.5 mg/mL. The acetone extract from *T. candida* inhibited all the tested microorganisms, but at higher concentrations. In related research, *Evernia prunastri* and *Pseudoevernia furfuracea* lichens were screened for their antimicrobial effects by Kosanić et al. (2013) who found varying antimicrobial success in inhibition of Gram-positive, Gram-negative bacteria and fungi and *Pseudoevernia furfuracea* was found to be the most effective.

Kosanić et al. (2014b) were extracted with acetone the three *Cladonia* species (*C. furcata, C. rangiferina* and *C. pyxidata*) in order to investigate their antimicrobial effect. As test organisms in this study were used *Bacillus mycoides, B. subtilis, Staphylococcus aureus, Escherichia coli* and *Klebsiella pneumonia, Aspergillus flavus, A. fumigatus, Candida albicans, Penicillium purpurescens* and *P. verrucosum*. They obtained results showed that extracts from *C. furcata* and *C. rangiferina* showed similar antibacterial and antifungal activity. They inhibited the microorganisms tested at concentrations from 0.78 to 25 mg/mL, while extracts from *C. pyxidata* inhibited all the tested microorganisms, but at higher concentrations.

Lecanora muralis, Parmelia saxatilis, Parmeliopsis ambigua, Umbilicaria crustulosa and *Umbilicaria polyphylla* were tested for their antibacterial and antifungal activity (Kosanić et al. 2014b). The antimicrobial activity was estimated by determination of the minimal inhibitory concentration by the broth microdilution method against six species of bacteria and ten species of fungi, and it has been found that of the lichens tested *Umbilicaria polyphylla* had largest antimicrobial activity with minimum inhibitory concentration values ranging from 0.78 to 1.56 mg/mL.

Kosanić et al. (2014c) investigated antibacterial and antifungal properties of acetone extracts of the lichens *Parmelia arseneana* and *Acarospora fuscata*. The extract from *P. arseneana* showed stronger antibacterial and antifungal activity. They inhibited the microorganisms tested at concentrations from 0.078 to 10 mg/mL. Extracts from *A. fuscata* also inhibited all of the tested microorganisms, but at higher concentrations. *Stereocaulon paschale* lichen as possible antibacterial and antifungal agent was studied by same authors (Ranković et al. 2014a, b) against five bacteria and five fungi. The methanol extract of the tested lichen showed relatively similar antibacterial and antifungal activity. The MIC for extract related to the tested bacteria and fungi were 0.625–10 mg/mL. The lowest measured MIC value (0.625 mg/mL) was related to the *Bacillus subtilis* and *B. cereus* species. Antibacterial activity of methanol, acetone, petroleum ether and chloroform extracts of lichens *Heterodermia boryi*, *Parmotrema stuppeum*, *Usnea nilgirica*, *Pyxine* sp. and *Parmotrema melanothrix* has been screened in vitro against *E. coli*, *Pseudomonas* sp., *Klebsiella* sp., Coagulase-negative *Staphylococcus aureus*, *Streptococcus viridans*, *Staphylococcus aureus*, *Acinetobacter* sp. and MRSA by disk diffusion method (Senthil Prabhu and Sudha 2015). The extracts of all the lichens exhibit variable range of antibacterial activity against all types of bacterial pathogens. Compared with all the above samples *Parmotrema stuppeum* is found to have the highest activity against all the pathogens. It is particularly effective against *Staphylococcus aureus* and MRSA. However the petroleum ether and chloroform extracts were found to be ineffective against *Streptococcus viridans* and CT-ve *Staphylococcus*. All the tested extracts were highly effective against *E. coli*, *Pseudomonas* sp., *Klebsiella* sp. and *Acinetobacter* sp.

Antimicrobial activity of the ethanolic, isopropanolic, acetone, DMSO and aqueous extracts of the two lichen species from Ecuadorian highland, *Usnea* sp. and *Stereocaulon* sp. were explored in vitro against bacteria *Bacillus subtilis*, *Escherichia coli* and *Staphylococcus aureus* by the disk diffusion method (Matvieieva et al. 2015). Also the minimal inhibitory concentration (MIC) was determined. The strongest antimicrobial activity was found in DMSO extract of *Usnea* sp. compared to antibacterial activity of ciprofloxacin and cefazolin antibiotics. The inhibition zone was 28 mm, 30 mm and 31 mm (DMSO extract, ciprofloxacin and cefazolin, respectively) in case of *B. subtilis* usage as the test bacteria. MIC value for *Usnea* sp. and *Stereocaulon* sp. DMSO extracts was 0.4 mg/mL. The *S. aureus* sensitivity to lichen DMSO extracts was comparable to sensitivity of these microorganisms to tetracycline and vancomycin. Thereby, most kinds of extracts (ethanol, isopropanol, hexane, DMSO and acetone solvents) from Ecuadorian lichens *Usnea* sp. and *Stereocaulon* sp. with the exception of aqueous *Stereocaulon* sp. extracts possessed antibacterial activity against *B. subtilis*. *E. coli* was resistant to all kinds of extracts.

Sariozlu et al. (2016) investigated antibacterial, antifungal activity and MIC values of the acetone, methanol and chloroform extracts of the lichen *Bryoria capillaris* against Gram-positive and Gram-negative bacteria, yeasts and filamentous fungi using disk diffusion method. The obtained results were shown that the tested extracts have considerable antimicrobial effect to tested pathogenic microorganism.

In study described by Ristić et al. (2016a), antimicrobial activity of the acetone extracts of *Melanelia subaurifera* and *Melanelia fuliginosa lichens* was determined. Both tested species have selective antimicrobial activity. The MIC for tested extracts ranged from 1.25 to 20 mg/mL against bacteria and from 2.5 to 20 mg/mL against fungi. Kosanić et al. (2016) investigated antimicrobial activity of the *Lasallia pustulata* methanol extract against five bacteria (*Staphylococcus aureus, Bacillus cereus, B. subtilis, Escherichia coli* and *Proteus mirabilis*) and five fungi (*Aspergillus flavus, Candida albicans, Fusarium oxysporum, Penicillium purpurescens* and *Trichoderma harsianum*). The methanol extract from the lichen tested showed relatively similar antibacterial and antifungal activity. The MIC for the extract in the tested bacteria and fungi were 0.625–20 mg/mL. The lowest measured MIC value (0.625 mg/mL) was related to the *B. subtilis* species.

Two *Ramalina* lichens were explored for their antimicrobial effect by Ristić et al. (2016b) against five species of bacteria and 10 species of fungi. Extract of *R. fraxinea* exhibited no inhibition of growth against bacterial strain *E. coli*. This extract exhibited no inhibition of growth against fungal strain *A. flavus* as well. Acetone extract of lichen species *R. fastigiata* showed significant antimicrobial properties. This lichen extract had an impact on all tested organisms. Similarly, antibacterial and antifungal activity of an extract of three *Ramalina* species (Ramalinaceae) viz. *R. hossei, R. conduplicans* and *R. pacifica* obtained by maceration process were investigated by Ankith et al. (2017). The antibacterial and antifungal activity of lichen extracts was carried out by Agar well diffusion and Poisoned food technique, respectively. Overall, *B. cereus* and *E. coli* were inhibited to higher extent and least extent, respectively, by extracts of *Ramalina* species. *R. pacifica* and *R. hossei* inhibited bacteria to highest and least extent, respectively. In the case of antifungal activity, marked and least inhibitory activity was shown by an extract of *R. hossei* and *R. pacifica*, respectively. Among fungi, *Alternaria* sp. and *Fusarium* sp. were inhibited to highest and least extent, respectively. Prior to this research, Cansaran et al. (2007) tested antimicrobial potential of five *Ramalina* species: *R. fastigiata, R. capitata, R. polymorpha, R. pollinaria* and *R. fraxinea* from Turkey. The agar disk diffusion method was used for screening of antimicrobial effect. Among the tested *Ramalina species, R. fastigiata* showed the maximum antibacterial influence. The extract of *R. fastigiata* showed the highest antimicrobial effect towards *Bacillus subtilis* and *B. megaterium*.

The antimicrobial activity of the lichen *Pleurosticta acetabulum* was evaluated by Tomović et al. (2017). They found that the MIC for the acetone extract of *P. acetabulum* fluctuated in a range of 1.25–20 mg/mL for bacteria and 5–20 mg/mL for fungi. The extract did not show inhibitory activity against *E. coli* and *A. flavus* which have been shown to be the most resistant bacteria and fungi. Bondarenko et al. (2017) examined antimicrobial activity of *Parmelia vagans*. Extracts of *Parmelia vagans* were initially obtained by using various solvents to find the most efficient extraction of antimicrobial compound(s). The antimicrobial activity of the extracts was evaluated using the Resazurin cell viability assay; the microbes investigated were *Staphylococcus aureus*, *Bacillus subtilis*, *Streptococcus pneumoniae*, *Klebsiella pneumoniae* and *Escherichia coli*. It was observed that extracts had no inhibitory effect on the growth of *E.coli* and *K. pneumonia* at the tested concentrations, but inhibition of growth was observed in *S. aureus, B. subtilis and S. pneumoniae*. They found that extracts obtained using ethyl acetate have the highest antimicrobial activity.

Extracts of five *Cladonia* lichens were screened for their antimicrobial effects by the broth microdilution method (Kosanić et al. 2018). Their results showed that the higher antibacterial activity was in the acetone extracts of the lichen *C. fimbriata*, which had an inhibiting effect on all of the tested bacteria. Similar inhibitory activity but with a slightly higher concentrations showed an extract of *C. furcata*. The MIC for the acetone extracts of the lichen *C. subulata, C. foliacea* and *C. rangiferina* were within the range 1.25–20 mg/mL in relation to the bacteria. These extracts did not show inhibitory activity against *E. coli*, which was proved to be the most resistant bacteria on all of the tested samples. Extract of lichen *C. subulata* inhibited all tested fungi. Extracts of other lichen species did not exhibit inhibitory effect on the fungus *A. flavus*. The MIC fluctuated in the range from 1.25 to 20 mg/mL.

Dixit et al. (2018) evaluated the antimicrobial and antifungal properties of ichen extract (*Usnea* sp. and *Parmotrema* sp.) against some human pathogenic bacterial and fungal species. In this study, the lichen was extracted in acetone and methanol. The bacterial isolates examined in this study were *S. aureus*, *Staphylococcus* sp. and *E.coli*. The fungal isolates used in this study were *Aspergillus niger, A. flavus*, *Candida* sp. and *Tricophyton* sp. In vitro antimicrobial activity was tested initially by Kirby-Bauer technique of disk diffusion method and was confirmed by minimum inhibitory concentration using broth microdilution method. The acetone and methanol extracts of *Parmotrema* sp. had a potential towards antibacterial activity more as compared to *Usnea* sp. Methanol extracts of *Parmotrema* sp. shows better effect against *Candida* sp., *A. flavus* and *Tricophyton* sp. fungal pathogens.

3.5 Studies on Antimicrobial Activities of Lichen Secondary Metabolites

As mentioned above, there are many studies on the antimicrobial activity of crude lichen extracts. However, studies on antimicrobial activity of lichen compounds are scarce and scattered. Lichens have been found to contain a variety of secondary lichen substances with strong antimicrobial activity (Table 3.1). Literature sources mentioning data for numerous lichen substances with examined antimicrobial activity are mentioned below.

Atranorin (from *Physcia aipolia*), fumarprotocetraric acid (from *Cladonia furcata*), gyrophoric acid (from *Umbilicaria polyphylla*), lecanoric acid (from *Ochrolechia androgyna*), physodic acid (from *Hypogymnia physodes*), protocetraric acid (from *Flavoparmelia caperata*), stictic acid from (*Parmelia conspersa*) and usnic acid (from *Flavoparmelia caperata*) showed relatively strong antimicrobial effects against six bacteria and ten fungi, among which were human, animal and plant pathogens, mycotoxin producers and food-spoilage organisms (Ranković and Mišić 2008; Ranković et al. 2008). Usnic acid was found to be the strongest

Table 3.1 List of lichen secondary metabolites used to evaluate antimicrobial activity

Lichen compounds	References
Lecanoric acid	Gomes et al. (2003), Ranković and Mišić (2008), Honda et al. (2010), Ristić et al. (2016a)
Atranorin	Kumar and Müller (1999), Yilmaz et al. (2004), Turk et al. (2006), Ranković et al. (2008, 2014a, b), Kosanić et al. (2014a)
Zeorin	Kosanić et al. (2010)
Gyrophoric acid	Candan et al. (2006), Ranković et al. (2008), Kosanić et al. (2014c)
Stenosporic acid	Candan et al. (2006)
Protocetraric acid	Tay et al. (2004), Ranković and Mišić (2008), Manojlović et al. (2012)
Fumarprotocetraric acid	Yilmaz et al. (2004), Ranković et al. (2008), Kosanic et al. (2014)
Stictic acid	Ranković and Mišić (2008)
Salazinic acid	Candan et al. (2007), Manojlović et al. (2012)
Usnic acid	Lauterwein et al. (1995), Perry et al. (1999), Yilmaz et al. (2004), Ivanova et al. (2004), Tay et al. (2004), Ranković et al. (2008, 2012, 2014a, b), Schmeda-Hirschmann et al. (2008), Paudel et al. (2010), Ramos and Almeida da Silva (2010)
Vulpinic acid	Whiton and Lawrey (1982), Lawrey (1986), Lauterwein et al. (1995)
Evernic acid	Whiton and Lawrey (1982), Lawrey 1986; Halama and Van Haluwin (2004), Kosanić et al. (2013)
Lobaric acid	Ingolfsdottir et al. (1998), Piovano et al. (2002), Sundset et al. (2008).
Physodic acid	Turk et al. (2006), Ranković et al. (2008, 2014a, b), Kosanić et al. (2013)
Protolichesterinic acid	Ingolfsdottir et al. (1998), Türk et al. (2003)
Norstictic acid	Tay et al. (2004), Honda et al. (2010), Ranković et al. (2014a, b)
Ramalin	Paudel et al. (2008, 2010)
Barbatic acid	Martins et al. (2010)
Divaricatic acid	Piovano et al. (2002), Kosanić et al. (2010), Oh et al. (2018)
Difractaic acids	Piovano et al. (2002), Honda et al. (2010)
Umbilicaric acid	Buçukoglu et al. (2013).
Homosekikaic acid	Sisodia et al. (2013)
Sekikaic acid	Sisodia et al. (2013)
Parietin	Manojlović et al. (2002, 2005)
Parietinic acid	Manojlović et al. (2002)
Emodin	Manojlović et al. (2002)
Fallacinal	Manojlović et al. (2002)
Fallacinol	Manojlović et al. (2002)
Isodivaricatic acid	Schmeda-Hirschmann et al. (2008)
Divaricatinic	Schmeda-Hirschmann et al. (2008)
Hirtusneanoside	Rezanka and Sigler (2007)
Neuropogonines A, B and C	Ivanova et al. (2002)
Hypostictic acid	Honda et al. (2010)
Norstictic acid	Honda et al. (2010)
Secalonic acid	Honda et al. (2010)

(continued)

Table 3.1 (continued)

Lichen compounds	References
Psoromic acid	Tasdemir and Franzblau (2007)
Vulpic acid	Tasdemir and Franzblau (2007)
Parietin	Basile et al. (2015)
Obtusatic acid	Ristić et al. (2016a)
Methyl evernate	Ristić et al. (2016a)
2′-O-methyl anziaic acid	Ristić et al. (2016a)
Barbatolic acid	Sariozlu et al. (2016)
Usimine A, B and C	Paudel et al. (2010)

antimicrobial agent (comparable to streptomycin), and physodic and stictic acids the weakest.

The antifungal activity of ten depsidones and five depsides was evaluated, as well as the antibacterial of these compounds and three additional depsides and the one diaryl ether; all of them were isolated from lichens growing in Chile (Piovano et al. 2002). Obtained results showed, in general, negative activity against yeast and filamentous fungi at concentrations of 250 mg/mL. However, divaricatic and difractaic acids, and to a lesser degree lobaric acid, presented activity against *Microsporum gypseum, Trichophyton mentagrophytes, T. rubrum* and *Epidermophyton floccosum*, all of them being dermatophyte fungi which cause skin infections. Regarding antibacterial activity, results indicated that against Gram-negative bacteria the 19 compounds are inactive. In contrast against Gram-positive bacteria, a marked action can be observed for seven compounds.

Anthraquinones (parietin, parietinic acid, emodin, fallacinal and fallacinol) from *Caloplaca schaereri* were tested for antimicrobial activity using *Bacillus subtilis, Staphylococcus aureus, Escherichia coli, Pseudomonas fluorescens, Candida albicans, Trichoderma harzianum, Aspergillus niger* and *Penicillium verrucosum* (Manojlović et al. 2002). All the anthraquinones tested showed potent antibacterial activity against *B. subtilis, S. aureus* and *P. fluorescens* (MIC 20–320 μg/mL), but only parietinic acid showed any activity against *E. coli* (MIC 160 μg/mL). Their effects are generally most potent on *B. subtilis* and *P. fluorescens*. Fallacinol was most potent against *S. aureus*. Fallacinol was the most active (potent) of the isolated compounds against all the fungi tested but was particularly active against *T. harzianum, A. niger* and *P. verrucosum* (MIC 10–40 μg/mL). Parietinic acid also showed potent antifungal effects on the fungi tested (MIC 20–80 μg/mL), while parietin had MIC values of 80, 40 and 20 μg/mL for *C. albicans, P. verrucosum, A. niger* and *T. harzianum*, respectively. Emodin showed MIC values of 20–40 μg/mL for *A. niger, T. harzianum* and *P. verrucosum* but was much less effective against *C. albicans* (MIC 80 μg/mL).

According to Schmeda-Hirschmann et al. (2008), isodivaricatic acid, divaricatinic acid and usnic acid, the main lichen metabolites in *Protousnea poeppigii*, displayed antifungal action against *Microsporum gypseum*, *Trichophyton mentagrophytes* and *T. rubrum*, usnic acid being less active. Divaricatic acid and zeorin constituents of *Lecanora frustulosa* and *Parmeliopsis hyperopta* has been screened in vitro against the *Bacillus mycoides, Bacillus subtilis, Staphylococcus aureus, Enterobacter cloacae, Escherichia coli, Klebsiella pneumoniae, Aspergillus flavus, Aspergillus fumigatus, Botrytis cinerea, Candida albicans, Fusarium oxysporum, Mucor mucedo, Paecilomyces variotii, Penicillium purpurescens, Penicillium verrucosum* and *Trichoderma harsianum.* Divaricatic acid and zeorin showed strong activity against both bacteria and fungi (Kosanić et al. 2010). *Bacillus subtilis, B. cereus, Staphylococcus aureus, Streptococcus faecalis, Proteus vulgaris, Listeria monocytogenes, Aeromonas hydrophila, Candida albicans* and *Candida glabrata* growth were inhibited by usnic acid, atranorin, fumarprotocetraric acid constituents of *Cladonia foliacea* (Yilmaz et al. 2004). Similarly, anti-Gram-positive activities have been reported for evernic acid, vulpinic acid and hirtusneanoside (Rezanka and Sigler 2007).

Antimicrobial activity of salazinic acid isolated from *Parmelia sulcata* has been screened against 28 food-borne bacteria and fungi and it has been found that this compounds showed activity against *Pseudomonas aeruginosa* and *Salmonella typhimurium* as well (Candan et al. 2007). Antimicrobial activity of fumarprotocetraric acid, lecanoric acid, protocetraric acid and stictic acid isolated from the lichen of *Cladonia furcata, Ochrolechia androgyna, Parmelia caperata* and *Parmelia conspersa* was studied in relation to *Bacillus mycoides, Bacillus subtilis, Staphylococcus aureus, Enterobacter cloacae, Escherichia coli, Klebsiella pneumoniae, Aspergillus flavus, Aspergillus fumigatus, Botrytis cinerea, Candida albicans, Fusarium oxysporum, Mucor mucedo, Paecilomyces variotii, Penicillium purpurescens, Penicillium verrucosum* and *Trichoderma harsianum* (Ranković and Mišić 2008). The antimicrobial activity was estimated by determining the minimal inhibited concentration by the Broth tube Dilution method. The researched lichen components inhibited the growth of all the tested microorganisms. The lowest MIC value (0.031 mg/mL) was measured for the fumarprotocetraric acid related to the *Klebsiella pneumoniae* species. The weakest antimicrobial activity was found in stictic acid, which inhibited most of the tested microorganisms at significantly higher concentrations.

Whiton and Lawrey (1982) reported that ascospore germination of *Sordaria fimicola* was significantly inhibited by evernic and vulpinic acids. Usnic acid, evernic acid and vulpinic acid inhibited the growth of the Gram-positive bacteria *Staphylococcus aureus, Bacillus subtilis* and *Bacillus megaterium*, but had no effect on the Gram-negative bacteria *Escherichia coli* or *Pseudomonas aeruginosa* (Lawrey 1986). Three new depsidones (neuropogonines A, B and C) isolated from a *Neuropogon* showed a moderate activity on a *Mycobacterium vaccae* strain (Ivanova et al. 2002). Two years later, the same authors from this lichen were isolated usnic acid and further investigated for antimicrobial activity against pathogenic Gram-positive and Gram-negative bacteria as *Streptococcus pyogenes, Staphylococcus epidermidis, Bacillus cereus, Listeria monocytogenes, Yersinia enterocolitica, Klebsiella pneumoniae* and *Salmonella typhimurium* (Ivanova et al. 2004).

Manojlović et al. (2005) reported antifungal activity of the anthraquinone parietin isolated from *Caloplaca cerina*. A potent fungitoxic compound, lecanoric acid, was isolated from *Parmotrema tinctorum* lichen and tested against the fungus *Cladosporium sphaerospermum* (Gomes et al. 2002). In addition, MICs values obtained from monoaromatic phenols (methyl beta-orsellinate, methyl and ethyl orsellinates) derived from various Icelandic lichen species were found equal or higher than usual preservatives (methyl and propyl-p-hydroxybenzoates, o-cresol) (Ingólfsdóttir et al. 1985).

The antimicrobial activity of gyrophoric acid and stenosporic acid constituents of the *Xanthoparmelia pokornyi* lichen has been screened against some food-borne bacteria and fungi. Both acids showed antimicrobial activity against *Aeromonas hydrophila, Bacillus cereus, Bacillus subtilis, Listeria monocytogenes, Proteus vulgaris, Staphylococcus aureus, Streptococcus faecalis, Yersinia enterocolitica, Candida albicans* and *Candida glabrata*, while were inactive against the tested filamentous fungi (Candan et al. 2006).

Tasdemir and Franzblau (2007) investigated four commercially available lichen metabolites, (+)-usnic acid, evernic acid, psoromic acid and vulpic acid, for their in vitro antituberculotic effects. $H_{37}Rv$ strain of *Mycobacterium tuberculosis* and the well-established microplate alamar blue assay were used for the determination of the MIC values. (+)-Usnic acid proved to be the most potent antituberculotic agent (MIC = 5.2 μg/mL), followed by psoromic acid (MIC = 44 μg/mL) and vulpic acid (MIC = 140 μg/mL). Evernic acid was found to be inactive at highest concentrations tested (MIC > 200 μg/mL). Honda et al. (2010) describes the extraction and identification of several classes of phenolic compounds from the lichens *Parmotrema dilatatum, Parmotrema tinctorum, Pseudoparmelia sphaerospora* and *Usnea subcavata* and determined their anti-tubercular activity. The depsides (atranorin, diffractaic and lecanoric acids), depsidones (protocetraric, salazinic, hypostictic and norstictic acids), xanthones (lichexanthone and secalonic acid) and usnic acid, as well seven orsellinic acid esters, five salazinic acid 8′,9′-O-alkyl derivatives and four lichexanthone derivatives, were evaluated for their activity against *Mycobacterium tuberculosis*. Diffractaic acid was the most active compound (MIC value 15.6 mg/mL, 41.6 mM), followed by norstictic acid (MIC value 62.5 mg/mL, 168 mM) and usnic acid (MIC value 62.5 mg/mL, 182 mM). Hypostictic acid (MIC value 94.0 mg/mL, 251 mM) and protocetraric acid (MIC value 125 mg/mL, 334 mM) showed moderate inhibitory activity. The other compounds showed lower inhibitory activity on the growth of *M. tuberculosis*, varying from MIC values of 250 to 1370 mM.

Cladonia verticillaris lichen lectin was evaluated by Ramos Dalia et al. (2014) for its antimicrobial potential and it showed activity against Gram-positive (*Bacillus subtilis, Staphylococcus aureus* and *Enterococcus faecalis*) and Gram-negative (*Escherichia coli* and *Klebsiella pneumoniae*) assayed strains, with greater inhibitory effect on growth of *E. coli* (MIC of 7.18 μg/mL). The lowest minimum bactericidal concentration (MBC, 57.4 μg/mL) was detected against *E. faecalis*. The antifungal assay performed with *Trichophyton mentagrophytes, Microsporum gypseum, Trichophyton rubrum, Trichosporon cutaneum* and *Trichosporon asahii.*

Cladonia verticillaris lichen lectin was the most active against *T. rubrum* with an inhibition percentage of 35% compared to negative control.

A total of five compounds, usnic acid, usimine A, usimine B, usimine C and ramalin, were isolated by bioactivity guided-fractionation of the methanol extract of *Ramalina terebrata* (Paudel et al. 2010). The qualitative antibacterial activities of the isolated compounds were determined by the disk diffusion method while the minimum inhibitory concentration (MIC) determination assay gave the quantitative strength of the test samples. All the test samples showed antibacterial activity against *Bacillus subtilis*, and usnic acid showed antibacterial activity against *Staphylococcus aureus*. The MIC values of the isolated compounds against *B. subtilis* were in the range of 1–26 μg/mL.

Manojlović et al. (2012) were investigated antibacterial and antifungal activities for protocetraric acid from *Parmelia caperata* lichen and depsidone salazinic acid from *Parmelia saxatilis* species. Antioxidant activities of this isolated metabolites was evaluated against *Bacillus mycoides, Bacillus subtilis, Staphylococcus aureus, Escherichia coli, Klebsiella pneumoniae, Aspergillus flavus, Aspergillus fumigatus, Candida albicans, Penicillium purpurescens* and *Penicillium verrucosum.* As a result of the study salazinic acid and protocetraric acid showed similar antimicrobial activity, but antibacterial activity was stronger than antifungal activity for both components.

Lichen compounds norstictic acid isolated from *Toninia candida* and usnic acid from *Usnea barbata* demonstrated very strong antimicrobial activity (Ranković et al. 2012). The MIC for different components relative to the tested microorganisms ranged from 0.0008 to 1 mg/mL. The strongest antimicrobial activity was found in usnic acid, which in extremely low amounts inhibited all species of bacteria and fungi. Similarly, 2 years later, antimicrobial activities of major lichen metabolites in *Hypogymnia physodes* lichen (physodic acids, atranorin and usnic acid) were studied by Ranković et al. (2014a, b). They found that usnic acid and physodic acid showed very strong and similar antimicrobial activity, followed by atranorin. Antibacterial activity was stronger than antifungal activity for all compounds.

Evernic acid and physodic acid from *Evernia prunastri* and *Pseudoevernia furfuraceae* lichens were screened for their antimicrobial effects by the broth microdilution method (Kosanić et al. 2013), and physodic acid was found to be the most effective. One year later, in related experiment, Kosanić et al. (2014a) investigated antimicrobial activity for atranorin and fumarprotocetraric acid isolated from *Cladonia* lichen. Antimicrobial activity was studied in relation to five species of bacteria and five species of fungi. The isolated lichen components demonstrated very strong antimicrobial activity. The MIC for different components relative to the tested microorganisms ranged from 0.015 to 1 mg/mL. The strongest antimicrobial activity was found in fumarprotocetraric acid, which in extremely low amounts inhibited all the species of bacteria and fungi.

Kosanić et al. (2014c) studied antimicrobial capacity for gyrophoric acid from *Acarospora fuscata* lichen: Antimicrobial activity was estimated by determination of the minimal inhibitory concentration by the broth microdilution method against five bacterial and five fungal species. The isolated lichen component demonstrated strong antibacterial and antifungal activity inhibiting all of the species of bacteria and fungi in extremely low amounts (MIC values were 0.019–1.25 mg/mL).

Basile et al. (2015) investigated the antibacterial and antifungal activity of parietin, major secondary metabolite from lichen *Xanthoria parietina.* The parietin was tested for antimicrobial activity against nine American Type Culture Collection standard and clinically isolated bacterial strains, and three fungal strains. The isolated compound showed strong antibacterial activity against all bacterial strains and matched clinical isolates, particularly against *Staphylococcus aureus* from standard and clinical sources. Among the fungi tested, *Rhizoctonia solani* was the most sensitive.

Two depsidones obtusatic acid isolated from *Ramalina fraxinea* and methyl evernate isolated from *R. fastigiata* were studied for antimicrobial activity (Ristić et al. 2016b). Antimicrobial capacity was tested towards five bacterial and ten fungal species using broth microdilution method to determine the minimum inhibitory concentration. Both isolated lichen components demonstrated relatively strong antimicrobial influence (showed inhibition against all tested microorganisms) and their MIC values ranged from 0.125 to 1 mg/mL. Methyl evernate showed the stronger antimicrobial capacity, inhibiting the tested microorganisms at low doses. In this research, *B. cereus* was the most sensitive bacterial strain and *C. albicans* was the most sensitive fungal strain. The same authors tested antimicrobial properties for 2′-O-methyl anziaic acid which is isolated from lichen *M. fuliginosa* and lecanoric acid which is isolated from lichen *M. subaurifera.* Both tested compounds inhibited all tested bacteria and fungi. The higher antimicrobial activity was found in the 2′-O-methyl anziaic acid which in relatively low concentrations inhibited the tested microorganisms. The MIC for tested compound ranged from 0.0625 to 1 mg/mL (Ristić et al. 2016a).

Antibacterial, antifungal activity and MIC values of barbatolic acid constituent of lichen *Bryoria capillaris* have been investigated against Gram-positive and Gram-negative bacteria, yeasts and filamentous fungi by Sariozlu et al. (2016). The lichen compound was shown considerable antimicrobial effect to tested pathogenic microorganisms but it should be emphasized that it significantly affected against the infectious pathogen *Mycobacterium tuberculosis* with a MIC value of 31.25 μg/mL.

In study described by Oh et al. (2018), the purified divaricatic acid isolated from the lichen *Evernia mesomorpha* was found to be effective against *Staphylococcus aureus* (32 μg/mL) and *Enterococcus faecium* (16 μg/mL) and against other Gram-positive bacteria, such as *Bacillus subtilis*, *Staphylococcus epidermidis*, *Streptococcus mutans* and *Enterococcus faecium* with MICs ranging from 7.0 to 64.0 μg/mL, whereas vancomycin had MICs ranging from 0.78 to 25.0 μg/mL. Interestingly, the antibacterial activity of divaricatic acid was higher than that of vancomycin against *S. epidermidis* and *E. faecium*. Furthermore, divaricatic acid was active against *Candida albicans*, whereas as vancomycin did not have antifungal properties, but slightly less effective than vancomycin against *S. aureus*. However, like vancomycin, divaricatic acid was not effective against Gram-bacteria.

Divaricatic acid was more effective than cefotaxime, a broad-spectrum antibiotic, against *E. faecium* and the *S. aureus*.

The results obtained in studies related to the antimicrobial activity of lichens indicate differences in activity between extracts depending on the species of lichen and as a function of the type of extracting solvent. These results are in agreement with the suggestion of Oloke and Kolawole (1998) that bioactive components of any medical plant have different solubility in different extracting solvents. Numerous researchers found lower antimicrobial activity of aqueous extracts in comparison with acetone, methanol or other organic solvent extracts (Land and Lundstrom 1998; Madamombe and Afolajan 2003; Ranković et al. 2008; Kosanić et al. 2010). The reason for weak activity of aqueous extracts is that active substances present in the thalli of lichens are insoluble or poorly soluble in water (Kinoshita et al. 1994).

In general, in studies regarding antimicrobial activity of lichens, fungi and Gram-negative bacteria were more resistant than Gram-positive ones. The difference of sensitivity between Gram-positive, Gram-negative bacteria and fungi can be ascribed to morphological differences between these microorganisms, above all to differences in permeability of the cell wall (Nostro et al. 2000). The cell walls of Gram-positive bacteria are made of peptidoglycans and teichoic acids, while those of Gram-negative bacteria are made of peptidoglycans, lipopolysaccharides and lipoproteins (Heijenoort 2001; Ranković et al. 2008). The cell walls of fungi are poorly permeable and consist of polysaccharides such as hitchin and glucan (Farkaš 2003).

Since microorganisms have developed resistance to many antibiotics, pharmacologists need to pursue new sources for antimicrobial agents. All these results suggest that lichens and their metabolites yield significant new bioactive substances for the treatment of various diseases caused by microorganisms. New compounds are to be described from poorly studied lichens, and even in species that are considered chemically well known, new chemical strains are still detected (e.g. Stocker-Worgotter et al. 2004). Many lichens are known which contain structurally unknown lichen products. Here, we opened the entrance door of a vast and interesting field of research.

References

Abdullah ST, Hamid H, Ali M et al (2007) Two new terpenes from the lichen *Parmelia perlata*. Ind J Chem 46B:173–176

Abo-Khatwa AN, al-Robai AA, al-Jawhari DA (1996) Lichen acids as uncouplers of oxidative phosphorylation of mouse-liver mitochondria. Nat Toxins 4:96–102

Abraham EP, Chain E (1940) An enzyme from bacteria able to destroy penicillin. Nature 146:837

Ankith GN, Rajesh MR, Karthik KN et al (2017) Antibacterial and antifungal activity of three Ramalina species. J Drug Deliv Ther 7:27–32

Bačkor M, Hudák J, Repčák M et al (1998) The influence of pH and lichen metabolites (vulpinic acid and (+) usnic acid) on the growth of lichen photobiont *Trebouxia irregularis*. Lichenologist 30:577–582

Balaji P, Hariharan GN (2007) In vitro antimicrobial activity of Parmotrema prasorediosum thallus extracts. Res J Bot 2:54–59

Balouiri M, Sadiki M, Ibnsouda SK (2016) Methods for *in vitro* evaluating antimicrobial activity: a review. J Pharm Anal 6:71–79

Basile A, Rigano D, Loppi S et al (2015) Antiproliferative, antibacterial and antifungal activity of the lichen Xanthoria parietina and its secondary metabolite parietin. Int J Mol Sci 16:7861–7875

Behera BC, Verma N, Sonone A et al (2005) Antimicrobial activity of various solvent extracts of lichen *Usnea ghattensis*. Agarkar Research Institute, Pune

Bondarenko V, Korczynski M, Techathaveewat W (2017) The antimicrobial properties of extracts isolated from lichen Parmelia vagans [supplemental material]. FASEB J 31(1):939.13

Bouaid K, Vicente C (1998) Chlorophyll degradation effected by lichen substances. Ann Bot Fenn 35:71–74

Buçukoglu TZ, Albayrak S, Gökhan Halici M et al (2013) Antimicrobial and antioxidant activities of extracts and lichen acids obtained from some *Umbilicaria* species from Central Anatolia, Turkey. J Food Process Preserv 37:1103–1110

Burkholder PR, Evans AW (1945) Further studies on the antibiotic activity of lichens. Bull Torrey Bot Club 72:157–164

Burkholder PR, Evans AW, McVeigh I et al (1944) Antibiotic activity of lichens. Proc Natl Acad Sci 30:250–255

Byron F, Brehm S, Eric AJ (2003) Sensitization of *Staphylococcus aureus* and *Escherichia coli* to antibiotics by the sesquiterpenoids. Antimicrob Agents Chemother 47:3357–3360

Candan M, Yilmaz M, Tay T et al (2006) Antimicrobial activity of extracts of the lichen *Xanthoparmelia pokornyi* and its gyrophoric and stenosporic acid constituents. Z Naturforsch 61:319–323

Candan M, Yýlmaz M, Tay T et al (2007) Antimicrobial activity of extracts of the lichen *Parmelia sulcata* and its Salazinic acid constituent. Z Naturforsch 62:619–621

Cansaran D, Atakol O, Halici M et al (2007) HPLC analysis of usnic acid in some Ramalina species from Anatolia and investigation of their antimicrobial activities. Pharm Biol 45:77–81

Chauhan R, Abraham J (2013) In vitro antimicrobial potential of the lichen *Parmotrema* sp. extracts against various pathogens. Iran J Basic Med Sci 16:882–885

Clinical and Laboratory Standards Institude (2009) Methods for dilution antimicrobial susceptibility tests for bacteria that grow aerobically; approved standard-eighth edition M07-A8. National Committee for Clinical Laboratory Standards 29

Cowan MM (1999) Plants products as antimicrobial agents. Clin Microbiol Rev 12:564–582

Culberson CF (1970) Supplement to chemical and botanical guide to lichen products. Bryologist 73:1–28

Dickert H, Machka K, Braveny I (1981) The uses and limitations of disc diffusion in the antibiotic sensitivity testing of bacteria. Infection 9:18–24

Dixit A, Mishra M, Singh AN et al (2018) Antimicrobial and antifungal activity of Indian Lichen (Usnea sps. And Parmotrema sps.) against human phatogenic bacterial and fungal sps. World J Pharm Pharm Sci 7:711–720

Elo H, Matikainen J, Pelttari E (2007) Potent activity of the lichen antibiotic (+)-usnic acid against clinical isolates of vancomycin-resistant enterococci and methicillin-resistant *Staphylococcus aureus*. Naturwissenschaften 94:465–468

Epand RF, Savage PB, Epand RM (2007) Bacterial lipid composition and the antimicrobial efficacy of cationic steroid compounds. Biochim Biophys Acta 1768:2500–2509

Esimone CO, Adikwn MU (1999) Antimicrobial activity of the cytotoxicity of *Ramalina farinacea*. Fitoterapia 7:428–431

Farkaš V (2003) Structure and biosynthesis of fungal cell walls: methodological approaches. Folia Microbiol 48:469–478

Francolini I, Norris P, Piozzi A et al (2004) Usnic acid, a natural antimicrobialagent able to inhibit bacterial biofilm formation on polymer surfaces. Antimicrob Agents Chemother 48:4360–4365

Goel M, Dureja P, Rani A et al (2011) Isolation, characterization and antifungal activity of major constituents of the Himalayan lichen *Parmelia reticulate* Tayl. J Agric Food Chem 59: 2299–2307

Gomes AT, Honda NK, Roese FM et al (2002) Bioactive derivatives obtained from lecanoric acid, a constituent of the lichen *Parmotrema tinctorum* (Nyl.). Hale (*Parmeliaceae*). Rev Bras Farmacogn 12:74–75

Gomes AT, Smania A Jr, Seidel C et al (2003) Antibacterial activity of orsellinates. Braz J Microbiol 34(3):194–196

Gulluce M, Aslan A, Sokmen M et al (2006) Screening the antioxidant and antimicrobial properties of the lichens *Parmelia saxatilis, Platismatia glauca, Ramalina pollinaria, Ramalina polymorpha* and *Umbilicaria nylanderiana*. Phytomedicine 13:515–521

Halama P, van Haluwin C (2004) Antifungal activity of lichen extracts and Lichenic acids. BioControl 49:95–107

Hauck M, Jurgens SR (2008) Usnic acid controls the acidity tolerance of lichens. Environ Pollut 156:115–122

Heijenoort J (2001) Formation of the glycan chains in the synthesis of bacterial peptidoglycan. Glycobiology 11:25–36

Honda NK, Pavan FR, Coelho RG et al (2010) Antimycobacterial activity of lichen substances. Phytomedicine 17:328–332

Ingólfsdóttir K, Bloomfield SF, Hylands PJ (1985) In vitro evaluation of the antimicrobial activity of lichen metabolites as potential preservatives. Antimicrob Agents Chemother 28:289–292

Ingolfsdottir K, Chung GAC, Skulason VG et al (1998) Antimycobacterial activity of lichen metabolites in vitro. Eur J Pharm Sci 6:141–144

Ivanova V, Aleksieva K, Kolarova M et al (2002) Neuropogonines A, B and C, new depsidon-type metabolites from *Neuropogon* sp., an Antarctic lichen. Pharmazie 57:73–74

Ivanova V, Graefe U, Schlegel B et al (2004) Usnic acid, metabolite from Neuropogon Sp., an Antarctic lichen isolation, structure elucidation and biological activity. Biotechnol Biotechnol Equip 18:66–71

Javeria S, Shahi SK, Shahi MP et al (2013) *Parmotrema nilgherrense*: potential antimicrobial activity against drug resistant pathogens. Int J Microb Res Technol 2:36–40

Karagoz A, Dogruoz N, Zeybek Z et al (2009) Antibacterial activity of some lichen extracts. J Med Plants Res 3:1034–1039

Karthikaidevi G, Thirumaran G, Manivannan K et al (2009) Screening of the antibacterial properties of lichen *Roccella belangeriana* (Awasthi) from Pichavaram mangrove (*Rhizophora* sp.). Adv Biol Res 3:127–131

Kekuda PTR, Vinayaka KS, Swathi D et al (2011) Mineral composition, total phenol content and antioxidant activity of a macrolichen *Everniastrum cirrhatum* (Fr.) Hale (*Parmeliaceae*). E-Journal of Chemistry 8:1886–1894

Kim JS, Kim Y (2007) The inhibitory effect of natural bioactives on the growth of pathogenic bacteria. Nutr Res Pract 1(4):273–278

Kinoshita K, Matsubara H, Koyama K et al (1994) Topics in the chemistry of lichen compounds. J Hattori Bot Lab 76:227–233

Klancnik A, Piskernik S, Jersek B et al (2010) Evaluation of diffusion and dilution methods to determine the antibacterial activity of plant extracts. J Microbiol Methods 81:121–126

Kosanić M, Ranković B (2011) Antioxidant and antimicrobial properties of some lichens and their constituents. J Med Food 14:1624–1630

Kosanić M, Ranković B, Slobodan S (2010) Antimicrobial activity of the lichen *Lecanora frustulosa* and *Parmeliopsis hyperopta* and their divaricatic acid and zeorin constituents. Afr J Microbiol Res 4:885–890

Kosanić M, Manojlović N, Janković S et al (2013) *Evernia prunastri* and *Pseudoevernia furfuraceae* lichens and their major metabolites as antioxidant, antimicrobial and anticancer agents. Food Chem Toxicol 53:112–118

Kosanić M, Ranković B, Stanojković T et al (2014a) *Cladonia* lichens and their major metabolites as possible natural antioxidant, antimicrobial and anticancer agents. LWT – Food Sci Technol 59:518–525

Kosanić M, Seklic D, Markovic S et al (2014b) Antimicrobial and anticancer properties of selected lichens from Serbia. Dig J Nanomater Bios 9:273–287

Kosanić M, Ranković B, Stanojković T et al (2014c) Biological activities and chemical composition of lichens from Serbia. EXCLI J 13:1226–1238

Kosanić M, Ranković B, Stanojković T et al (2016) *Lasallia pustulata* lichen as possible natural antigenotoxic, antioxidant, antimicrobial and anticancer agent. Cytotechnology 68:999–1008

Kosanić M, Ristić S, Stanojković T et al (2018) Extracts of five Cladonia lichens as sources of biologically active compounds. Farmacia 66:644–651

Kumar KC, Müller K (1999) Lichen Metabolites. 2. Antiproliferative and cytotoxic activity of gyrophoric, usnic, and diffractaic acid on human keratinocyte growth. J Nat Prods 62(6): 821–823

Kumar PSV, Kekuda PTR, Vinayaka KS et al (2010) Studies on proximate composition,antifungal and anthelmintic activity of macrolichen *Ramalina hoss* H. Magn and G.Awasthi. Int J Biotech Biochem 6:193–203

Land CJ, Lundstrom J (1998) Inhibition of fungal growth by water extracts from the lichen *Nephroma arcticum*. Lichenologist 30:259–262

Lauterwein M, Oethinger M, Belsner K et al (1995) In vitro activities of the lichen secondary metabolites vulpinic acid, (+)-usnic acid, and (−)-usnic acid against aerobic and anaerobic microorganisms. Antimicrob Agents Chemother 39(11):2541–2543

Lawrey JD (1986) Biological role of lichen substances. The Byrologist 89:111–122

Maciąg-Dorszyńska M, Węgrzyn G, Guzow-Krzemińska B (2014) Antibacterial activity of lichen secondary metabolite usnic acid is primarily caused by inhibition of RNA and DNA synthesis. FEMS Microbiol Lett 353:57–62

Madamombe IT, Afolajan AJ (2003) Evaluation of antimicrobial activity of extracts from south African *Usnea barbata*. Pharm Biol 41:199–202

Manojlović N, Solujić S, Sukdolak S (2002) Antimicrobial activity of an extract and anthraquinones from *Caloplaca schaereri*. Lichenologist 34:83–85

Manojlović NT, Solujic S, Sukdolak S et al (2005) Antifungal activity of *Rubia tinctorum*, *Rhamnus frangula* and *Caloplaca cerina*. Fitoterapia 76:244–246

Manojlović NT, Vasiljevic PJ, Markovic ZS (2010) Antimicrobial activity of extracts and various fractions of chloroform extract from the lichen *Laurera benguelensis*. J Biol Res-Thessalon 13:27–34

Manojlović N, Ranković B, Kosanić M et al (2012) Chemical composition of three *Parmelia* lichens and antioxidant, antimicrobial and cytotoxic activities of some their major metabolites. Phytomedicine 19:1166–1172

Martins MCB, Gonçalves de Lima MJ, Silva FP et al (2010) *Cladia aggregata* (lichen) from Brazilian northeast, chemical characterization and antimicrobial activity. Braz Arch Biol Technol 53:115–122

Matvieieva NA, Pasichnyk LA, Zhytkevych NV et al (2015) Antimicrobial activity of extracts from Ecuadorian lichens. Mikrobiol Z 77:23–27

Micheletti AC, Beatriz A, de Lima DP et al (2009) Constituintes quımicos de Parmotrema lichexanthonicum Eliasaro & Adler – Isolamento, Modificacoes estruturais e avaliacao das atividades antibiotica e citotoxica. Quim Nova 32:12–20

Mitrović T, Stamenković S, Cvetković V et al (2011) Antioxidant, antimicrobial and antiproliferative activities of five lichen species. Int J Mol Sci 12:5428–5448

Mohammed SG (2013) Comparative study of *in vitro* antibacterial activity of miswak extracts and different toothpastes. Am J Agric Biol Sci 8:82–88

Moreno S, Scheyer T, Romano CS et al (2006) Antioxidant and antimicrobial activities of rosemary extracts linked to their polyphenol composition. Free Radic Res 40:223–231

Nostro MP, Germano V, D'Angelo A et al (2000) Extraction methods and bioautography for evaluation of medical plant antimicrobial activity. Lett Appl Microbiol 30:379–384

Odabasoglu F, Aslan A, Cakir A et al (2004) Comparison of antioxidant activity and phenolic content of three lichen species. Phytother Res 18:938–941

Ofokansi KC, Esimone CO (2005) Evaluation of the in vitro antimicrobial activity and release behaviour of ointments and applications containing extract of lichen *Ramalina farinaceae*. Plant Prod Res J 9:6–10

Oh JM, Kim YJ, Gang HS et al (2018) Antimicrobial activity of divaricatic acid isolated from the lichen *Evernia mesomorpha* against methicillin-resistant *Staphylococcus aureus*. Molecules 23: 3068

Oloke JO, Kolawole DO (1998) The antibacterial and antifungal activities of certain components of *Aframomum meleguetea* fruits. Fitoterapia 59:384–388

Paudel B, Bhattarai HD, Lee JS et al (2008) Antioxidant activity of polar lichens from King George Island (Antarctica). Polar Biol 31:605–608

Paudel B, Bhattarai HD, Lee HK et al (2010) Antibacterial activities of Ramalin, usnic acid and its three derivatives isolated from the Antarctic lichen *Ramalina terebrata*. Z Naturforsch C 65: 34–38

Perry NB, Benn MH, Brennan NJ et al (1999) Antimicrobial, antiviral and citotoxic activity of New Zealand lichens. Lichenologist 31:627–636

Piovano M, Garbarino JA, Giannini FA et al (2002) Evaluation of antifungal and antibacterial activities of aromatic metabolites from lichens. Bol Soc Chil Quím 47:235–240

Ramos DF, Almeida da Silva PE (2010) Antimycobacterial activity of usnic acid against resistant and susceptible strains of and non-tuberculous mycobacteria. Pharm Biol 48(3): 260–263

Ramos Dalia BM, Gomes Francis S, Napoleo Thiago H et al (2014) Antimicrobial preparations on bacteria and fungi of medical importance. Chin J Biol 2014:219392

Ranković B, Mišić M (2007) Antifungal activity of extract of the lichens *Alectoria sarmentosa* and *Cladonia rangiferina*. Mikol Fitopatol 41:276–281

Ranković B, Mišić M (2008) The antimicrobial activity of the lichen substances of the lichens *Cladonia furcata, Ochrolechia androgyna, Parmelia caperata* and *Parmelia conspresa*. Biotechnol Biotech Equip 22:1310–2818

Ranković B, Mišić M, Sukdolak S et al (2007a) Antimicrobial activity of the lichens *Aspicilia cinerea, Collema cristatum, Ochrolechia androgyna, Physcia aipolia* and *Physcia caesia*. Ital J Food Sci 19:461–469

Ranković B, Mišić M, Sukdolak S (2007b) Evaluation of antimicrobial activity of the lichens *Lasallia pustulata, Parmelia sulcata, Umbilicaria crustulosa* and *Umbilicaria cylindrical*. Microbiology 76:723–727

Ranković B, Mišić M, Sukdolak S (2008) The antimicrobial activity of substances derived from the lichens *Physcia aipolia, Umbilicaria polyphylla, Parmelia caperata* and *Hypogymnia physodes*. World J Microbiol Biotechnol 24:1239–1242

Ranković B, Mišić M, Sukdolak S (2009) Antimicrobial activity of extracts of the lichens *Cladonia furcata, Parmelia caperata, Parmelia pertusa, Hypogymnia physodes* and *Umbilicaria polyphylla*. Biologia 64:53–58

Ranković B, Ranković D, Kosanić M et al (2010) Antioxidant and antimicrobial properties of the lichen *Anaptychya cilaris, Nephroma parile, Ochrolechia tartarea* and *Parmelia centrifuga*. Cent Eur J Biol 5:649–665

Ranković B, Kosanić M, Stanojković T (2011) Antioxidant, antimicrobial and anticancer activity of the lichens *Cladonia furcata, Lecanora atra* and *Lecanora muralis*. BMC Complement Altern Med 11:97. https://doi.org/10.1186/1472-6882-11-97

Ranković B, Kosanić M, Stanojković T et al (2012) Biological activities of *Toninia candida* and *Usnea barbata* together with their norstictic acid and usnic acid constituents. Int J Mol Sci 13:14707–14722

Ranković B, Kosanić M, Stanojković T (2014a) Stereocaulon pashale lichen as antioxidant, antimicrobial and anticancer agent. Farmacia 62:306–317

Ranković B, Kosanić M, Manojlović N et al (2014b) Chemical composition of *Hypogymnia physodes* lichen and biological activities of some its major metabolites. Med Chem Res 23:408–416

Rezanka T, Sigler K (2007) Hirtusneanoside, an unsymmetrical dimeric tetrahydroxanthone from the lichen *Usnea hirta*. J Nat Prod 70:1487–1491
Ristić S, Ranković B, Kosanić M et al (2016a) Phytochemical study and antioxidant, antimicrobial and anticancer activities of *Melanelia subaurifera* and *Melanelia fuliginosa* lichens. J Food Sci Technol 53:2804–2816
Ristić S, Ranković B, Kosanić M et al (2016b) Biopharmaceutical potential of two Ramalina lichens and their metabolites. Curr Pharm Biotechnol 17:651–658
Rundel PW (1978) The ecological role of secondary lichen substances. Biochem Syst Ecol 6: 157–170
Saenz MT, Garcia MD, Rowe JG (2006) Antimicrobial activity and phytochemical studies of some lichens from south of Spain. Fitoterapia 77:156–159
Santiago KA, Borricano JN, Canal JN et al (2010) Antibacterial activities of fruticose lichens collected from selected sites in Luzon Island, Philippines. Philipp Sci Lett 3:18–28
Santos PS, Lat B, Palo M (1964) The antibiotic activities of some Philippine lichens. Philipp J Sci 93:325–335
Sariozlu NY, Cankilic MY, Candan M et al (2016) Antimicrobial activity of lichen *Bryoria capillaris* and its compound barbatolic acid. Biomed Res India 27:419–423
Sasidharan S, Chen Y, Saravanan D et al (2011) Extraction, isolation and characterization of bioactive compounds from plants' extracts. Afr J Tradit Complement Altern Med 8:1–10
Schmeda-Hirschmann G, Tapia A, Lima B et al (2008) A new antifungal and antiprotozoal depside from the Andean lichen *Protousnea poeppigii*. Phytother Res 22:349–355
Senthil Prabhu S, Sudha SS (2015) Evaluation of the antibacterial properties of some lichen species against human pathogens. Int J Adv Res Biol Sci 2:177–181
Sharma BC, Kalikotay S, Rai B (2012) Assessment of antimicrobial activity of extracts of few common lichens of Darjeeling hills. Ind J Fund Appl Life Sci 2:120–126
Sheldon AT (2005) Antibiotic resistance: a survival strategy. Clin Lab Sci Summer 18:170–180
Shrestha G, Raphael J, Leavitt SD et al (2014) In vitro evaluation of the antibacterial activity of extracts from 34 species of north American lichens. Pharm Biol 27:1–5
Sinha SN, Biswas M (2011) Evaluation of antibacterial activity of some lichen from Ravangla, Sikkim, India. Int J Pharm Bio Sci 2:B23–B28
Sisodia R, Geol M, Verma S et al (2013) Antibacterial and antioxidant activity of lichen species *Ramalina roesleri*. Nat Prod Res 27:2235–2239
Srivastava P, Logesh AR, Upreti DK et al (2013) In-vitro evaluation of some Indian lichens against human pathogenic bacteria. Mycosphere 4:734–743
Stocker-Worgotter E, Elix JA, Grube M (2004) Secondary chemistry of lichen-forming fungi: chemosyndromic variation and DNA-analyses of cultures and chemotypes in the *Ramalina farinacea* complex. Bryologist 107(2):152–162
Stoll A, Brack A, Renz J (1950) The effect of lichenic substances on the tubercle bacillus and certain other microorganisms; seventh article on antibacterial substances. Schweiz Z Pathol Bakteriol 13:729–751
Sundset MA, Kohn A, Mathiesen SD (2008) Eubacterium rangiferina, a novel usnic acid-resistant bacterium from the reindeer rumen. Naturwissenschaften 95(8):741–749
Tasdemir D, Franzblau SG (2007) In vitro antituberculotic activity of several lichen metabolites. Planta Med 73:174
Tay T, Özdemir Türk A, Yılmaz M et al (2004) Evaluation of the antimicrobial activity of the acetone extract of the lichen Ramalina farinacea and its (+)-usnic acid, norstictic acid, and protocetraric acid constituents. Z Naturforsch C 59(5–6):384–388
Tiwari P, Rai H, Upreti DK et al (2011) Assessment of antifungal activity of some Himalayan foliose lichen against plant pathogenic fungi. Am J Plant Sci 2:841–846
Tomasi S, Picard S, Lainé C et al (2006) Solid-phase synthesis of polyfunctionalized natural products: application to usnic acid, a bioactive lichen compound. J Comb Chem 8:11–14
Tomović J, Kosanić M, Ristić S et al (2017) Chemical composition and bioactive properties of the lichen, *Pleurosticta acetabulum*. Trop J Pharm Res 16:2977–2984

Türk OA, Yılmaz M, Kıvanc M et al (2003) The antimicrobial activity of extracts of the lichen Cetraria aculeate and its protolichesterinic acid constituent. Z Naturforsch C 58:850–854

Turk H, Yılmaz M, Tay T et al (2006) Antimicrobial activity of extracts of chemical races of the lichen Pseudoevernia furfuracea and their physodic acid, chloroatratorin, atratorin and olivetoric acid constituents. Z Naturforsch C 61(7–8):499–507

Vartia KO (1973) Antibiotics in lichens. In: Ahmadjian V, Hale ME Jr (eds) The lichens. Academic, New York, NY, pp 547–561

Verma N, Behera BC, Parizadeh H et al (2011) Bactericidal activity of some lichen secondary compounds of *Cladonia ochrochlora, Parmotrema nilgherrensis* and *Parmotremasancti-angelii*. Int J Drug Dev Res 3:222–232

Vivek MN, Kambar Y, Manasa M et al (2014) Radical scavenging and antibacterial activity of three *Parmotrema* species from Western Ghats of Karnataka, India. J Appl Pharm Sci 4: 086–091

Whiton JC, Lawrey JD (1982) Inhibition of *Cladonia cristatella* and *Sordaria fimicola* ascospore germination by lichen acids. Bryologist 85:222–226

Yilmaz M, Turk AO, Tay T et al (2004) The antimicrobial activity of extracts of the lichen Cladonia foliacea and its (-)-usnic acid, atranorin and fumaprocetraric acid constituents. Z Naturforsch C 59:149–254

Studies on Antioxidant Properties of Lichen Secondary Metabolites

4

Marijana Kosanić and Branislav Ranković

Abstract

At present time, it is suspected that much used synthetic antioxidants have toxic and carcinogenic effects. Consequently, there is a growing interest towards finding new antioxidants of natural resources without any undesirable effect. Numerous in vitro studies on plants, micro- and macroalgae, macromycetes and lichens strongly support the fact that their constituents with antioxidant capacity are capable of exerting protective effects against oxidative stress in biological systems. Therefore, it is of prime importance to utilize natural antioxidants for their protective effect against oxidative stress and physiological dysfunctions. In the quest for novel natural antioxidant sources, our prime interest has focused on lichens. In recent time, numerous studies point to the importance of lichens in the neutralization of free radicals. Lichens are rich in secondary metabolites, primarily phenols, which are well known for its antioxidant properties. Because of that, the present chapter focuses on the role of lichens and their secondary metabolites in combating danger posed by overproduced free radicals.

4.1 Free Radicals, Oxidative Stress and Antioxidants

Free radicals (reactive oxygen species (ROS), such as hydroxyl radical, superoxide anion and hydrogen peroxide, and reactive nitrogen species (RNS), such as nitric oxide and peroxynitrite) (Fig. 4.1) play an important role in many chemical processes in the cells. At low or moderate concentrations, reactive oxygen species (ROS) and reactive nitrogen species (RNS) are necessary for the maturation process

M. Kosanić (✉) · B. Ranković
Department of Biology, Faculty of Science, University of Kragujevac, Kragujevac, Serbia

B. Ranković (ed.), *Lichen Secondary Metabolites*,
https://doi.org/10.1007/978-3-030-16814-8_4

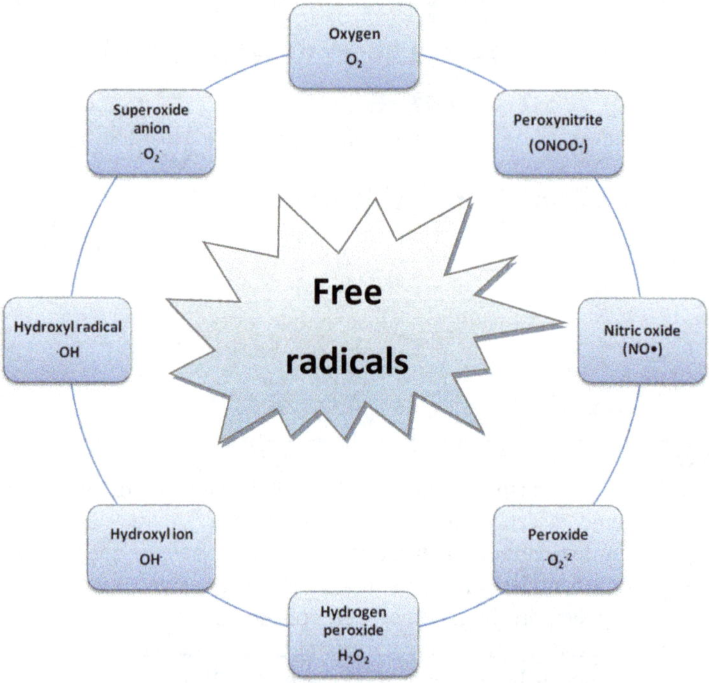

Fig. 4.1 Free radicals (reactive oxygen species and reactive nitrogen species)

of cellular structures and can act as weapons for the host defence system (Halliwell 1996; Squadriato and Pelor 1998; Young and Woodside 2001; Droge 2002; Sangameswaran et al. 2009; Huda-Faujan et al. 2009). Under normal conditions, the balance between the generation and diminution of ROS is controlled by the antioxidant defence system. However, under certain pathological conditions such as drought, salinity, chilling, metal toxicity and UV-B radiation as well as pathogens, when ROS are not effectively eliminated by the antioxidant defence system, the dynamic balance between the generation and diminution of ROS is broken (Lobo et al. 2010).

When produced in excess, free radicals and oxidants generate a phenomenon called oxidative stress, a deleterious process that can seriously alter the cell membranes and other structures such as proteins, lipids, lipoproteins and deoxyribonucleic acid (Betteridge 2000). If not regulated properly, oxidative stress can induce a variety of chronic and degenerative diseases such as Alzheimer's disease, atherosclerosis, emphysema, hemochromatosis, many forms of cancer (e.g. melanoma, Parkinson's disease and schizophrenia) as well as the ageing process (Fig. 4.2) (Sangameswaran et al. 2009; Sachindra et al. 2010).

Antioxidants are substances which possess the ability to protect the body from damage caused by oxidative stress (Souri et al. 2008). The body has several

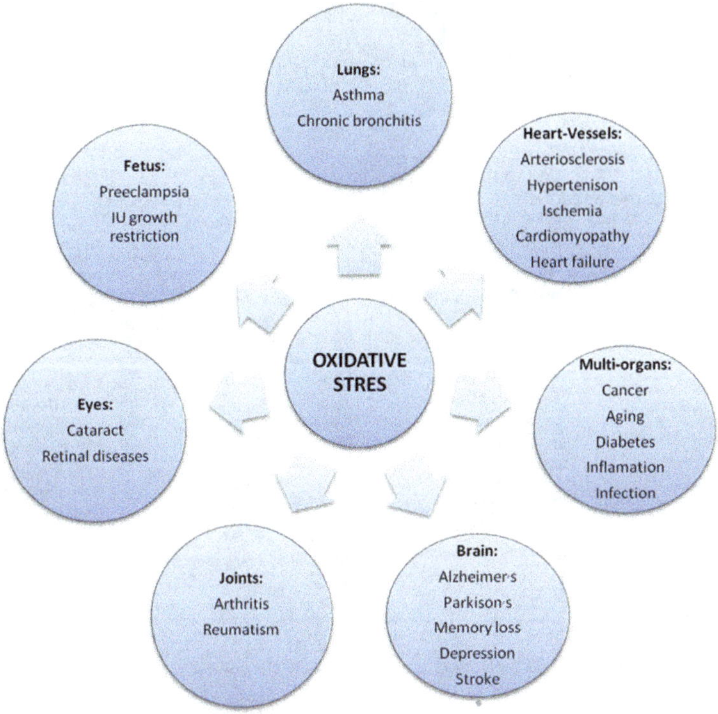

Fig. 4.2 Diseases induced by oxidative stress

mechanisms to counteract oxidative stress by producing antioxidants, either naturally generated in situ (endogenous antioxidants) or externally supplied through foods (exogenous antioxidants) (Fig. 4.3).

In addition, in order to protect biomolecules against the attack of ROS and/or to suppress the resultant damage, synthetic antioxidants have been used for industrial processing in recent years. The most extensively used synthetic antioxidants are butylated hydroxyanisole (BHA), butylated hydroxytoluene (BHT), tert-butylhydroquinone (TBHQ) and propyl gallate (PG) (Ramachandra Shajyothi and Padmalatha 2012). They are used widely in the food industry because of their effectiveness and generally being less expensive than natural antioxidants. However, since synthetic antioxidants are often carcinogenic, finding natural substitutes is of great interest (Zhang et al. 2009). Concerns regarding toxicological effects and carcinogenic potential of synthetic antioxidants have prompted the need for natural alternatives in the last few decades (Lu and Foo 2001; Aligiannis et al. 2003; Vagi et al. 2005; Es-Safi et al. 2006). Therefore, the importance of search for the exploitation of effective natural antioxidants has greatly increased in recent years (Pokorny et al. 2001; Gulcin et al. 2002; Naveena et al. 2008).

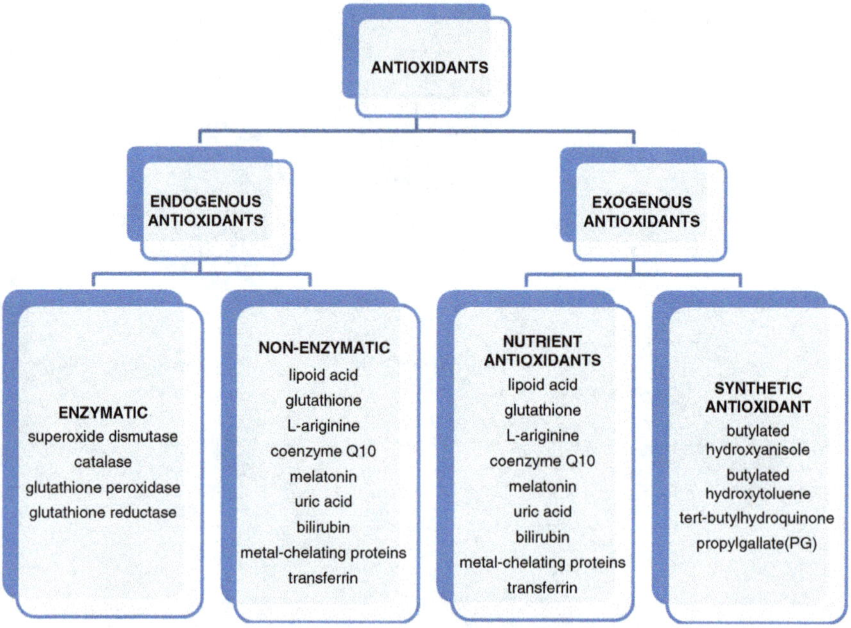

Fig. 4.3 Endogenous antioxidants and exogenous antioxidants for protecting the body from damage caused by oxidative stress

4.2 Antioxidant Properties of Lichens

Lichens are a good source of natural antioxidants. Although mainly hydrophobic, the phenolic nature of the major secondary metabolites of lichens is expected to afford antioxidant properties. Phenolics are the major secondary metabolites from lichens that play crucial function in the regulation of lichen growth and development in stressful and unfavourable climatic conditions. Lichens grow in extreme environmental conditions which in turn could cause up-regulated pathways of secondary metabolite synthesis and increased production of polyphenolics that have protective effects against oxidative stress owing to their antioxidant capacities (Kumar et al. 2014). Lichens produce many phenolic compounds including depsides, depsidones, dibenzofurans and pulvinic acid derivatives (Paudel et al. 2008).

Phenolic compounds are widespread products of secondary metabolism of lichen, and antioxidant activity is most frequently associated with their presence. Lichen phenolics are mainly depsides, depsidones and dibenzofurans, whereas vascular plant phenolics include tannins, lignins and flavonoids. Orsellinic acid is the basic unit in the biosynthesis of lichen phenolics. Lichen phenolics are generally secreted by the fungal partner and deposited as crystals on the surface of the cell wall of the fungal hyphae. Lichen phenolics are primarily acetate-polymalonate-derived with the exception of pulvinic acid derivates which are synthesized via the shikimic acid pathway. Lichen phenolics are composed of two monocyclic phenols joined either by an ester bond as in depsides or by both ester and ether bonds in depsidones or a furane heterocycle bond as found in dibenzofurans, such as usnic acid (Ross Watson 2014).

Phenolic compounds contain in its structure an aromatic ring by one or more hydroxyl groups. It is believed that the antioxidant activity of the phenol is primarily a result of their ability to be the donor of hydrogen atoms and eliminate the free radicals to form less reactive phenoxyl radicals (Sawa et al. 1999).

$$\mathrm{Ph-OH+ROO\rightarrow Ph-O\bullet+ROOH}$$

$$\mathrm{ArOH+OH\rightarrow H2O+ArO\bullet}$$

It is known that with the increase of the number of hydroxyl groups in the molecule, as well as the extension of the side chain of an antioxidant increases the activity of these compounds, probably due to the possibility of stabilization free radical forms the side chain conjugation (Yanishlieva et al. 1999). Because of the large number of potential reactive centres, the phenol molecule has the ability to react with the radicals of the same time.

Phenolics possess ideal structural chemistry for antioxidant activity and have been shown to be more effective in vitro than vitamins E and C on molar basis (Michalak 2006). As described by Bors et al. (1990), there are three structural features that are important determinants for the antioxidant potential of phenols:

1. The *ortho* 3′,4′-dihydroxy structure in the B ring
2. The 2,3-double bond in conjunction with the 4-oxo group in the C ring (which allows conjunction between A and B rings or electron delocalization)
3. The presence of a 3-OH group in C ring and a 5-OH group in the A ring

Among them the 3-OH group is the most significant determinant of electron-donating activity.

Mainly, the antioxidant activity of phenolic compounds is affected by their chemical structure. The antioxidant activity of phenolics also depends on the type and polarity of the extracting solvent, the isolation procedures, purity of active compounds as well as the test system and substrate to be protected by the antioxidant (Moure et al. 2001).

When antioxidant capacities of the lichens are compared with their phenolic constituents, it could be concluded that antioxidative nature of the lichens might depend on their phenolics. There are many reports which correlate the total phenolic content of lichens and their antioxidant activity. For example, for the methanol extract of *Lobaria pulmonaria*, there was a strong correlation between antioxidant activity and total phenolic content (Odabasoglu et al. 2004). Kekuda et al. (2012) found highly positive relationship between total phenolics and antioxidant activity in methanol extract of a foliose macrolichen *Everniastrum cirrhatum.* Pavithra et al. (2013) found a high correlation between antioxidant efficacy of a macrolichen *Usnea pictoides* and its phenolic content. A positive correlation was seen between the phenolic content and total antioxidant activity of *Anaptychia ciliaris*, *Nephroma parile*, *Ochrolechia tartarea* and *Parmelia centrifuga* having correlation coefficient values (r) of 0.990 (Ranković et al. 2010a, b). Kosanić and Ranković (2011a, b) found that total phenolic content of *Cetraria islandica*, *Lecanora atra*, *Parmelia pertusa*, *Pseudevernia furfuracea* and *Umbilicaria cylindrica* extracts was strongly related with DPPH radical scavenging activity ($r = 0.966$), with reducing power ($r = 0.944$) and with superoxide anion radical scavenging ($r = 0.823$). A certain correlation also was established between the antioxidant activity and the total phenolic content for the *Toninia candida* methanol, chloroform and petrol ether extracts (Manojlović et al. 2012).

Many other studies have shown a direct correlation between the phenolic content and the antioxidant activity (Tilak et al. 2004; Coruh et al. 2007; Rekha et al. 2012; Poornima et al. 2012).

However, the total phenolic content and antioxidant potency did not always correlate. Odabasoglu et al. (2005) determined the total antioxidant activity, total phenolic content and reducing power of methanol and water extracts of four lichen species, *Bryoria fuscescens*, *Dermatocarpon intestiniformis*, *Peltigera rufescens* and *Pseudevernia furfuracea*. Water and methanol extracts of *P. rufescens* showed the highest antioxidant activity. However, there was no correlation between antioxidant activity and total phenolic content of the extracts. Although the methanol extract of *P. furfuracea* had the highest total phenolic contents, it exhibited low antioxidant activity. In contrast, there was a strong correlation between reducing power and total antioxidant activity of the extracts. Stanly et al. (2011) evaluated the antioxidant activity and total phenolic content of four lichen species belonging to the genus *Ramalina*, *Parmotrema*, *Bulbothrix* and *Cladia* collected from Malaysia. There was no correlation between total phenolic content and radical scavenging activity of methanol extracts of all the tested species. For example, the antioxidant properties of some isolated phenols are not so impressive. For instance, two depsidones were found only slightly more active than the commercial quercetin in a superoxide scavenging assay with IC50 # 600 lM (Lohézic-Le Dévéhat et al. 2007), and none of the orcinol or orsellinate derivatives were as active as the commercial gallic acid to reduce DPPH (Lopes et al. 2008).

These results suggest that the antioxidant activity of some tested extracts might be attributed to the presence of non-phenolic compounds. Nevertheless, it should be taken into consideration that individual phenolics may have distinct antioxidant activities; there may be antagonistic or synergistic interactions between phenolics and other compounds like carbohydrates, proteins, etc. (Odabasoglu et al. 2005).

4.3 Assays to Determine Antioxidant Capacities of Lichens

Methods to assess the antioxidant activity of lichens are based on hydrogen atom transfer (HAT) and others on electron transfer (ET).

HAT-based assays measure the capability of phenolic antioxidants to quench free radicals (generally, peroxyl radicals considered to be biologically more relevant) by

H-atom donation. The HAT mechanisms of antioxidant action in which the hydrogen atom (H) of a phenol (ArOH) is transferred to a ROO• radical can be summarized by the reaction

$$\mathrm{ROO\bullet} + \mathrm{AH/ArOH} \rightarrow \mathrm{ROOH} + \mathrm{A\bullet/ArO\bullet}$$

where the aryloxy radical (ArO•) formed from the reaction of antioxidant phenol with peroxyl radical is stabilized by resonance. The AH and ArOH species denote the protected biomolecules and phenolic antioxidants, respectively. Effective phenolic antioxidants need to react faster than biomolecules with free radicals to protect the latter from oxidation. Since in HAT-based antioxidant assays, both the fluorescent probe and antioxidants react with ROO•, the antioxidant activity can be determined from competition kinetics by measuring the fluorescence decay curve of the probe in the absence and presence of antioxidants, integrating the area under these curves and finding the difference between them (Huang et al. 2005; Prior et al. 2005).

HAT-based assays include oxygen radical absorbance capacity (ORAC) assay, TRAP assay using R-phycoerythrin as the fluorescent probe, crocin bleaching assay using 2,2′-azobis(2-amidinopropane) dihydrochloride (AAPH) as the radical generator and β-carotene bleaching assay, although the latter bleaches not only by peroxyl radical attack but by multiple pathways (Huang et al. 2005; Prior et al. 2005).

In most *ET-based assays*, the antioxidant action is simulated with a suitable redox-potential probe, namely, the antioxidants react with a fluorescent or coloured probe (oxidizing agent) instead of peroxyl radicals. Spectrophotometric ET-based assays measure the capacity of an antioxidant in the reduction of an oxidant, which changes colour when reduced. The degree of colour change (either an increase or decrease of absorbance of the probe at a given wavelength) is correlated to the concentration of antioxidants in the sample. 2,2′-Azino-bis-(3-ethylbenzothiazoline-6-sulfonic acid) (ABTS)/Trolox-equivalent antioxidant capacity (TEAC) and 2,2-di (4-*tert*-octylphenyl)-1-picrylhydrazyl (DPPH) (Brand-Williams et al. 1995; Bondet et al. 1997; Sanchez-Moreno et al. 1998) are decolourisation assays, whereas in Folin total phenols assay (Folin and Ciocalteu 1927, Singleton et al. 1999), ferric reducing antioxidant power (FRAP) (Benzie and Strain 1996; Benzie and Szeto 1999) and cupric reducing antioxidant capacity (CUPRAC) (Apak et al. 2004), there is an increase in absorbance at a prespecified wavelength as the antioxidant reacts with the chromogenic reagent (i.e. in the latter two methods, the lower valencies of iron and copper, namely, Fe(II) and Cu(I), form charge transfer complexes with the corresponding ligands, respectively).

ET-based assays generally set a fixed time for the concerned redox reaction and measure thermodynamic conversion (oxidation) during that period. Although the reducing capacity of a sample is not directly related to its radical scavenging capability, it is a very important parameter of antioxidants (Apak et al. 2013).

Except for the most commonly used spectrophotometric methods, the antioxidant capacity of various compounds is also frequently evaluated using electrochemical methods. Commonly used electrochemical methods are cyclic voltammetric methods, amperometric technique and bioamperometric and biosensor technique.

Cyclic voltammetric technique: This method is based on the potentiodynamic electrochemical evaluation. In this process the working electrode is ramped linearly to the time. The working electrode potential is measured at different intervals in order to calculate the current intensity at respective electrode. On attaining the preset potential, the potential ramp of working electrode is reversed. In a single experiment, this reversal can take place many times. Then the cyclic voltammogram is made by plating working electrode's current to the applied voltage. A cyclic voltammogram gives the following significant parameters such as cathodic oxidation potential, anodic oxidation potential, anodic peaks and cathodic peaks (Chevion et al. 2000; Ahmad et al. 2014).

Amperometric technique: It is based on the principle that the intensity of the current flowing between a reference and a working electrode is measured at a fixed potential. A current is produced by the oxidation/reduction reaction of an analyte whose value is fixed at a preset value as compared to the reference electrode. The measurement of the antioxidant activity by an amperometric method is based on the reduction of the DPPH using carbon electrode (Milardovic et al. 2006; Ahmad et al. 2014).

Biamperometric technique: In biamperometry, between two same working electrodes that is polarized at the minute difference in potential and immersed in a solution that contains a reversible redox couple. The current flows in this process are measured. Commonly used redox couple in biamperometric measurement is Fe^{3+}/Fe^{2+}, I^2/I^-, $Fe(CN)6^3$-/$Fe(CN)6^{4-}$. Controlled parameters in biamperometry is potential difference between two same kind of working electrodes while the potential value of reference electrode is not controlled with respect to two electrodes (Tougas et al. 1985; Ahmad et al. 2014).

Biosensor technique: Because of their electron giving nature, oxidoreductases are very commonly used biosensors. These are very stable in nature and do not require any other cofactor or chemical for their working. They have been broadly studied and discussed in the literature. In broader sense most of the antioxidant properties of the extract is due to their polyphenolic constituents. These polyphenols are detected by the enzyme-based biosensors. These enzymes include tyrosinase and lacease (Barroso et al. 2011; Ahmad et al. 2014).

Chromatographic techniques are also employed for the assessment of the antioxidant capacity of the extracts. Of the chromatographic techniques, gas chromatography is used for these purposes. Gas chromatography is one of the primary types of chromatography that is extensively applied for the separation and analysis of stable but vaporizable compounds. The process is carried out by passing the mixture through the stationary phase which is liquid, and the mobile phase in this type of chromatography is gas. Gases used as mobile phase are helium and nitrogen that are considered to be inert, whereas the stationary phase is a microfilm of a liquid incorporated on to a non-reactive solid support. The retention time comparison gives its value. Usually thermal conductivity and flame ionization detectors are used in gas chromatography (Ahmad et al. 2014).

4.4 Antioxidant Activities of Lichen Extracts

Lichens are proven to be a good source of antioxidants, and a plenty of literatures have supported the antioxidant action of these organisms. Few reports concerning the antioxidative nature of pure lichen metabolites are available in the literature; most of the publications describe the antioxidant activities of crude lichen extracts.

There was a first report in 1993 (Yamamoto et al. 1993) on anti-oxidation activity in lichens by the method using SOD. Thereafter, antioxidant activities of many lichen species were assessed by numerous researchers. For example, Aslan et al. (2006) investigated the antioxidant activity of methanol extract of *Evernia divaricata*, *Evernia prunastri*, *Cladonia foliacea*, *Dermatocarpon miniatum* and *Neofuscella pulla* by scavenging free radical DPPH and the inhibition of linoleic acid oxidation. They found that extracts of *Cladonia foliacea*, *Evernia divaricata*, *Evernia prunastri* and *Neofuscella pulla* did not exert any activity in both assays, whereas those of *Dermatocarpon miniatum* provided 50% inhibition at 396.1 μg/ml concentration in the former and gave 49% inhibition in the latter. Mitrović et al. (2011) also studied antioxidant activity of methanol extract of *Parmelia sulcata*, *Flavoparmelia caperata*, *Evernia prunastri*, *Hypogymnia physodes* and *Cladonia foliacea* and found that they exhibited DPPH radical scavenging activity.

The antioxidant activity of ethanol extract of *Sanionia uncinata* was evaluated by Bhattarai et al. (2008a, b) by analysing its reducing power, superoxide scavenging activity, ABTS (2,2′-azino-bis(3-ethylbenzothiazoline-6-sulfonic acid)) cation scavenging activity and DPPH free radical scavenging activity. They found that tested lichen species could be an important source of natural antioxidant agents. Kumar et al. (2010a, b) reported significant antioxidant activity for methanol extracts of *Ramalina hossei* and *Ramalina conduplicans*; Yucel et al. (2007) for chloroform, methanol and aqueous extracts of *Cladonia rangiformis*; and Odabasoglu et al. (2005) for methanol and aqueous extracts of *Bryoria fuscescens*, *Dermatocarpon intestiniformis*, *Peltigera rufescens* and *Pseudevernia furfuracea*. A strong antioxidant effect was also found in the methanol extract of *Usnea ghattensis* (Verma et al. 2008a), in methanol extract of *Parmotrema pseudotinctorum* (Kumar et al. 2010a, b) as well as in acetone, methanol and aqueous extracts of lichens *Cladonia furcata*, *Hypogymnia physodes*, *Lasallia pustulata*, *Parmelia caperata* and *Parmelia sulcata* (Kosanić et al. 2011).

Sisodia et al. (2013) evaluated antioxidant activity for different extracts of *Ramalina roesleri* species and found that the DPPH radical scavenging activity of extracts ranged from 29.42 to 87.90%. Sharma and Kalikotay (2012) examined the antioxidant activity of two common lichens, namely, *Parmotrema reticulatum* and *Usnea* sp., from Darjeeling hills. The antioxidant assay of different concentrations of ethanolic and methanolic extracts of lichens was determined with respect to five parameters, i.e. DPPH radical scavenging activity, total antioxidant activity, reducing power ability and flavonoid and phenolic content. The DPPH radical scavenging ranged from 10 to 31.5% for methanol extracts of *Parmotrema reticulatum* and *Usnea* sp., respectively, and for reducing power measured values of absorbance varied from 0.376 to 0.514. In addition, total phenolic content of the extracts was high, and total flavonoid content was moderate.

The methanol extracts of 24 lichen species were tested for antioxidant activities in vitro. It was found that in DPPH assay, three species, *Peltigera* sp., *Cladonia* sp. and *Canoparmelia* sp., showed comparable activity with commercial standard, BHA. In ABTS+ assay, extracts of *Parmoterma* sp., *Ramalina* sp., *Peltigera* sp. and *Cladonia* sp. showed stronger activity than ascorbic acid. The observed data indicated that the high altitude lichens contain stronger antioxidant constituents (Paudel et al. 2008).

In the study described by Mastan et al. (2014), the lichens *Cladomia fimbriata*, *Permilopsis ambigua*, *Punctelia subrudecta* and *Evernia mesomorpha* secondary metabolites were extracted in the two solvents methanol and water. The lichen extracts showed comparable and strong antioxidant activity and exhibited higher DPPH and hydroxyl radical scavenging activity. Among the tested lichen extracts, water extract of *Evernia mesomorpha* gave highest reducing power, although the reducing activity was lower than the standard ascorbic acid. These findings provided evidence that crude aqueous and organic solvent extracts of lichens contain antioxidant important compounds.

Gulcin et al. (2002) found that the aqueous extracts of *Cetraria islandica* had a strong inhibition on peroxidation of linoleic acid, reducing power, superoxide anion radical scavenging and free radical scavenging activities. Similar results were reported by Behera et al. (2005) for different extracts from the lichen *Usnea ghattensis* which was found their strong effect on DPPH radical, superoxide anion radical, nitric oxide radical and strong lipid peroxidation inhibition. Kekuda et al. (2009) found strong antioxidant activity for the extracts of the lichen *Parmotrema pseudotinctorum* and *Ramalina hossei*. Manojlovic et al. (2010) explored antioxidant properties of *Laurera benguelensis* by scavenging of DPPH radical and found that this lichen provided a weak radical scavenging activity.

Anaptychya ciliaris, *Nephroma parile*, *Ochrolechia tartarea* and *Parmelia centrifuga* were screened for antioxidant activity by Ranković et al. (2010a, b). They found that the methanol extract of the *Parmelia centrifuga* showed a strong antioxidant activity, in comparison with the extracts from *Anaptychya ciliaris*, *Nephroma parile* and *Ochrolechia tartarea* which were relatively weaker. *Lecanora muralis*, *Parmelia saxatilis*, *Parmeliopsis ambigua*, *Umbilicaria crustulosa* and *Umbilicaria polyphylla* were tested for DPPH radical scavenging, superoxide anion radical scavenging and reducing power (Kosanić et al. 2014a), and it has been found that of the lichens tested, *Umbilicaria polyphylla* had the largest antioxidant activity.

Antioxidant activity of petroleum ether, chloroform, ethyl acetate and methanol extracts of *Usnea pictoides* lichen was determined by DPPH free radical scavenging assay and ferric reducing assay (Pavithra et al. 2013). The scavenging potential of methanol extract was higher than other extracts, and, also, in ferric reducing assay, methanol extract showed stronger reducing power than other extracts.

Usnea pictoides was found to be a strong antioxidant agent by Pavithra et al. (2013). The lichen was powdered and extracted sequentially using solvents of increasing polarity, viz. petroleum ether, chloroform, ethyl acetate and methanol. Antioxidant activity of solvent extracts was determined by DPPH free radical scavenging assay and ferric reducing assay. A dose-dependent scavenging of DPPH radicals by solvent extracts was observed. The scavenging potential of methanol extract was higher than other extracts. In ferric reducing assay, methanol extract showed stronger reducing power than other extracts. Overall, extracts containing high phenolic contents exhibited stronger antioxidant activity. A positive correlation was observed between total phenolic content and the antioxidant activity of lichen extracts.

In a related study, Odabasoglu et al. (2004) determined the antioxidant activities, reducing powers and total phenolic contents of methanol and water extracts of three lichen species, *Usnea longissima*, *Usnea florida* and *Lobaria pulmonaria*. Of the extracts tested, the methanol extracts of *Lobaria pulmonaria* and *Usnea longissima* showed potent antioxidant activities. The methanol extract of *Lobaria pulmonaria* also had the highest total phenolic contents (87.9 mg/g lyophylisate). For the methanol extract of this species, there was also a strong correlation between antioxidant activity and total phenolic contents. However, a similar correlation was not observed for *Usnea longissima*. Although the methanol extract of *Usnea longissima* had a lower phenolic content (38.6 mg/g lyophylisate), it exhibited potent antioxidant activity. On the other hand, there was a strong correlation between the reducing powers and the total phenolic contents of the extracts. The highest reducing power was determined for the methanol extract of *Lobaria pulmonaria*.

Ranković et al. (2010a, b) evaluated methanol extracts of the lichens *Cetraria pinastri*, *Cladonia digitata*, *Cladonia fimbriata*, *Fulgensia fulgens*, *Ochrolechia parella* and *Parmelia crinite* for their antioxidant activity. They found that the methanol extract of the *Cetraria pinastri* showed a strong antioxidant activity, whereas the extracts of the species *Fulgensia fulgens*, *Cladonia fimbriata* and *Parmelia crinita* showed the moderate one and the extract of the species *Ochrolechia parella* and *Cladonia digitata* the weak one. The methanol extract of the lichen *Cetraria pinastri* had the biggest total phenol content (32.9 mg/g of the dry extract). A certain correlation was established between the antioxidant activity and the total phenol content for the researched lichen extracts.

Acetone, methanol and aqueous extracts of the lichen *Cetraria islandica*, *Lecanora atra*, *Parmelia pertusa*, *Pseudevernia furfuracea* and *Umbilicaria cylindrica* were found to possess effective antioxidant activities (Kosanić and Ranković 2011a). Antioxidant activities of the tested extracts were studied by DPPH radical scavenging, superoxide anion radical scavenging and reducing power. The DPPH radical scavenging activity for studied species ranged from 32.68 to 94.70%. For reducing power, measured values of absorbance varied from 0.016 to 0.109. The superoxide anion scavenging activity for different extracts was 7.31–84.51%. In addition, the high contents of total phenolic compounds suggest that phenols might be the major antioxidant compounds in tested extracts. The same authors also found relatively strong antioxidant activity for acetone, methanol and aqueous extracts of the lichens *Cladonia furcata*, *Hypogymnia physodes* and *Umbilicaria polyphylla* (Kosanić and Ranković 2011b).

The antioxidant effect of methanol, chloroform and petrol ether extracts from the lichen *Toninia candida* was assayed by Manojlović et al. (2012). The lichen extracts showed comparable and strong antioxidant activity and exhibited higher DPPH and hydroxyl radical scavenging, chelating activity and inhibitory activity towards lipid peroxidation.

Stereocaulon paschale, *Parmeliopsis ambigua*, *Parmelia pertusa*, *Parmelia caperata*, *Parmelia sulcata*, *Parmelia saxatilis*, *Hypogymnia physodes*, *Cladonia furcata*, *Lecanora atra*, *L. muralis*, *Umbilicaria crustulosa*, *Umbilicaria cylindrica* and *Umbilicaria polyphylla* were studied for their antioxidant capacity by Serbian group of scientists (Ranković et al. 2011, 2014a, b; Kosanić et al. 2012a, b). They found that the tested lichens have effective free radical scavenging activity, reducing power and superoxide anion radical scavenging, and based on strong relationships between total phenolic contents and the antioxidant effect of tested species, it could be concluded that antioxidative nature of these lichens depends on their phenolics.

The n-hexane, methanol and water extracts of 14 saxicolous lichens from trans-Himalayan Ladakh region were evaluated for their antioxidant capacities by Kumar et al. (2014). The ferric reducing antioxidant power (FRAP), 2,29-azino-bis-(3-ethylbenzothiazoline-6-sulfonic acid) diammonium salt (ABTS), 1,1-diphenyl-2-picrylhydrazyl (DPPH) and nitric oxide (NO) radical scavenging capacities and b-carotene-linoleic acid bleaching property exhibited analogous results where the lichen extracts showed high antioxidant action. The lichen extracts were also found to possess good amount of total proanthocyanidin, flavonoid and polyphenol as the main antioxidant components in extracts. The methanolic extract of *Lobothallia alphoplaca* exhibited highest FRAP value. Methanolic extract of *Xanthoparmelia stenophylla* showed the highest ABTS radical scavenging capacity. The n-hexane extract of *Rhizoplaca chrysoleuca* exhibited highest DPPH radical scavenging capacity. Highest antioxidant capacity in terms of b-carotene linoleic acid bleaching property was observed in the water extract of *Xanthoria elegans*. Similarly, *Melanelia disjuncta* water extract showed highest NO scavenging capacity. Among n-hexane, methanol and water extracts of all lichens, the methanolic extract of *Xanthoparmelia mexicana* showed highest total proanthocyanidin, flavonoid and polyphenol content.

In the study by Vivek et al. (2014), the radical scavenging potential of three *Parmotrema* species, viz. *Parmotrema tinctorum*, *Parmotrema grayanum* and *Parmotrema praesorediosum*, from Maragalale and Guli Guli Shankara, Western Ghats of Karnataka, India, was reported. The powdered lichen materials were extracted using methanol. Radical scavenging activity of lichen extracts was determined by DPPH free radical scavenging assay. Total phenolic content of lichen extracts was estimated by Folin-Ciocalteau reagent method. Scavenging of DPPH radicals by lichen extracts was concentration dependent. Among the lichen species, *Parmoterma grayanum* showed higher scavenging potential as indicated by lower IC50 value. Total phenolic content was also high in *Parmoterma grayanum*.

Kosanić et al. (2014c) investigated antioxidant properties of acetone extracts of the lichens *Parmelia arseneana* and *Acarospora fuscata*. Antioxidant activity was evaluated by measuring the scavenging capacity of tested samples on DPPH and superoxide anion radicals and reducing the power of extracts. As a result of the study, *Parmelia arseneana* extract was found to have the larger DPPH radical scavenging activity, superoxide anion radical scavenging and reducing power than *Acarospora fuscata* extract.

In a study described by Ristić et al. (2016a), DPPH radical scavenging activity and reducing power of the acetone extracts of *Melanelia subaurifera* and *Melanelia fuliginosa* lichens were determined. The tested extracts showed a dose-dependent activity. The results of this assay present that acetone extract of *M. subaurifera* exhibits larger DPPH radical scavenging activity (IC_{50} = 165.13 μg/ml) than extract of *M. fuliginosa* (IC_{50} = 266.32 μg/ml) as well as showed higher reducing power. The same authors explored two *Ramalina* lichens for their antioxidant capacity. Extracts of the studied lichens showed a moderate DPPH radical scavenging activity. The IC_{50} values for the studied extracts were 285.45 and 423.51 μg/mL for *R. fraxinea* and *R. fastigiata*, respectively. Acetone extract of lichen *R. fraxinea* also showed the higher reducing power (Ristić et al. 2016b). Kosanić et al. (2016) investigated antioxidant activity of the *Lasallia pustulata* methanol extract by measuring free radical scavenging activity, superoxide anion scavenging activity and reducing power. Their results revealed that the methanol extract had moderate free radical scavenging activity with IC_{50} values of 395.56 μg/mL. Moreover, the extract tested had effective reducing power and superoxide anion radical scavenging.

Similarly, the antioxidant potential of the lichen *Pleurosticta acetabulum* was evaluated by Tomović et al. (2017). The antioxidant capacity of the acetone lichen extract was evaluated by determining the radical scavenging capacity on 2,2-diphenyl-1-picrylhydrazyl (DPPH) radicals and reducing power. They found that *P. acetabulum* extract exhibited moderate free radical scavenging activity (IC_{50} of 151.7301 μg/mL), while spectrophotometric absorbance for reducing power varied from 0.035 to 0.127.

The study performed by Aoussar et al. (2017) focused on screening of antioxidant activity of dichloromethane, acetone and methanol extracts of *Evernia prunastri* and *Pseudevernia furfuracea* growing in Morocco. The antioxidant activity was performed in vitro by DPPH (1,1-diphenyl-2-picrylhydrazyl) radical scavenging method and ferric reducing antioxidant power assay. The highest DPPH radical scavenging activity was obtained from the acetone extract of *P. furfuracea.* Also, for *P. furfuracea*, the absorbance of the acetone extract was higher than those of other extracts and increased from 0.043 ± 0.005 to 0.890 ± 0.036.

Five lichen species of *Cladonia* genus (*C. fimbriata*, *C. furcata*, *C. subulata*, *C. foliacea* and *C. rangiferina*) were examined for their antioxidant effect by Kosanić et al. (2018). They found that all tested extracts showed relatively high DPPH radical scavenging activity (IC_{50} values of lichen extracts range from 183.41 to 312.56 μg/mL). The extract of *C. furcata* showed the highest DPPH radical scavenging activity. Measured values of absorbances for reducing power varied in range from 0.010 to 0.1178. The extract of *C. furcata* showed the highest reducing power, followed by *C. foliacea*, *C. fimbriata*, *C. rangiferina* and *C. subulata*. The extract of *C. fimbriata* contains hypoprotocetraric acid, fumarprotocetraric acid, usnic acid (USN) and a small amount of the atranorin for which is a proven antioxidant activity, so that such greater activity was expected.

Ganesan et al. (2017) evaluated the antioxidant activities of *Parmotrema reticulatum* obtained from Yercaud hills of southern India. *P. reticulatum* was extracted with different solvents such as petroleum ether, ethyl acetate, ethanol, methanol, acetone and water based on their polarity. Then, it was subjected to antioxidant assays such as 1,1-diphenyl-2-picrylhydrazyl and ferric reduction activity potential. The DPPH free radical scavenging activity of lichen extracts using different solvents was found to increase while increasing the concentration of extracts. Methanol extracts of *P. reticulatum* showed highest activity of capturing hydrogen-free radicals than the other extracts. The IC_{50} value was found to be 19 μg/mL indicating the presence of bioactive molecules. The free radical capturing activity was found to be increased in the order of methanol > petroleum ether > water > acetone > ethanol > ethyl acetate extracts of *P. reticulatum*. The result revealed that methanol extract has the similar potential as that of standard ascorbic acid. In contrast, very less antioxidant activity was observed in the extract of ethyl acetate, and it was measured with an inhibition of 32.8% at 100 μg/ml. For ferric reduction activity potential, the absorbance was found to be increasing with an increase in the concentration of various extracts of lichen, and the activity was higher in the ethanol extracts among all other extracts. The order of activity was methanol > ethyl acetate > petroleum ether > ethanol > acetone > water. The same authors studied lichen species such as *Parmotrema austrosinense*, *P. hababianum* and *P. tinctorum* collected from the Eastern Ghats of India. From this species, the extracts were obtained by using various solvents; they were then subjected to various antioxidants assays, such as 2,2-diphenyl-1-picrylhydrazyl, ferric reducing antioxidant power and hydrogen peroxide. *P. tinctorum* showed very good antioxidant activity followed by *P. hababianum* and *P. austrosinense*. With respect to IC_{50}, acetone extracts of *P. tinctorum* were found to be superior (18.41 μg/mL) followed by methanol (60.3 μg/mL) and petroleum ether (74.5 μg/mL) extracts. The obtained results in this study indicated that selected lichen species showed moderate levels of reducing the ferric ions. In this regard, acetone extracts of *P. tinctorum* showed the highest reducing power as 3.77 at 1000 μg of concentration followed by petroleum ether extracts (1.46), whereas the least activity was measured for methanol extracts (0.125). In case of *P. hababianum*, methanol extracts showed highest reducing power (0.687) followed by water extracts (0.45). In contrast, water extracts of *P. austrosinense* showed better activity (0.48). Similar effects were seen for both methanol and petroleum ether extracts of *P. austrosinense*. For hydrogen peroxide scavenging assays, methanol and benzene extracts of *P. hababianum* showed the highest potential of scavenging, whereas other extracts did not show any activity. Benzene extract of *P. tinctorum* showed highest scavenging power as 26.5 followed by acetone extracts (35.32). Similar effects were observed in both methanol and benzene extracts of *P. austrosinense* for scavenging power of 80.99 and 89.85, respectively (Ganesan et al. 2015).

Antioxidant capacity of many other lichen extracts was confirmed by other researchers (Ranković et al. 2012; Kosanić et al. 2013a, b, 2014b).

4.5 Antioxidant Activities of Lichen Secondary Metabolites

Lichens have been found to contain a variety of secondary lichen substances with strong antioxidant activity. There are reports concerning the antioxidative nature of pure lichen metabolites that are available in the literature (Table 4.1).

Methanol-water (90:10 v/v) extracts of five polar lichen species, namely, *Stereocaulon alpinum*, *Ramalina terebrata*, *Caloplaca* sp., *Lecanora* sp. and

Table 4.1 Literature sources mentioning data for lichen substances responsible for antioxidant activity of lichens

Lichen substances	References
Lecanoric acid	Jayaprakasha and Rao (2000), Lopes et al. (2008), Thadhani et al. (2011), Verma et al. (2008b), Buçukoglu et al. (2013), Ristić et al. (2016a)
Atranorin	Hidalgo et al. (1994), Jayaprakasha and Rao (2000), Melo et al. (2011), Sisodia et al. (2013). Ranković et al. (2014a, b), Kosanić et al. (2014a, b, c)
2,4-Dihydroxy-6-propyl	Sisodia et al. (2013)
Ramalin	Paudel et al. (2008)
Barbatic acid	Verma et al. (2008b)
Divaricatic acid	Hidalgo et al. (1994)
Pannarin	Hidalgo et al. (1994), Russo et al. (2008)
Zeorin	Verma et al. (2008b)
Gyrophoric acid	Buçukoglu et al. (2013), Kosanić et al. (2014c)
Umbilicaric acid	Buçukoglu et al. (2013)
Protocetraric acid	Manojlović et al. (2012)
Fumarprotocetraric acid	de Barros Alves et al. (2014), Kosanić et al. (2014a, b, c)
Stictic acid	Lohézic-Le Dévéhat et al. (2007)
Salazinic acid	Verma et al. (2012), Manojlović et al. (2012), Selvaraj et al. (2015)
Usnic acid	Odabasoglu et al. (2006), Manojlović et al. (2012), Ranković et al. (2012, 2014a, b), Verma et al. (2012), Sisodia et al. (2013), Ananthi et al. (2015)
Variolaric acid	Brisdelli et al. (2013)
2-Hydroxy-4-methoxy-6-propylbenzoic acid	Sisodia et al. (2013)
Evernic acid	Kosanić et al. (2013a, b)
Erythrin	Choudhary et al. (2009)
2,4-Dihydroxy-3,6-dimethylbenzoate	Sisodia et al. (2013)
Lobaric acid	Thadhani et al. (2011); Brisdelli et al. (2013)
Physodic acid	Kosanić et al. (2013a, b), Ranković et al. (2014a, b)
Protolichesterinic acid	Sisodia et al. (2013)
Norstictic acid	Ranković et al. (2012)
Cuculloquinone	Stepanenko et al. (2002)
Homosekikaic acid	Sisodia et al. (2013)
Sekikaic acid	Thadhani et al. (2011), Verma et al. (2012), Sisodia et al. (2013)
Benzoic acid	Sisodia et al. (2013)
Diffractaic acid	Brisdelli et al. (2013)
Vicanicin	Brisdelli et al. (2013)
Sphaerophorin	Russo et al. (2008)
Isidiophorin	Atalay et al. (2011)
Rhizonaldehyde	Atalay et al. (2011)
Rhizonyl alcohol	Atalay et al. (2011)

(continued)

Table 4.1 (continued)

Lichen substances	References
Pulmonarianin	Atalay et al. (2011)
Hypotrachynic acid	Papadopoulou et al. (2007)
Deoxystictic acid	Papadopoulou et al. (2007)
Cryptostictinolide	Papadopoulou et al. (2007)
8′-Methylconstictic acid	Papadopoulou et al. (2007)
8′-Methylstictic acid	Papadopoulou et al. (2007)
8′-Methylmenegazziaic acid	Papadopoulou et al. (2007)
Ethylstictic acid	Papadopoulou et al. (2007)
2′-O-methyl anziaic acid	Ristić et al. (2016a)
Obtusatic acid	Ristić et al. (2016b)
Methyl evernate	Ristić et al. (2016b)

Caloplaca regalis from King George Island were analysed using thin layer chromatography (TLC) followed by a DPPH (2,2-diphenyl-1-picrylhydrazyl) spray technique. The experimental data showed that 33–50% of the major constituents of the test extracts were active antioxidants (Bhattarai et al. 2008a, b).

Hidalgo et al. (1994) reported the antioxidant activity of some depsides, such as atranorin (isolated from *Placopsis* sp.) and divaricatic acid (isolated from *Protousnea malacea*), and depsidones, such as pannarin (isolated from *Psoroma pallidum*) and 1′-chloropannarin (isolated from *Erioderma chilense*). All of these secondary compounds inhibited rat brain homogenate autoxidation and β-carotene oxidation, and depsidones were found to be the most effective. Russo et al. (2008) found that both sphaerophorin (depside) and pannarin (depsidone) inhibited superoxide anion formation in vitro, pannarin being more efficient, confirming Hidalgo et al. (1994). Similarly, de Barros Alves et al. (2014) found high antioxidant power of fumarprotocetraric acid produced by the lichen *Cladonia verticillaris* evaluated using the thiobarbituric acid reactive species assay in mouse lung tissue.

Thadhani et al. (2011) assessed the antioxidant activity of several classes of lichen metabolites in the in vitro superoxide radical (SOR), nitric oxide radical and 2,2-diphenyl-1-picrylhydrazyl radical scavenging assays. The depsides sekikaic acid and lecanoric acid showed promising antioxidant activity in SOR assay with IC_{50} values of 82.0 ± 0.3 μmol and 91.5 ± 2.1 μmol, respectively, while the depsidone lobaric acid exhibited an IC_{50} value of 97.9 ± 1.6 μmol, all relative to the standard, propyl gallate ($IC_{50} = 106.0 \pm 1.7$ μmol). One of the most abundant mononuclear phenolic compounds, methyl-β-orcinol carboxylate, was found to be a potent NO scavenger ($IC_{50} = 84.7 \pm 0.1$ μmol), compared to the standard rutin ($IC_{50} = 86.8 \pm 1.9$ μmol).

Brisdelli et al. (2013) investigated the effects of six lichen metabolites (diffractaic acid, lobaric acid, usnic acid, vicanicin, variolaric acid, protolichesterinic acid) on reactive oxygen species (ROS) level towards three human cancer cell lines, MCF-7 (breast adenocarcinoma), HeLa (cervix adenocarcinoma) and HCT-116 (colon carcinoma). All tested lichen compounds did not exhibit free radical scavenging activity using the 1,1-diphenyl-2-picrylhydrazyl (DPPH) assay. The lichen metabolites did not significantly increase the intracellular ROS level and did not prevent oxidative injury induced by t-butyl hydroperoxide in tested cells.

Choudhary et al. (2009) studied the superoxide radical scavenging activity assay. They found that the para-depsides lecanoric acid and erythrin and the meta-depside sekikaic acid showed exceptionally high percentage of radical scavenging activity in the SOR assay along with the depsidone lobaric acid. The structural feature in all of the above compounds is having two aromatic rings connected by an ester linkage and ortho to the carbonyl bearing carbon of ring A, an oxygen atom which may act as the electron acceptor from the antibonding orbitals of superoxide radical leading to molecular oxygen. The electron thus obtained could be stabilized due to extended conjugation available in such compounds. In the case of depsidone lobaric acid, the electron accepted by C2O could be stabilized by both aromatic rings. The SOR activity of both the depsides lecanoric acid and erythrin was lost on permethylation suggesting that when C2O is methylated, the molecule loses its ability to accept electrons. Importantly, the IC_{50} values of the sekikaic acid, lecanoric acid and lobaric acid were lower than the propyl gallate standard.

Atalay et al. (2011) determined the lipid peroxidation inhibition potential for nine lichen compounds in two lipid peroxidation test systems (liposome and emulsion systems). They found that isidiophorin, rhizonaldehyde, rhizonyl alcohol and pulmonarianin retarded lipid peroxidation in both test systems. However, stictic acid and ergosterol peroxide exhibited antioxidant activity in only the liposome test system. Usnic acid and diffractaic acid were not antioxidants in either system, while stictic acid was not lipid peroxidation inhibitor in the emulsion test system. All compounds, which inhibited lipid peroxidation in both test systems, were also DPPH radical scavengers.

Odabasoglu et al. (2006) investigated gastroprotective effect of usnic acid isolated from *Usnea longissima* in the indomethacin-induced gastric ulcers in rats. They found that the gastroprotective effect of usnic acid can be attributed to its reducing effect on the oxidative damage. All tested doses of usnic acid a significant increase in the levels of superoxide dismutase (SOD), glutathione peroxidase (GPx) and reduced glutathione (GSH), and a reduction in the lipid peroxidation (LPO) level in tissues.

For some stictic acid derivates, Lohézic-Le Dévéhat et al. (2007) found moderate antiradical activity in the 1,1-diphenyl-2-picrylhydrazyl (DPPH) assay and very high superoxide anion scavenging activity.

Methanolic extracts and lichen acids (gyrophoric acid, lecanoric acid and umbilicaric acid) obtained from six *Umbilicaria* species were tested for their antioxidant activity (Buçukoglu et al. 2013). The antioxidant ability was measured using a free radical scavenging activity assay using 2,2-diphenyl-1-picrylhydrazyl (DPPH). The methanolic extracts showed moderate DPPH radical scavenging activity. Among the lichen acids, umbilicaric acid showed the highest antioxidant activity with 68.14% inhibition.

Jayaprakasha and Rao (2000) extracted with benzene and acetone the lichen *Parmotrema stuppeum* in order to investigate its antioxidant effect. Both the extracts were fractionated on 1% oxalic acid impregnated silica gel column to obtain methyl orsenillate, orsenillic acid, atranorin and lecanoric acid respectively. Antioxidant activities of benzene extract, acetone extract and isolated compounds were evaluated in a carotene-linoleate model system. The obtained results showed that the pure compounds and extracts have moderate antioxidant activity.

Four new β-orcinol metabolites, hypotrachynic acid, deoxystictic acid, cryptostictinolide acid and 8′-methylconstictic acid along with the metabolites 8-′-methylstictic acid, 8′-methylmenegazziaic acid, stictic acid, 8′-ethylstictic acid and atranorin, which have been previously described, were isolated for the first time from the tissue extracts of the lichen *Hypotrachyna revoluta* (Papadopoulou et al. 2007). The structures of the new metabolites were elucidated on the basis of extensive spectroscopic analyses. Radical scavenging activity of the metabolites isolated in adequate amounts was evaluated using luminol chemiluminescence. The evaluation of the lichen metabolites by the chemiluminescence method showed most of them possess noteworthy antioxidant activity, with the highest levels being exhibited by compound 8′-methylmenegazziaic acid, which was only seven times less potent than the standard antioxidant Trolox® that was used for comparison reasons. The results showed that compounds 8′-methylmenegazziaic acid and atranorin that possess an additional hydroxyl group on the aromatic ring are the most active ones, and the activity is reduced by half when the hydroxyl of C-3 is replaced by an aldehyde moiety. Finally the scavenging activity of the metabolites possessing an aldehyde group on C-3 seems to be drastically reduced when the methylene of the γ-lactone ring is substituted by a hydroxy or methoxy moiety, as observed in the cases of metabolites deoxystictic acid, stictic acid and 8′-methylstictic acid.

Cuculloquinone, a bisnaphthoquinone of *Flavocetraria cucullata* inactivated DPPH to a 80% extent while the BHT that was used as a standard antioxidant was twofold less active (Stepanenko et al. 2002).

Lopes et al. (2008) found that lecanoric acid and some depside isolated from a *Parmotrema tinctorum* were very active DPPH radical scavengers.

Various antioxidant assays were used by Paudel et al. (2008) to evaluate the antioxidant capacities of the ramalin isolated from the methanol-water extract of the Antarctic lichen *Ramalina terebrata.* The experimental data showed that ramalin was five times more potent than commercial butylated hydroxyanisole (BHA) in scavenging l-diphenyl-2-picrylhydrazyl (DPPH) free radicals, 27 times more potent in scavenging 2,2′-azino-bis(3-ethylbenzothiazoline-6-sulfonic acid free radicals (ABTS*+) than the vitamin E analogue, Trolox, and 2.5 times more potent than BHT in reducing Fe3+ to Fe2+ ions. Similarly, ramalin was 1.2 times more potent than ascorbic acid in scavenging superoxide radicals and 1.25 times more potent than commercial kojic acid in inhibiting tyrosinase enzyme activity, which ultimately leads to whitening of skin cells. Furthermore, ramalin was assessed to determine its antioxidant activity in vivo. One microgram per millilitre ramalin significantly reduced the released nitric oxide (NO), and 0.125 μg/ml ramalin reduced the produced hydrogen peroxide in LPS (lipopolysaccharide)-stimulated murine macrophage Raw 264.7 cells. Considering all the data together, ramalin can be a strong therapeutic candidate for controlling oxidative stress in cells.

Melo et al. (2011) evaluated free radical scavenging activities and antioxidant potential of atranorin using various in vitro assays for scavenging activity against hydroxyl radicals, hydrogen peroxide, superoxide radicals and nitric oxide. Besides, the total reactive antioxidant potential and total antioxidant reactivity indexes and in vitro lipoperoxidation were also evaluated. They found that atranorin exerts differential effects towards reactive species production, enhancing hydrogen peroxide and nitric oxide production and acting as a superoxide scavenger; no activity towards hydroxyl radical production/scavenging was observed. Also, total reactive antioxidant potential and total antioxidant reactivity analysis indicated that atranorin acts as a general antioxidant, although it demonstrated to enhance peroxyl radical-induced lipoperoxidation in vitro.

Atranorin, protolichesterinic acid, usnic acid, 2-hydroxy-4-methoxy-6-propyl benzoic acid, homosekikaic acid, sekikaic acid, benzoic acid, 2,4-dihydroxy-6-propyl and 2,4-dihydroxy-3,6-dimethyl benzoate isolated from the hexane extract from *Ramalina roesleri* were assayed for antioxidant activity by 1,1-diphenyl-2-picrylhydrazyl (Sisodia et al. 2013). Maximum DPPH radical scavenging activity was exhibited by sekikaic acid followed by homosekikaic acid.

Verma et al. (2012) found high free radical scavenging activities of salazinic acid, sekikaic acid and usnic acid isolated from three terrestrial natural lichen species *Ramalina celastri*, *Ramalina nervulosa* and *Ramalina pacifica.* Contrary, Brisdelli et al. (2013) investigated the effects of six lichen metabolites (diffractaic acid, lobaric acid, usnic acid, vicanicin, variolaric acid, protolichesterinic acid) on reactive oxygen species (ROS) using the 1,1-diphenyl-2-picrylhydrazyl (DPPH) assay. They found that lichen compounds did not exhibit DPPH radical scavenging activity and did not significantly increase the intracellular ROS level.

Manojlović et al. (2012) investigated antioxidant activities for protocetraric and usnic acids from *Parmelia caperata* lichen and depsidone salazinic acid from *Parmelia saxatilis* species. Antioxidant activities of these isolated metabolites were evaluated by free radical scavenging, superoxide anion radical scavenging and reducing power. As a result of the study, usnic acid had stronger antioxidant activity than salazinic acid and protocetraric acid in all used tests.

In the study described by Ranković et al. (2012), lichen compounds norstictic acid isolated from *Toninia candida* and usnic acid from *Usnea barbata* exhibited high antioxidant potential in vitro, but it should be noted that the norstictic acid had a larger antioxidant capacity.

Evernia prunastri and *Pseudevernia furfuracea* lichens and their major metabolites evernic acid and physodic acid were screened for their antioxidant effects by Kosanić et al. (2013a, b) who found varying antioxidant success in free radical scavenging, superoxide anion radical scavenging and reducing power, with physodic acid being the most effective.

Antioxidant activities of major lichen metabolites in *Hypogymnia physodes* lichen (physodic acids, atranorin and usnic acid) were studied by Ranković et al. (2014a, b). An physodic acid was found to be the most effective antioxidant in free radical and superoxide anion scavenging, as well as in reducing power assays among tested lichen metabolites.

Kosanić et al. (2014a, b, c) investigated the antioxidant activities of acetone extracts of the lichens *Cladonia furcata*, *Cladonia pyxidata* and *Cladonia rangiferina* and their atranorin and fumarprotocetraric acid constituents. Antioxidant activities were evaluated by free radical scavenging, superoxide anion radical scavenging and reducing power. As a result of the study, isolated components had larger antioxidant activity than tested extracts. Atranorin had largest free radical scavenging activity with IC_{50} values 131.48 μg/mL. Moreover, atranorin had the most effective reducing power and superoxide anion radical scavenging.

Kosanić et al. (2014c) studied antioxidant capacity for gyrophoric acid from *Acarospora fuscata* lichen. The isolated lichen component demonstrated strong DPPH radical scavenging activity. The IC_{50} value for gyrophoric acid was 105.75 μg/mL. The isolated component also expressed the high superoxide anion radical scavenging activity (the IC_{50} was 196.62 μg/mL) and reducing power. Ristić et al. (2016a) tested antioxidant properties for 2′-O-methyl anziaic acid from lichen *M. fuliginosa* and lecanoric acid from lichen *M. subaurifera*. Both tested compounds showed a dose-dependent activity. For DPPH assay, the IC_{50} value of lecanoric acid was higher (IC_{50} = 424.51 μg/mL) compared to another compound 2′-*O*-methyl anziaic acid (IC_{50} = 121.52 μg/mL). Compound 2′-*O*-methyl anziaic acid also showed higher reducing power than lecanoric acid. The same group of authors have done similar research with two *Ramalina* lichens from which isolated obtusatic acid (from *R. fraxinea*) and methyl evernate (from *R. fastigiata*). Both compounds exhibited lower DPPH radical scavenging activity. For reducing power it is interesting that the isolated components showed higher effect than lichen extracts from which they are isolated (Ristić et al. 2016b).

Study performed by Selvaraj et al. (2015) was to estimate the antioxidant activity of salazinic acid and its derivate hexaacetyl salazinic acid isolated from *Parmotrema reticulatum*. The antioxidant properties were studied under the categories DPPH, FRAP, metal chelating activity, hydroxyl scavenging activity, lipid peroxidation activity, phosphomolybdenum activity and superoxide dismutase activity. But among the two test compounds, hexaacetyl salazinic acid gives lower IC_{50} value and shows higher inhibition activity than salazinic acid for DPPH, superoxide dismutase activity and hydroxyl radical scavenging activity. It was already known that the lower the IC_{50} value, the higher is the inhibition activity. In lipid peroxidation, both the test compounds show nearly equivalent activity. In phosphomolybdenum assay, salazinic acid shows better activity than hexaacetyl salazinic acid. In FRAP assay, hexaacetyl salazinic acid shows better activity than salazinic acid. The metal chelating capacity of both the test compounds was found to be less.

Ananthi et al. (2015) repeated the same experiment with usnic acid and its derivative usnic acid diacetate isolated from *Usnea luridorufa*. They concluded that the lichen secondary metabolite usnic acid and its derivative usnic acid diacetate have possessed efficient antioxidant properties and are reported to be dose dependent. Among these two test compounds usnic acid gives lower IC_{50} value for hydroxyl scavenging activity, superoxide dismutase scavenging activity and shows better activity than usnic acid diacetate, for DPPH radical scavenging while usnic acid diacetate gives lower IC_{50} value and possesses better activity than usnic acid for lipid peroxidation inhibiting activity both the test compounds shows lesser activity. Also usnic acid diacetate possesses higher total antioxidant activity via phosphomolydenum assay than usnic acid. In FRAP assay usnic acid diacetate shows higher activity than usnic acid. On the other hand, metal chelating activity of both usnic acid and usnic acid diacetate was found to be very less.

This work reveals that the lichens can be an interesting source of new antioxidative agents, with a potential use in different fields (food, cosmetics, pharmaceutical). Further work should be focused on the isolation of new pure compounds from lichens and investigation of their antioxidant activity.

References

Ahmad S, Arshad MA, Ijaz S et al (2014) Review on methods used to determine antioxidant activity. Int J Multidiscip Res Dev 1:35–40

Aligiannis N, Mitaku S, Tsitsa-Tsardis E et al (2003) Methanolic extract of *Verbascum macrurum* as a source of natural preservatives against oxidative rancidity. J Agric Food Chem 51: 7308–7312

Ananthi R, Tinabaye A, Selvaraj G (2015) Antioxidant study of usnic acid and its derivative usnic acid diacetate. Int J Adv Res Technol 4:356–366

Aoussar N, Manzali R, Nattah I et al (2017) Chemical composition and antioxidant activity of two lichens species (*Pseudevernia furfuracea* L. and *Evernia prunastri* L.) collected from Morocco. JMES 8:1968–1976

Apak R, Güçlü K, Özyürek M et al (2004) A novel total antioxidant capacity index for dietary polyphenols, vitamin C and E, using their cupric ion reducing capability in the presence of neocuproine: the CUPRAC method. J Agric Food Chem 52:7970–7981

Apak R, Gorinstein S, Bohm V et al (2013) Methods of measurement and evaluation of natural antioxidant capacity/activity (IUPAC technical report). Pure Appl Chem 85:957–998

Aslan A, Güllüce M, Sökmen M et al (2006) Antioxidant and antimicrobial properties of the lichens *Cladonia foliacea, Dermatocarpon miniatum, Evernia divaricata, Evernia prunastri and Neofuscella pulla*. Pharm Biol 44:247–252

Atalay F, Halici MB, Mavi AA et al (2011) Antioxidant phenolics from *Lobaria pulmonaria* (L.) Hoffm. and *Usnea longissima* Ach. Lichen species. Turk J Chem 35:647–661

Barroso MF, De-los-Santos-Alvarez N, Delerue-Matos C et al (2011) Towards a reliable technology for antioxidant capacity and oxidative damage evaluation: electrochemical (bio) sensors. Biosens Bioelectron 30:1–12

Behera BC, Verma N, Sonone A et al (2005) Antimicrobial activity of various solvent extracts of Lichen Usnea ghattensis. Agarkar Research Institute, Pune, India. Biotechnol Lett 27:991–995

Benzie IFF, Strain JJ (1996) The ferric reducing ability of plasma (FRAP) as a measure of "antioxidant power": the FRAP assay. Anal Biochem 239:70–76

Benzie IFF, Szeto YT (1999) Total antioxidant capacity of teas by the ferric reducing/antioxidant power assay. J Agric Food Chem 47:633–636

Betteridge DJ (2000) What is oxidative stress? Metabolism 49:3–8

Bhattarai HD, Paude B, Lee HS et al (2008a) Antioxidant activity of *Sanionia uncinata*, a polar mass species from King George Island, Antarctica. Phytother Res 22:1635–1639

Bhattarai HD, Paudel B, Hong SG et al (2008b) Thin layer chromatography analysis of antioxidant constituents of lichens from Antarctica. J Nat Med 62:481–484

Bondet V, Brand-Williams W, Berset C (1997) Kinetics and mechanisms of antioxidant activity using the DPPH free radical method. Food Sci Technol 30:609–615

Bors W, Heller W, Michel C et al (1990) Flavonoids as antioxidants: determination of radical-scavenging efficiencies. Methods Enzymol 186:343–355

Brand-Williams W, Cuvelier ME, Berset C (1995) Use of a free radical method to evaluate antioxidant activity. LWT Food Sci Technol 28:25–30

Brisdelli F, Perilli M, Sellitri D et al (2013) Cytotoxic activity and antioxidant capacity of purified lichen metabolites: an in vitro study. Phytother Res 27:431–437

Buçukoglu TZ, Albayrak S, Halici MG et al (2013) Antimicrobial and antioxidant activities of extracts and lichen acids obtained from some *Umbilicaria* species from Central Anatolia, Turkey. J Food Process Preserv 37:1103–1110

Chevion S, Roberts MA, Chevion M (2000) The use of cyclic voltammetry for the evaluation of antioxidant capacity. Free Radic Biol Med 28:860–870

Choudhary MI, Ali S, Thadhani VM et al (2009) Natural novel antioxidants, Patent Application Number: 11/838567, Publication Date 02/19/2009

Coruh N, Celep AGS, Ozgokçe F et al (2007) Antioxidant capacities of *Gundelia tournefortii* L. extracts and inhibition on glutathione-S-transferase activity. Food Chem 100(3):1249–1253

de Barros Alves GM, de Sousa Maia MB, de Souza FE et al (2014) Expectorant and antioxidant activities of purified fumarprotocetraric acid from *Cladonia verticillaris* lichen in mice. Pulm Pharmacol Ther 27:139–143

Droge W (2002) Free radicals in the physiological control of cell function. Physiol Rev 82(1):47–95

Es-Safi N, Khlifi S, Kerhoas L et al (2006) Iridoid glucosides from the aerial parts of *Globularia alypum* L. (*Globulariaceae*). Chem Pharm Bull 54:85–88

Folin O, Ciocalteu V (1927) Tyrosine and tryptophan determination in proteins. J Biol Chem 73: 627–650

Ganesan A, Thangapandian M, Ponnusamy P et al (2015) Antioxidant and antibacterial activity of Parmeliod lichens from Shevaroy hills of Eastern Ghats, India. Int J Pharm Tech Res 8:13–23

Ganesan A, Thangapandian M, Ponnusamy P et al (2017) Antibacterial and antioxidant activity of *Parmotrema reticulatum* obtained from Eastern Ghats, Southern India. Biomed Res 28: 1593–1597

Gulcin I, Oktay M, Kufrevioglu OI et al (2002) Determination of antioxidant activity of lichen *Cetraria islandica* (L) Ach. J Ethnopharmacol 79:325–329

Halliwell B (1996) Antioxidants in human health and disease. Annu Rev Nutr 16(1):33–50

Hidalgo ME, Fernández E, Quilhot W et al (1994) Antioxidant activity of depsides and depsidones. Phytochemistry 37:1585–1587

Huang D, Ou B, Prior RL (2005) The chemistry behind antioxidant capacity assays. J Agric Food Chem 53:1841–1856

Huda-Faujan N, Noriham A, Norrakiah AS et al (2009) Antioxidant activity of plants methanolic extracts containing phenolic compounds. Afr J Biotechnol 8:484–489

Jayaprakasha GK, Rao LJ (2000) Phenolic constituents from the lichen *Parmotrema stuppeum* (Nyl.) Hale and their antioxidant activity. Z Naturforsch 55:1018–1022

Kekuda PTR, Vinayaka KS, Kumar PSV et al (2009) Antioxidant and antibacterial activity of lichen extracts, honey and their combination. J Pharm Res 2:1875–1878

Kekuda TRP, Raghavendra HL, Swathi D et al (2012) Antifungal and cytotoxic activity of Everniastrum cirrhatum (Fr.) Hale. Chiang Mai J Sci 39:76–83

Kosanić M, Ranković B (2011a) Lichens as possible sources of antioxidants. Pak J Pharm Sci 24: 165–170

Kosanić M, Ranković B (2011b) Antioxidant and antimicrobial properties of some lichens and their constituents. J Med Food 14:1624–1630

Kosanić M, Ranković B, Vukojević J (2011) Antioxidant properties of some lichen species. J Food Sci Technol 48:584–590

Kosanić M, Ranković B, Stanojković T (2012a) Antioxidant, antimicrobial, and anticancer activities of three *Parmelia* species. J Sci Food Agric 92:1909–1916

Kosanić M, Ranković B, Stanojković T (2012b) Antioxidant, antimicrobial and anticancer activity of 3 *Umbilicaria* species. J Food Sci 77:T20–T25

Kosanić M, Ranković B, Stanojković T (2013a) Investigation of selected Serbian lichens for antioxidant, antimicrobial and anticancer properties. J Anim Plant Sci 23:1628–1633

Kosanić M, Manojlović N, Janković S et al (2013b) *Evernia prunastri* and *Pseudevernia furfuracea* lichens and their major metabolites as antioxidant, antimicrobial and anticancer agents. Food Chem Toxicol 53:112–118

Kosanić M, Šeklić D, Marković S et al (2014a) Evaluation of antioxidant, antimicrobial and anticancer properties of selected lichens from Serbia. Dig J Nanomater Bios 9:273–287

Kosanić M, Ranković B, Stanojković T et al (2014b) *Cladonia* lichens and their major metabolites as possible natural antioxidant, antimicrobial and anticancer agents. LWT Food Sci Technol 59: 518–525

Kosanić M, Ranković B, Stanojković T et al (2014c) Biological activities and chemical composition of lichens from Serbia. EXCLI J 13:1226–1238

Kosanić M, Ranković B, Stanojković T et al (2016) *Lasallia pustulata* lichen as possible natural antigenotoxic, antioxidant, antimicrobial and anticancer agent. Cytotechnology 68:999–1008

Kosanić M, Ristić S, Stanojković T et al (2018) Extracts of five Cladonia lichens as sources of biologically active compounds. Farmacia 66:644–651

Kumar SVP, Kekuda TRP, Vinayaka KS et al (2010a) Studies on antibacterial, anthelmintic and antioxidant activities of a macrolichen *Parmotrema pseudotinctorum* (des. Abb.) Hale (parmeliaceae) from Bhadra wildlife sanctuary, Karnataka. Int J Pharm Tech Res 2:1207–1214

Kumar SVP, Prashith Kekuda TR, Vinayaka KS et al (2010b) Anthelmintic and antioxidant efficacy of two macrolichens of *Ramalinaceae*. Phcog J 1:4

Kumar J, Dhar P, Tayade AB et al (2014) Antioxidant capacities, phenolic profile and cytotoxic effects of saxicolous lichens from trans-himalayan cold desert of Ladakh. PLoS One 9(6):e98696. https://doi.org/10.1371/journal.pone.0098696

Lobo V, Patil A, Phatak A et al (2010) Free radicals, antioxidants and functional foods: impact on human health. Pharmacogn Rev 4:118–126

Lohézic-Le Dévéhat F, Tomasi S, Elix JA et al (2007) Stictic acid derivatives from the lichen *Usnea articulate* and their antioxidant activities. J Nat Prod 70:1218–1220

Lopes TIB, Coelho RG, Yoshida NC et al (2008) Radical scavenging activity of orsellinates. Chem Pharm Bull 56:1551–1554

Lu Y, Foo LY (2001) Antioxidant activities of polyphenols from sage (*Salvia officinalis*). Food Chem 75:197–202

Manojlovic NT, Vasiljevic PJ, Gritsanapan W et al (2010) Phytochemical and antioxidant studies of *Laurera benguelensis* growing in Thailand. Biol Res 43:169–176

Manojlović N, Ranković B, Kosanić M et al (2012) Chemical composition of three *Parmelia* lichens and antioxidant, antimicrobial and cytotoxic activities of some their major metabolites. Phytomedicine 19:1166–1172

Mastan A, Sreedevi B, Kumari J et al (2014) Evaluation of the in vitro antioxidant and antibacterial activities of secondary metabolites produced from lichens. Asian J Pharm Clin Res 7:193–198

Melo MG, dos Santos JP, Serafini MR et al (2011) Redox properties and cytoprotective actions of atranorin, a lichen secondary metabolite. Toxicol In Vitro 25:462–468

Michalak A (2006) Phenolic compounds and their antioxidant activity in plants growing under heavy metal. Polish J Environ Stud 15:523–530

Milardovic S, Ivekovic D, Grabaric BS (2006) A novel amperometric method for antioxidant activity determination using DPPH free radical. Bioelectrochemistry 68:175–180

Mitrović T, Stamenković S, Cvetković V et al (2011) Antioxidant, antimicrobial and antiproliferative of five lichen species. Int J Mol Sci 12:5428–5448

Moure A, Cruz JM, Franco D et al (2001) Natural antioxidants from residual sources. Food Chem 72:145–171

Naveena BM, Sen AR, Kingsly RP et al (2008) Antioxidant activity of pomegranate rind powder extract in cooked chicken patties. Int J Food Sci Technol 43(10):1807–1812

Odabasoglu F, Aslan A, Cakir A et al (2004) Comparison of antioxidant activity and phenolic content of three lichen species. Phytother Res 18:938–941

Odabasoglu F, Aslan A, Cakir A et al (2005) Antioxidant activity, reducing power phenolic content of some lichen species. Fitoterapia 76:216–219

Odabasoglu F, Cakir A, Suleyman H et al (2006) Gastroprotective and antioxidant effects of usnic acid on indomethacin-induced gastric ulcer in rats. J Ethnopharmacol 103:59–65

Papadopoulou P, Tzakou O, Vagias C et al (2007) Beta-orcinol metabolites from the lichen *Hypotrachyna revolute*. Molecules 12:997–1005

Paudel B, Bhattarai HD, Lee JS et al (2008) Antioxidant activity of polar lichens from King George Island (Antarctica). Polar Biol 31:605–608

Pavithra GM, Vinayaka KS, Rakesh KN et al (2013) Antimicrobial and antioxidant activities of a macrolichen *Usnea pictoides* G. Awasthi (*Parmeliaceae*). J Appl Pharm Sci 3:154–160

Pokorny J, Yanishlieva N, Gordan M (2001) Antioxidants in food: practical applications. Woodhead, Cambridge

Poornima G, Kekuda PTR, Vinayaka KS (2012) Antioxidant efficacy of Olea dioica Roxb (Oleaceae) leaves. Biomedicine 32:506–510

Prior RL, Wu X, Schaich K (2005) Standardized methods for the determination of antioxidant capacity and phenolics in foods and dietary supplements. J Agric Food Chem 53:4290–4302

Ramachandra Shajyothi C, Padmalatha RS (2012) Antioxidant activity of *Alstonia scholaris* extracts containing flavonoids and phenolic compounds. Int J Pharm Pharm Sci 4:424–426

Ranković B, Ranković D, Marić D (2010a) Antioxidant and antimicrobial activity of some lichen species. Microbiology 79:809–815

Ranković B, Ranković D, Kosanić M et al (2010b) Antioxidant and antimicrobial properties of the lichen *Anaptychya ciliaris, Nephroma parile, Ochrolechia tartarea* and *Parmelia centrifuga*. Cent Eur J Biol 5:649–655

Ranković B, Kosanić M, Stanojković T (2011) Antioxidant, antimicrobial and anticancer activity of the lichens *Cladonia furcata, Lecanora atra* and *Lecanora muralis*. BMC Complement Altern Med. https://doi.org/10.1186/1472-6882-11-97

Ranković B, Kosanić M, Stanojković T et al (2012) Biological activities of *Toninia candida* and *Usnea barbata* together with their norstictic acid and usnic acid constituents. Int J Mol Sci 13: 14707–14722

Ranković B, Kosanić M, Stanojković T (2014a) *Stereocaulon paschale* Lichen as antioxidant, antimicrobial and anticancer agent. Farmacia 62:306–317

Ranković B, Kosanić M, Manojlovic N et al (2014b) Chemical composition of *Hypogymnia physodes* lichen and biological activities of some its major metabolites. Med Chem Res 23: 408–416

Rekha C, Poornima G, Manasa M et al (2012) Ascorbic acid, total phenol content and antioxidant activity of fresh juices of four ripe and unripe citrus fruits. Chem Sci Trans 1(2):303–310

Ristić S, Ranković B, Kosanić M et al (2016a) Phytochemical study and antioxidant, antimicrobial and anticancer activities of *Melanelia subaurifera* and *Melanelia fuliginosa* lichens. J Food Sci Technol 53:2804–2816

Ristić S, Ranković B, Kosanić M et al (2016b) Biopharmaceutical potential of two Ramalina lichens and their metabolites. Curr Pharm Biotechnol 17:651–658

Ross Watson R (2014) Polyphenols in plants: isolation, purification and extract preparation. Academic Press, San Diego, CA

Russo A, Piovano M, Lombardo L et al (2008) Lichen metabolites prevent UV light and nitric oxide-mediated plasmid DNA damage and induce apoptosis in human melanoma cells. Life Sci 83:468–474

Sachindra NM, Airanthi MKWA, Hosokawa M et al (2010) Radical scavenging and singlet oxygen quenching activity of extracts from Indian seaweeds. J Food Sci Technol 47:94–99

Sanchez-Moreno C, Larrauri JA, Saura-Calixto FA (1998) A procedure to measure the antiradical efficiency of polyphenols. J Sci Food Agric 76:270–276

Sangameswaran B, Balakrishnan BR, Chumbhale D et al (2009) In vitro antioxidant activity of roots of *Thespesia lampas* dalz and gips. Pak J Pharm Sci 22:368–372

Sawa T, Nakao M, Akaike T et al (1999) Alkylperoxyl radical-scavenging activity of various flavonoids and other phenolic compounds: implications for the anti-tumor-promoter effect of vegetables. J Agric Food Chem 47:397–402

Selvaraj G, Tinabaye A, Ananthi R (2015) In vitro antioxidant activities of salazinic acid and its derivative hexaacetyl salazinic acid. Int J Adv Res Technol 4:345–355

Sharma BC, Kalikotay S (2012) Screening of antioxidant activity of lichens *Parmotrema reticulatum* and *Usnea* sp. from Darjeeling Hills, India. IOSR J Pharm 2:54–60

Singleton VL, Orthofer R, Lamuela-Raventos RM (1999) Analysis of total phenols and other oxidation substrates and antioxidants by means of Folin-Ciocalteu reagent. Methods Enzymol 299:152–178

Sisodia R, Geol M, Verma S et al (2013) Antibacterial and antioxidant activity of lichen species *Ramalina roesleri*. Nat Prod Res 27:2235–2239

Souri E, Amin G, Farsam H et al (2008) Screening of thirteen medicinal plant extracts for antioxidant activity. Iran J Pharm Res 7:149–154

Squadriato GL, Pelor WA (1998) Oxidative chemistry of nitric oxide: the roles of superoxide, peroxynitrite, and carbon dioxide. Free Radic Biol Med 25(4–5):392–403

Stanly C, Ali DMH, Keng CL et al (2011) Comparative evaluation of antioxidant activity and total phenolic content of selected lichen species from Malaysia. J Pharm Res 4:2824–2827

Stepanenko LS, Krivoshchekova OE, Skirina IF (2002) Functions of phenolic secondary metabolites in lichens from Far East Russia. Symbiosis 32:119–131

Thadhani VM, Choudhary MI, Ali S et al (2011) Antioxidant activity of some lichen metabolites. Nat Prod Res 25:1827–1837

Tilak JC, Adhikari S, Devasagayam TPA (2004) Antioxidant properties of *Plumbago zeylanica*, an Indian medicinal plant and its active ingredient, plumbagin. Redox Rep 9:220–227

Tomović J, Kosanić M, Ristić S et al (2017) Chemical composition and bioactive properties of the lichen, *Pleurosticta acetabulum*. Trop J Pharm Res 16:2977–2984

Tougas T, Jannetti J, Collier W (1985) Theoretical and experimental response of a biamperometric detector for flow injection analysis. Anal Chem 57:1377–1381

Vagi E, Rapavi E, Hadolin M et al (2005) Phenolic and triterpenoid antioxidants from *Origanum majorana* L. herb and extracts obtained with different solvents. J Agric Food Chem 53:17–21

Verma N, Behera BC, Sonone A et al (2008a) Lipid peroxidation and tyrosinase inhibition by lichen symbionts grown in vitro. Afr J Biochem Res 2:225–231

Verma N, Behera BC, Sonone A et al (2008b) Cell aggregates derived from natural lichen thallus fragments: antioxidant activities of lichen metabolites developed in vitro. Nat Prod Commun 3: 1911–1918

Verma N, Behera BC, Joshi A (2012) Studies on nutritional requirement for the culture of lichen Ramalina nervulosa and Ramalina pacifica to enhance the production of antioxidant metabolites. Folia Microbiol 57(2):107–114

Vivek MN, Kambar Y, Manasa M et al (2014) Radical scavenging and antibacterial activity of three *Parmotrema* species from Western Ghats of Karnataka, India. J Appl Pharm Sci 4: 086–091

Yamamoto Y, Miura Y, Higuchi M et al (1993) Using lichen tissue cultures in modern biology. Bryologists 96:384–393

Yanishlieva NV, Marinova EM, Gordon HM et al (1999) Antioxidant activity and mechanism of action of thymol and carvacrol in two lipid systems. Food Chem 64:59–66

Young I, Woodside J (2001) Antioxidants in health and disease. J Clin Pathol 54:176–186

Yucel O, Odabasoglu F, Gulluce M et al (2007) Antioxidant and antimicrobial properties of a lichen species, Cladonia rangiformis growing in Turkey. Turk J Pharm Sci 4:101–109

Zhang XY, Chen da C, Mei Hong Xiu MH et al (2009) The novel oxidative stress marker thioredoxin is increased in first-episode schizophrenic patients. Schizophr Res 113(2–3):151–157

Investigations of Lichen Secondary Metabolites with Potential Anticancer Activity

5

Tatjana Stanojković

Abstract

Cancers are among the leading causes of morbidity and mortality worldwide. In the past half-century, natural products have served us well in combating cancer. The main sources of these compounds are microorganisms, plants, and marine organisms. Lichens as chemically significant biota represent a large group of symbiotic organisms of fungi (mycobionts) and algae (photobionts) constituting about 17,000 species and are a source of diverse secondary metabolites. This chapter focuses primarily on the anticancer properties of lichen secondary metabolites. We have reviewed various publications related to anticancer activity emphasizing results about specific lichen compounds. We have shown that various isolated lichen compounds often demonstrate significant inhibitory activity against various cancer cell lines at very low concentrations. Although lichens are a source for excellent anticancer active compounds, only a small number have been tested for their biological significance. Our effort is another attempt to expand and deepen research in this area, especially on compounds that have shown promising results.

5.1 Introduction

Throughout history, natural products have afforded a rich source of compounds that have found many applications in the fields of medicine, pharmacy, and biology. Within the sphere of cancer, a number of important new commercialized drugs have been obtained from natural sources, by structural modification of natural compounds, or by the synthesis of new compounds, designed following a

T. Stanojković (✉)
Institute of Onology and Radiology of Serbia, Belgrade, Serbia
e-mail: stanojkovict@ncrc.ac.rs

B. Ranković (ed.), *Lichen Secondary Metabolites*,
https://doi.org/10.1007/978-3-030-16814-8_5

natural compound as a model. The search for improved cytotoxic agents continues to be important in the discovery of modern anticancer drugs. The huge structural diversity of natural compounds and their bioactivity potential have meant that several products isolated from plants, marine flora, and microorganisms can serve as "lead" compounds for improvement of their therapeutic potential by molecular modification. More than 70% of anticancer compounds are either natural products or substances derived from natural products. On the other hand, conjugation of toxic natural products to monoclonal antibodies or polymeric carriers can lead to more efficacious targeted therapies. Because less than 15% of higher plants have been systematically investigated, the natural products research towards chemotherapy requires further attention and multi-scientific collaboration (Karikas 2010). Also, some herbal compounds have been subjected to clinical trials. This chapter focuses only on those herbal compounds originating from lichens whose anticancer effect has been investigated. The aim of this chapter is to highlight the importance of lichen secondary metabolites with potential anticancer activity.

5.2 Anticancer Secondary Metabolites of Lichens

5.2.1 Significance of Lichens and Lichen Secondary Metabolites

Lichen metabolites in general can be divided into two groups: primary and secondary. Primary metabolites are proteins, lipids, carbohydrates, and other organic compounds involved in lichen metabolism and structure. Secondary metabolites, also known as lichen substances, are produced mainly by the fungus and secreted onto the surface of the lichen hyphae either in amorphous forms or as crystals. They often accumulate in the upper cortex or in specialized structures such as fruiting bodies (Fahselt 1994; Elix 1996). Lichens synthesize a great variety of secondary metabolites, many of which are unique. Approximately 1050 secondary compounds have been identified to date (Stocker-Worgotter 2008). These compounds are produced by the mycobiont (Elix 1996; Huneck 1999) and accumulate in the cortex (such as atranorin, parietin, usnic acid, fungal melanins) or in the medullary layer (such as physodic acid, physodalic acid, protocetraric acid) as tiny extracellular crystals on the outer surfaces of the hyphae. The photobiont might also have an influence on the secondary metabolism of the mycobiont (Brunauer et al. 2006, 2007; Yamamoto et al. 1993, 1994; Molnár and Farkas 2010).

Lichens had to evolve diverse biosynthetic pathways to produce such complex arrays of secondary metabolites: polyketide, shikimic acid, and mevalonic acid pathways. Most of the lichen substances are phenolic compounds. Polyketide-derived aromatic compounds, the depsides, depsidones, dibenzofurans, xanthones, and naphthaquinones, are of great interest. Compounds from other pathways are esters, terpenes, steroids, terphenylquinones, and pulvinic acid (Fahselt 1994; Cohen and Neil Towers 1995; Elix and Stocker-Worgotter 2008; Muller 2001; Brunauer et al. 2006, 2007; Stocker-Worgotter and Elix 2002; Johnson et al. 2011;

Manojlovic et al. 2012). Thus, many lichens and lichen products have proved to be a source of important secondary metabolites for the food and pharmaceutical industries (Huneck 1999; Oksanen 2006), and still hold considerable interest as alternative treatments in various parts of the world (Richardson 1991). Also, we know more about these substances through experimental studies, but the functions of these compounds in lichen symbioses are still poorly understood (Hager et al. 2008): they may impact the biotic and abiotic interactions of lichens with their environment. In addition, the compounds may help to protect the thalli against herbivores, pathogens, competitors, and external abiotic factors, such as high UV irradiation. A wide spectrum of biological potential is shown by the lichens, but they have been long neglected because of their slow-growing nature and difficulties in their artificial cultivation, and have scarcely been studied from a biochemical perspective (Crittenden and Porter 1991; Yamamoto et al. 1995; Yamamoto 2000; Behera et al. 2003, 2004). Industrial-scale harvests of lichens are neither ecologically sensible nor sustainable and, for many species, are not feasible. The new molecular biology technologies are coming into sight for direct access to lichen genomes to reveal and eventually to harvest or produce novel secondary metabolites (Miao et al. 2001).

Furthermore, lichen substances exhibit a great diversity of biological effects, including antimicrobial, antiinflammatory, analgesic, antipyretic, and antiproliferative and cytotoxic activities (Boustie and Grube 2005). However, relatively few lichen substances have been screened in detail for biological activity and therapeutic potential, principally because of difficulties in obtaining them in quantities and purity sufficient for structural elucidation and pharmacological testing (Muggia et al. 2009).

5.2.2 Lichen Secondary Metabolites as Potential Anticancer Drugs: Some Earlier Studies

As already noted, more than 1050 secondary metabolites have been reported for lichens and aposymbiotically cultured mycobionts (Molnár and Farkas 2010). Among them, a relatively small number of these secondary products (50–60) occur in non-lichenized fungi or higher plants (Elix and Stocker-Worgotter 2008). One example is the anthraquinone parietina, which is present in other fungi such as *Aspergillus* and *Penicillium*, as well as in the vascular plant genera *Rheum*, *Rumex*, and *Ventilago* (Romagni and Dayan 2002). This metabolic diversity is largely the result of the symbiotic relationship between the lichen partners (Lawrey 1986). Lichen secondary products can constitute up to 20% of the dry thallus weight, but in most lichens the amount varies from 5% to 10%. Many lichen secondary metabolites exhibit cytotoxic properties and could be potential sources of pharmaceutically useful chemicals. The purpose of this study was to provide insights regarding the anticancer properties of lichen secondary metabolites and also to provide information regarding the mode of action of lichen compounds against cancer cells. However, so far only a limited number of studies have been published in which the mechanism of action against cancer cell lines was explored (Molnár and

Farkas 2010). The molecular mechanism of cell death by lichen compounds includes cell-cycle arrest, apoptosis, necrosis, and inhibition of angiogenesis (Brisdelli et al. 2013). There is clearly an urgent need for expanding research in this area of study, including studies of those compounds which have shown promising results as well as a strong focus on identifying specific mechanisms of action and extensive clinical trials using the most promising lichen-based drug therapies, followed by large-scale production of the best of those compounds. Although a large number of representatives of this group has already been tested and become the source of pharmaceutically important anticancer drugs, a vast potential reservoir of untapped possibilities still remains.

Among the more promising possibilities are lichenized fungi with their more than 1000 identified secondary chemicals. The use of lichen secondary products as anticancer drugs dates back to the late 1960s when the activity of lichen polysaccharides against tumor cells was initially explored (Fukuoka et al. 1968; Shibata et al. 1968). Similarly, in early studies Kupchan and Kopperman (1975) first reported the tumor inhibitor activity of usnic acid extracted from *Cladonia* sp. against Lewis lung carcinoma: they reported a 35–52% increase in the lifespan of treated mice versus the control group using a dose range of 20–200 mg/kg usnic acid. The butyrolactone, protolichesterinic acid, was also found active as an antiproliferative against leukemia cells K-562 (IC_{50} = 20 mg/ml) and against Ehrlich solid tumor, whereas nephrosteranic acid derivatives have poor activity (Hirayama et al. 1980). Polyporic acid (a terphenylquinone) and derivatives (Cain 1966), a depsidone, physodalic acid (Shibamoto and Wei 1984), and lichen glucans (Nishikawa et al. 1969, 1979; Hirayama et al. 1980; Nishikawa and Ohno 1981), including lichenin derivatives (Demleitner et al. 1992), have also been investigated in this way.

5.2.3 Lichen Secondary Metabolites as Potential Anticancer Drugs: Recent Data

In the past few years, the interest of researchers in active metabolites originating in the lichen that show anticancer potential has not diminished. On the contrary, there is an increasing need for natural compounds that can respond to the challenges of modern therapy and be potential anticancer drugs.

In particular, an interesting study is about the anticancer capacities of ramalin, and secondary metabolites from the Antarctic lichen *Ramalina terebrata*, in the human colorectal cancer cell line HCT116. Sung et al. (2017) showed that ralmalin possesses anticancer activity against HCT116 cells, significantly reducing survival and inducing apoptosis. Also, this study has shown that ramalin induces arrest in the G_2/M phase of the cell cycle and affects the modulation of the corresponding genes, such as tumor protein p53, cyclin-dependent kinase inhibitor 1A, cyclin-dependent kinase 1, and cyclin B1. Similarly, another metabolite, lobaric acid, also from the Antarctic lichen *Stereocaulon alpinum*, leads to a significant decrease in the survival of the treated Hela and HCT116 malignant cells (Ju-Mi et al. 2018). Further analysis showed a significant downregulation of the apoptosis and DNA repair regulators.

Similarly, it is important to note that interest in researching metabolites such as olivetoric, physodic, and psoromic acids is not reduced. Namely, in the study by a group from Turkey (Bugrahan et al. 2016), the human brain GBM cell line U87MG, as well as PRCC cultures obtained from six newborn Sprague–Dawley rats, was used to investigate the anticancer activity of these three metabolites of lichens. This research has shown that olivetolic acid and psoromic acid have good cytotoxic activity, which is promising for application in the treatment of glioblastoma multiforme (GBM).

Another group (Cardile et al. 2017) also investigated the cytotoxicity of physodic acid as well as the other two lichen secondary metabolites, atranorine and gyrophoric acid, on the A375 melanoma cancer cell line. The obtained results showed that the depsidone, physodic acid, possessed good cytotoxicity (dose–response relationship in the range of 6.25–50 mM) associated with apoptosis. According to the data of a new study (Rui et al. 2017), lichen secondary metabolites significantly inhibited atranorin tumorigenesis in human lung cancer cells by affecting the AP-1, Wnt, and STAT signaling and RhoGTPase-suppressing activity. We also mention the results of a particularly interesting study on the evaluation of in vitro anticancer activity of vulpinic acid in which Nil et al. (2018) reported that vulpinic acid significantly inhibited the proliferation of some treated cell lines (CaCo2, HepG2, Hep2C, RD, Wehi, Vero, and L929 cell lines). In addition, the IC_{50} value of vulpinic acid altered the mRNA levels of the Bax, Bcl-2, and P53 genes in all examined cancer cells when compared to untreated cells, which is the first evaluation of the apoptotic effect of vulpinic acid on mRNA levels.

Finally, the research group of Serbia (Ristic et al. 2016a, b) examined the anticancer effect of lecanoric acid and 2-*O*-methyl anziaic acid isolated from the lichens *Melanelia subaurifera* and *Melanelia fuliginosa*. Unfortunately, the investigated compounds showed significantly less activity than the extracts. The same authors (Ristic et al. 2016a, b) also examined the anticancer potential of the species *Ramalina fraxinea* and *R. fastigiata*, especially isolated obtusatic acid, which showed a good cytotoxic effect on all tested malignant cell lines.

Finally, it is clear how important is the investigation in this area of study, including the research of those compounds and their mechanisms of action. One of the important goals in the future is certainly clinical studies using the most promising lichen-based drug therapies.

5.3 Overview of the Most Investigated Lichen Secondary Metabolites

5.3.1 Usnic Acid

Usnic acid, the most extensively studied lichen metabolite since its first isolation in 1844, exhibited an antiproliferative effect on human leukemia cells (K562) and endometrial carcinoma (HEC-50) cells (Cardarelli et al. 1997; Ingólfsdóttir 2002;

Kristmundsdóttir et al. 2002). In addition, it is known that Kupchan and Kopperman (1975) first reported the tumor inhibitor activity of usnic acid extracted from *Cladonia* sp. against Lewis lung carcinoma. They reported a 35% to 52% increase in the lifespan of treated mice versus the control group using a dose range of 20–200 mg/kg usnic acid. Therefore, usnic acid is one of the most interesting lichen metabolites for the study of their antitumor effects. The cytotoxicity, the in vitro antitumor effects, and the mechanism of action of usnic acid need to be investigated in greater detail to reach clinical trials and to allow further applications (Table 5.1).

Usnic acid, and usnic acid-amine derivatives, showed in vitro antiproliferative/ cytotoxic activity against a wide variety of murine and human cancer cell lines (Takai et al. 1979; Cardarelli et al. 1997; Bézivin et al. 2004; Mayer et al. 2005; Bazin et al. 2008; Sahu et al. 2011; Bačkorová et al. 2011; Burlando et al. 2009). The toxicity of usnic acid was associated with increased P_{450} activity and oxidative stress in human hepatoblastoma cells (Sahu et al. 2011), with mitochondrial dysfunction in HepG2 cells (Sahu et al. 2011), in the breast cancer T-47D cell line, and in the pancreatic cancer Capan-2 cell line (Einarsdóttir et al. 2010), with apoptotic induction in murine leukemia L1210 cells (Bézivin et al. 2004; Bazin et al. 2008).

(+)-Usnic acid was found to be a strong hepatotoxic agent against monogastric murine hepatocytes (Han et al. 2004). Also, Correché et al. investigated the cytotoxic and apoptotic effects of usnic acid obtained from continental (Chilean) and Antarctic lichens in primary cultures of rat hepatocytes (Correché et al. 2004).

The (−)-enantiomer of usnic acid (isolated from *Cladonia convoluta*) was moderately cytotoxic to various cancer cell lines, such as murine Lewis lung carcinoma, human chronic myelogenous leukemia, human brain, metastasis of a prostate carcinoma, human breast adenocarcinoma, and human glioblastoma (Molnár and Farkas 2010; Bézivin et al. 2004). Usnic acid also decreased the proliferation of human breast cancer cells and human lung cancer cells without any DNA damage (Mayer et al. 2005). Finding cancer therapies that do not have DNA-damaging effects and that do not cause the development of secondary malignancies later in life is of great interest. Accordingly, usnic acid may represent a novel source for a natural non-genotoxic anticancer drug. Usnic acid from the lichen *Usnea barbata* (Rankovic et al. 2012) induced a significant cytotoxic effect on tested human melanoma Fem-x and human colon carcinoma LS174 cell lines that was stronger than the lichen extracts. Then, as shown by numerous data, there is significant antitumor activity by usnic acid in vitro. For example, usnic acid activated programmed cell death in A2780 and HT-29 cell lines, probably through the mitochondrial pathway (Bačkorová et al. 2012).

Lichen compounds showed differential sensitivity to various cancer cells. Usnic acid was highly effective against the whole spectrum of cell lines (HeLa, MCT-7, A2780, HT-29, Jurkat, SK-BR-3, and HCT-116). Similar to cytotoxicity, usnic acid also significantly inhibited the clonogenic ability of all the tested cell lines. Usnic acid also demonstrated strong pro-apoptotic action associated with the altered cell-cycle distribution and accumulation of cells in S-phase (Brisdelli et al. 2013; Bačkorová et al. 2011).

Table 5.1 Overview of recent literature related to the in vitro anticancer activity of lichen secondary metabolites (according to Shrestha and St. Clair 2013)

Lichen metabolites/lichen species	Cell lines tested	References
Usnic acid (commercial) Atranorin (commercial) Parietin (*X. parietina*) Gyrophoric acid (*Umbilicaria hirsuta*)	Human ovarian carcinoma A2780 Human colon adenocarcinoma HT-29	Bačkorová et al. (2012)
Diffractaic acid (*Protousnea magellanica*) Vicanicin (*Psoroma pallidum*) Lobaric acid (*Stereocaulon alpium*) Variolaric acid (*Ochrolechia deceptionis*) Protolichesterinic acid (*Cornicularia aculeate*) Usnic acid (*Cladonia lepidophora*)	Human breast adenocarcinoma MCF-7 Human colon adenocarcinoma HCT-116 Human cervix adenocarcinoma HeLa	Brisdelli et al. (2013)
Atranorin (*Bacidia stipata*) Diffractaic acid (*P. magellanica*) Divaricatic acid (*Protousnea malacea*) Vicanicin (*Psoroma dimorphum*) Protolichesterinic acid (*R. melanophthalma*)	Human prostate cancer androgen-responsive (LNCaP) Human prostate cancer androgen-non responsive DU-145	Russo et al. (2012)
Usnic acid (commercial) Atranorin (commercial) Parietin (*X. parietina*) Gyrophoric acid (*U. hirsuta*)	Human ovarian A2780 Human breast MCF-7 Human colon HT-29 Human T cells, Jurkat Human cervix HeLa Human breast SK-BR-3 Human colon wild type p53 HCT-116 p53+/+ Human colon p53 null HCT-116 p53−/−	Bačkorová et al. (2012)
Lecanoric acid and it's orsellinate derivatives	Larynx carcinoma HEP-2 Breast carcinoma MCF-7 Kidney carcinoma 786-0 Murine melanoma cell B16-F-I0 Vero cell	Bogo et al. (2010)
(+) Usnic acid (*C. arbuscula*) (−) Usnic acid (*Alectoria ochroleuca*)	Breast cancer cell line T-47D Pancreatic cancer cell line Capan-2	Einarsdóttir et al. (2010)
Retigeric acid A and Retigeric acid B (*Lobaria kurokawae*)	Human Pca LNCaP PC-3, DU 145 Human epidermoid KB and vincristine resistant KB (KB/VCR) Human ovarian 3-AO and cisplatin resistant 3-AO (3-AO/CDDP) Human benign prostate epithelial RWPE1 Human hTERT-RPEl Human breast MCF-7 Human osteosarcoma U20S and Saos2	Liu et al. (2010)

(continued)

Table 5.1 (continued)

Lichen metabolites/lichen species	Cell lines tested	References
Depsidones—Vicanicin, Pannarin, 1-chloropannarin, Salazinic acid, Stictic acid, variolaric acid, Psoromic acid, Fumarprotocetraric acid, Lobaric acid Depsides—Atranorin, Sphaerophorin, Divaricatic acid, Diffractaic acid, Gyrophoric acid, Usnic acid	Hepatocytes from rat	Correché et al. (2004)
Pannarin, 10-chloro pannarin, Salazinic acid, Psoromic acid, Fumarprotocetaric acid, Lobaric acid, Vicanicin, Stictic acid, Variolaric acid, Atranorin, Sphaerophorin, Divaricatic acid, Diffractaic acid, Gyrophoric acid	Lymphocytes from rat spleens	Correché et al. (2002)
(+)-usnic acid Methyl a-orcinolcarboxylate Ethyl hematommate Diffractaic acid Gyrophoric acid (+)-protolichesterinic acid	Human keratinocyte cell line HaCaT	Kumar and Muller (1999)
Lobaric acid (*S. alpinum*) Protolichesterinic acid (*C. islandica*)	Breast cancer cell T-47D and ZR-75-1 Erythro-leukemia K-563	Ögmundsdóttir et al. (1998)
Usnic acid derivatives	Lewis lung carcinoma L1210	Takai et al. (1979)

Somewhat earlier (Einarsdóttir et al. 2010) it was announced that both (+)- and (−)-usnic acids are effective inhibitors of DNA synthesis, with IC_{50} values of 4.2 and 4.0 μg/ml, respectively, against T-47D (a breast cancer cell line), and 5.3 and 5.0 μg/ml against Capan-2 (a pancreatic cancer cell line). There was a reduction in cell size, and both acids inhibit cell entry into the S-phase. Regarding the mechanism of action, staining with the mitochondrial dye JC-1 demonstrated a dose-dependent loss of mitochondrial membrane potential following treatment with usnic acid in both cell lines. Study of the effects of usnic acid on MCF7 (estrogen-dependent, wild-type p53) indicated no morphological changes in microtubules or increase in the mitotic index, suggesting that the antineoplastic activity of usnic acid is not related to alterations in the formation and stabilization of microtubules (O'Neill et al. 2010).

Usnic acid from *Xanthoparmelia somloensis*, Burlando et al. (2009) have investigated its cytotoxic effect towards malignant mesothelioma cells (MM98), vulvar carcinoma cells (A431), and keratinocytes (HaCaT). Usnic acid showed high cytoxicity for all three cell lines. Further, both types of usnic acid showed dose- and time-dependent cytotoxicity against V79 (Chinese hamster lung fibroblast) and A549 (human lung carcinoma) cell lines. Cytotoxicity was more pronounced in A549 than V79, with cell viability more diminished in A549 versus V79 after 2 days of treatment (Koparal et al. 2006). To investigate the mechanism of action of usnic acid, elevated levels of the p53 and p21 proteins following treatment with usnic acid were confirmed, but there was no p53 transcriptional activity, suggesting that the accumulation of p21 was not secondary to p53 transactivation (Mayer et al. 2005). They concluded that usnic acid has antiproliferative activity against wild-type p53 (MCF7) and nonfunctional p53 (MDA-MB-231) breast cancer cells, as well as against the H1299 lung cancer cell line, which is null for p53. Usnic acid is therefore a non-genotoxic anticancer agent that works in an p53-independent manner. In another study (Bézivin et al. 2004), usnic acid also induced L1210 (murine lymphocytic leukemia) in apoptosis in a dose- and time-dependent manner as fluorescence microscopy revealed condensation of nuclear chromatin, nuclear fragmentation, and formation of apoptotic bodies.

5.3.2 Depsides and Depsidones

There are several studies about antitumor activity of depsides and depsidones and especially on atranorin. Bačkorová and colleagues reported that antiproliferative/cytotoxic effects of atranorin efficiently induced apoptosis and inhibited cell proliferation in various cancer cell lines tested. Similar to usnic acid, atranorin demonstrated strong pro-apoptotic action. Moreover, the same authors reported on the sensitivity of up to nine human cancer cell lines (A2780, HeLa, MCF-7, SK-BR-3, HT-29, HCT-116, p53 (+/+), HCT-116, p53 (−/−), HL-60, and Jurkat) to the antiproliferative/cytotoxic effects of some typical secondary metabolites of lichens (parietin and gyrophoric acid). Further, the analysis of cell-cycle distribution also revealed an accumulation of cells in S-phase. This study has confirmed a differential sensitivity of cancer cell lines to lichen secondary metabolites (Bačkorová et al. 2011). In addition, atranorin, diffractaic acid, and divaricatic acid were found to be active against prostate cancer cells [human prostate cancer androgen-responsive (LNCaP) and human prostate cancer androgen-nonresponsive DU-145 cells] only in high concentration (Russo et al. 2006, 2012). This study for the first time showed that apoptosis induced by the compounds appeared to be mediated, at least in part, via the inhibition of Hsp70 expression. Also, the depsides atranorin, sphaerophorin, divaricatic acid, diffractaic acid, and gyrophoric acid, and the depsidones vicanicin, pannarin, 1-chloropannarin, salazinic acid, stictic acid, variolaric acid, psoromic acid, fumarprotocetraric acid, and lobaric acid, were evaluated for their cytotoxic activity towards hepatocytes from rat and lymphocytes from rat spleens (Correché et al. 2002, 2004). The research has shown that salazinic acid, stictic acid, and psoromic acid showed apoptosis of hepatocytes in a dose-dependent manner, with stictic acid showing the strongest apoptotic activity. Ogmundsdottir and associates showed that lobaric acid and protolichesterinic acid when used towards breast cancer cells T-47D and ZR-75-1 as erythro-leukemia K-563 cells caused a significant reduction in DNA synthesis. Significant cell deaths in all three cell lines were observed at concentrations of 20 and 30 μg/ml of protolichesterinic acid and lobaric acid, respectively (Ögmundsdóttir et al. 1998). Similarly, Kumar and Muller showed that gyrophoric acid, usnic acid, and diffractaic acids were reported as potent antiproliferative agents towards the human keratinocyte cell line HaCaT, inhibiting cell growth at IC_{50} values of 1.7, 2.1, and 2.6 μM (Kumar and Muller 1999). Also, results from the work of Pejin and associates suggested a moderate anticancer activity towards malignant HT29 (IC_{50} value, 29.29 μg/ml) and a low growth inhibition on nonmalignant MRC5 cells (IC_{50} value, 2478.40 μg/ml) of stictic acid (Pejin et al. 2013). These findings may indicate that stictic acid can be considered as a promising lead compound for the design of novel human colon adenocarcinoma drugs. Generally, depsidones showed stronger cytotoxic activity than depsides. The strong biological activity of some depsidones may result from the strong hydrogen bond between the aldehyde group at C3 and the hydroxyl group at C4. Similarly, the cytotoxic activity of depsides may result in part from the presence of a COOH group on C1 and an OH group on C2. Similarly, Manojlovic and associates reported the strong cytotoxic effect of the depsidone salazinic acid as well as the phenolic compound protocetraric acid against FEM-x (human melanoma) and LS174 (human colon carcinoma) cell lines (Manojlovic et al. 2012). Pannarin, a depsidone, was shown to inhibit the growth of DU-145 prostate carcinoma and M14 human melanoma cells (Russo et al. 2006; Brandão et al. 2013) Also, for the purpose of identifying novel agents with antigrowth and pro-apoptotic activity on prostate cancer cells, Russo et al. (2012) evaluated the effect of lichen secondary metabolites, the depsides atranorin, diffrattaic, and divaricatic acids, as well as the depsidone vicanicin, on cell growth in androgen-sensitive (LNCaP) and androgen-insensitive (DU 145) human prostate cancer cells. The depsides resulting from decarboxylation of baeomycesic and squamatic acids showed antiproliferative effects on PC-3 prostate cancer cells [50% growth inhibitory concentration (GI_{50}), 70.06 and 79.37 μm, respectively] (Guo et al. 2011). Also, olivetoric acid as a di-depside displayed dose-dependent antiangiogenic activities, inhibited cell proliferation, and disrupted endothelial tube formation in rat adipose tissue endothelial cells (Koparal et al. 2010).

Very good cytotoxic activity against malignant FEM-x and LS174 cells was also shown by depsidone physodic acids and depside atranorin, which were identified from the lichen *Hypogymnia physodes* growing in Serbia (Rankovic et al. 2014). Their IC_{50} values were in the range of 17.89–24.63 μg/ml, respectively, somewhat less than the value of the IC_{50} of usnic acid. These authors examined the antitumor activity of evernic acid, which belongs to depsides, and also depsidone physodic acid isolated from the lichens *Evernia prunastri* and *Pseudevernia furfuracea* (Kosanic et al. 2013). The obtained results show that the tested compounds exhibited high cytotoxic activity against the target cells in vitro. The best cytotoxic activity was exhibited by physodic acid. The effect of tested samples on cell-cycle progression was investigated also in FEM-x and LS174 cells. An increase in cells containing sub-G1 amounts of DNA was observed, indicating that the evernic and physodic acids were inducing cell death. Similarly, somewhat earlier, Russo et al. (2008) reported that the depside sphaerophorin (isolated from *Sphaerophorus globosus*) and the depsidone pannarin (isolated from *Psoroma pholidotoides*) inhibited the growth of M14 human melanoma cells, triggering apoptotic cell death. The data obtained from cell culture show that these lichen metabolites inhibit the growth of melanoma cells, inducing their apoptotic cell death, demonstrated by the fragmentation of genomic DNA and by a significant increase of caspase-3 activity, and correlated, at least in part, with the increase of reactive oxygen species (ROS) generation. The anticancer activities of these lichen metabolites are promising in the treatment of this aggressive, therapy-resistant skin tumor (Molnár and Farkas 2010). However, one new study was done (Brisdelli et al. 2013) of six lichen metabolites (diffractaic acid, lobaric acid, lips acid, vicanicin, variolaric acid, protolichesterinic acid) on proliferation, viability, and ROS level towards three human cancer cell lines: MCF-7 (breast adenocarcinoma), HeLa (cervix adenocarcinoma), and HCT-116 (colon carcinoma). In this comparative study, lichen metabolites showed various cytotoxic effects in a concentration-dependent manner. Moreover, not all tested lichen compounds exhibited free radical scavenging activity. The lichen metabolites did not significantly increase intracellular ROS level. Further, the cytotoxic activities of the depsidone variolaric acid and two other secondary metabolites of lichens, R-alectoronic acid and ergosterol peroxide, were evaluated against the murine B16 melanoma cell line. All the tested compounds showed a significant antitumor activity, especially variolaric acid and alectoronic acid, as compared to cisplatin as a positive control (Millot et al. 2007). Anziaic acid was also found to act as an inhibitor of human topoisomerase II but had little effect on human topoisomerase I (Cheng et al. 2013). This is the first report of a depside with activity as a topoisomerase poison inhibitor and demonstrates the potential of this class of natural products as a source for new antibacterial and anticancer compounds. Protolichesterinic acid showed an inhibitory effect against 12 cell lines, with IC_{50} values of 2.4–18.1 μg/ml (Russo et al. 2012). Also, in recent years (Bogo et al. 2010) it has been shown that lecanoric acid (para-depside), a secondary metabolite of the lichen *Parmotrema tinctorum*, has moderate antitumor activity against some malignant cell lines tested (MCF7 breast carcinoma, 786-0 kidney carcinoma, and B16-F10 murine melanoma). Similarly, some cytotoxic activity in vitro of lecanoric acid, orsellinic acid, methyl ester, orcinol, and usnic acid isolated from the lichen *Parmelia subrudecta* is shown by other authors (Ivanova et al. 2010). The depsides resulting from decarboxylation of baeomycesic and squamatic acids showed antiproliferative effects on PC-3 prostate cancer cells (Haraldsdóttir et al. 2004; Brandão et al. 2013). Also, Brandao and his research group have reported evaluation of some lichen compounds (the depsides atranorin and diffractaic, divaricatic, and perlatolic acids; the depsidones psoromic, protocetraric, and norstictic acids) tested against UACC-62 and B16-F10 melanoma cell lines, and 3 T3 normal fibroblast cells (Brandão et al. 2013).

5.3.3 Naphthoquinones

Naphthazarin and its derivates were isolated from *Cetraria islandica*. This naphthoquinone demonstrated, in in vitro experiments, a strong cytotoxic effect to human epidermal carcinoma cells. The dimer of this naphthoquinone, hybocarpone, was isolated from *Lecanora hybocarpa* (Babula et al. 2009).

5.3.4 Anthraquinones

5.3.4.1 Emodin

The anthraquinones are a large family of compounds having diverse biological properties. Emodin (1,3,8-trihydroxy-6-methylanthraquinone) is a naturally occurring anthraquinone present in numerous lichens (Cohen and Neil Towers 1995). Emodin, first assigned to be a specific inhibitor of the protein tyrosine kinase p65lck, has now a number of cellular targets interacting with it. Its inhibitory effect on mammalian cell-cycle modulation in specific oncogene overexpressed cells formed the basis of using this compound as an anticancer agent. Identification of apoptosis as a mechanism of elimination of cells treated with cytotoxic agents initiated new studies deciphering the mechanism of apoptosis induced by emodin. At present, its role in combination chemotherapy with standard drugs to reduce toxicity and to enhance efficacy is being pursued vigorously. Its additional inhibitory effects on angiogenic and metastasis regulatory processes make emodin a sensible candidate as a specific blocker of tumor-associated events (Srinivas et al. 2007).

5.3.4.2 Parietin

Parietin is derived from polyaromatic ring polyketides and is present in lichen of the genera *Xanthoria* and *Teloschites*, in particular in the lichens *Teloschites chrysothalmus*, *Teloschites spinosus*, and *Xanthoria parietina*. Among others, Bačkorová and the authors showed that certain cytotoxic potential of parietin by a series of cancer cell lines (Bačkorová et al. 2011).

5.3.5 Xanthones: Lichexanthone

Lichexanthone was one of the xanthones also tested in a recent study (Brandão et al. 2013), together with other secondary metabolites of lichens. This study showed that lichexanthone was the least active substance tested, delineating a very distinct response relative to the other compounds and the standard doxorubicin.

5.3.6 Other Compounds

Apart from compounds derived from common pathways, which are found throughout all major lichen groups, there are also some unusual compound classes among these organisms; for example, arthogalin, a cyclic depsipeptide (Huneck and Himmelreich 1995), and other amino acid-derived compounds such as the cytotoxic scabrosin esters isolated from *Xanthoparmelia scabrosa* (Ernst-Russell et al. 1999). Thus, Magaya et al. tested the effects of arthogalin, a secondary metabolite of the lichen *Caloplaca inclinans,* on the growth of murine malignant prostate sarcoma cells in vitro (Magaya et al. 2013). The results of this study showed that arthogalin is a potent inhibitor of growth of tested cancer cells. They also showed that arthogalin increases sensitivity of cells to radiation, and this effect is significant at a radiation intensity lower than the standard intensity of cancer radiotherapy. On the basis of this study, arthogalin shows promise for combined-modality cancer treatment. Also, earlier studies have shown that scabrosin esters (SEs), which have been isolated from the lichen *Xanthoparmelia scabrosa*, belong to the epipolythiodioxopiperazine (ETP) class of secondary metabolites characterized by possession of a reactive disulfide bond. Colony-forming assays that were used in these studies have shown that these compounds are active against human tumor cell lines at nanomolar concentrations. Colony-forming assays also show that these toxins are active against human tumor cell lines at nanomolar concentrations (Moerman et al. 2003). These authors show that the typical scabrosin ester acetate butyrate induces early mitochondrial membrane hyperpolarization in accompanied by apoptotic cell death. Here, we mention retigeric acid A (RA) and retigeric acid (RB), both pentacyclic triterpenoids from the lichen species *Lobaria kurokawae*. Liu and associates (Liu et al. 2010) showed cytotoxicity towards malignant cells at lower concentrations ($<100\ \mu M$) of these compounds but RB is more potent than RA. Specially, investigation on the effect of RB on PC-3 cells showed that RB caused a dose-dependent accumulation of cells in the S phase accompanied with decreases in cyclin B and in cyclin E and cyclin A. Both caspase-dependent and caspase-independent pathways were responsible for apoptosis in PC-3 cells. It should also be noted that 16-*O*-acetyl-leucotylic acid, a new triterpenic acid, exhibited potent antiproliferative activity against HL-60 with an EC_{50} value of 21 μM although the leucotylic acid, the derivative of 16-*O*-acetly-leucotylic acid has a higher EC_{50} value (72 μM) (Tokiwano et al. 2009).

5.4 Overview of Existing In Vivo Studies

Without in vivo evaluation and clinical trials, no real efficacy in cancer therapy can be claimed for any of the potential agents. First, it is known that the most of the antitumor in vivo studies on lichen extracts have been performed by Japanese scientists in the 1970s (Fujikawa et al. 1972; Hirayama et al. 1974). The allogeneic tumor S-180-forming ascites implanted into albino mice is generally used as the basic screening model. Few antitumor assays have been conducted on Ehrlich carcinomas. In each model, samples dissolved in distilled water were administered by i.p. injection for 10 consecutive days, starting 24 h after tumor implantation. After 30 days, the antitumor effect was evaluated through the inhibition ratio (IR), which is linked to the reduction of tumor weight, and the complete regression rate (CR), which is linked to complete recovery of grafted animals. However, with advances in the isolation and characterization of secondary metabolites of lichens, in addition to in vitro testing of their antitumor effects, there is growing interest for the study in vivo (Ribeiro-Costa et al. 2004). Only scarce in vivo assays in mice have been attempted for some low molecular weight lichen compounds. One of the first lichen acids found to have some activity on L-1210 and S-180 models in mice was polyporic acid, a dihydroxyquinone isolated from *Sticta coronata.* It was given in a dose of 60 mg/kg administered by intraperitoneal (i.p.) injection (Burton and Cain 1959). The well-known (−)-usnic acid, a dibenzofuran, was proved to have a weak, if any, antitumoral effect against Lewis lung carcinoma and P-388 leukemia cells (Kupchan et al. 1975; cited in Takai et al. 1979). A series of 20 lichen compounds have been tested against Ehrlich carcinomas in mice, revealing some potential for the butyrolactones (+)-protolichesterinic acid and nephrosterinic acid, with 50% and 70% tumor growth inhibition, respectively (Hirayama et al. 1980). A significant in vivo antineoplastic activity (murine leukemia P-388, tested/control × 100 (T/C) = 40% at 160 μg/kg) is reported for ambewelamide A, an original diketopiperazine dione (Williams et al. 1998). This scabrosin ester and derivatives isolated from two lichen species have shown potent in vitro cytotoxic activity (IC_{50} within the μM to nM range for P=388, P-815, and MCF-7) (Williams et al. 1998; Ernst-Russell et al. 1999). Another rare compound, hybocarpone, isolated from a mycobiont culture of *Lecanora hybocarpa* (IC_{50} = 0.27 μM) (Ernst-Russell et al. 1999), is also a relevant compound to be investigated further in terms of anticancer studies. In 2004 Ribeiro-Costa and colleagues investigated the in vitro and in vivo properties of usnic acid encapsulated into poly(lactic-co-glycolic acid (PLGA) microspheres. Microparticles will probably have a promising role in the future of chemotherapy. These polymeric delivery systems are capable of maximizing therapeutic activity while reducing side effects of anticancer agents. In this study, PLGA microspheres contained usnic acid from *Cladonia substellata*. The antitumor assay was performed in mice against sarcoma-180 tumor (UA, 15 mg/ kg weight body/ day) during 7 days. Animals were then killed and tumors and organs were excised for histopathological analysis. A maximum release of 92% was achieved at the fifth day. The IC_{50} values for free and encapsulated usnic acid were 12 and 14 μg /ml, respectively. The encapsulation of usnic acid into microspheres promoted an increase of 21% in tumor inhibition as compared with the free usnic acid treatment. In summary, usnic acid was efficiently encapsulated into PLGA microspheres and the microencapsulation improved its antitumor activity (Ribeiro-Costa et al. 2004).

5.5 Lichen Secondary Metabolites as Potential Anticancer Drugs: Prospects and Promise

Chemotherapy is still the method most commonly used and most promising in the treatment of cancer patients. Also, chemotherapy holds the most promise for selectively eradicating cancer cells while at the same time minimizing collateral damage to surrounding tissues. Many of these chemical agents owe their origins to natural sources in the environment, whereas other anticancer chemotherapeutics are wholly designed by pharmaceutical scientists based upon current knowledge of cancer-onset mechanisms. Selectivity for cancer cell destruction without harming healthy cells is the central focus of these treatment protocols, and the chemotherapy's well-known side effects of chemotherapy (hair loss, nausea, immunocompromise, etc.) are a continuing reminder that much room for progress yet remains. Natural products and their derivatives represent more than 50% of all the drugs in clinical use in the world. Almost 60% of drugs approved for cancer treatment are of natural origin (Fakim 2006). Whether the promise of fully selective anticancer medicines will be realized in our lifetime remains unknown, but exciting developments from the investigation of lichen secondary metabolites lend credibility to the proposition that the best is yet to come. Therefore, increasing research on the lichen natural resources may provide good results for exploiting and developing valuable natural products which benefit humans. For the past 20–30 years, some studies with lichen, even with limited screening effort, have indicated the frequent occurrence of metabolites with antitumor properties (Ren et al. 2009; Rankovic et al. 2011, 2014; Kosanic et al. 2012a, b). All this evidence supports efforts for the development of new anticancer drugs with lichen secondary metabolites as a starting point. The potential of lichen secondary metabolites as a possible source of anticancer drugs is certainly large and visible. As we know, the structures of more than 700 lichen substances are available, but because lichens grow slowly, their availability is insufficient in quantity and difficult in large-scale industrial production, so lichens were frequently ignored by pharmaceutical industries. However, the secondary metabolites of lichen that are deposited on the surface of the mycelium were usually produced by fungi; therefore, it became possible that cultured mycobionts could replace the lichen. Although not many of the lichen metabolites are likely to become therapeutics, the information gained from studying them is likely to lead the development and understanding of novel molecular targets and chemical synthesis or chemical modification of natural metabolites, which in turn may lead to the development of new classes of therapeutic agents. On the other hand, powerful new technologies such as combinatorial chemistry, high-throughput screening, bioinformatics, proteomics, and genomics have emerged and are being integrated widely in the field of pharmaceutical discovery research. These technologies have enormous potential to make use of the chemical diversity of natural products (Lahlou 2013). All this, including compound library design, protein 3D structures, nuclear magnetic resonance (NMR)-based screening, 3D quantitative structure–activity relationship (QSAR) in modern drug design, and computer-aided prediction of drug toxicity and metabolism, may help in the development of new agents modeled on the basis of the secondary metabolites of lichens. Finally, a multidisciplinary collaboration among lichenologists, chemists, pharmacologists, and biologists is expected to be critical in the development of potential anticancer drugs from secondary metabolites of lichens.

References

Babula P, Adam V, Havel L, Kizek R (2009) Noteworthy secondary metabolites naphthoquinones–their occurrence, pharmacological properties and analysis. Curr Pharm Anal 5:47–68

Bačkorová M, Bačkor M, Mikeš J, Jendželovský R, Fedoročko P (2011) Variable responses of different human cancer cells to the lichen compounds parietin, atranorin, usnic acid and gyrophoric acid. Toxicol In Vitro 25:37–44

Bačkorová M, Jendželovský R, Kello M, Bačkor M, Mikeš J, Fedoročko P (2012) Lichen secondary metabolites are responsible for induction of apoptosis in HT-29 and A2780 human cancer cell lines. Toxicol In Vitro 26:462–468

Bazin MA, Le Lamer AC, Delcros JG, Rouaud I, Uriac P, Boustie J, Corbel JC, Tomasi S (2008) Synthesis and cytotoxic activities of usnic acid derivatives. Bioorg Med Chem 16:6860–6866

Behera BC, Adawadkar B, Makhija U (2003) Inhibitory activity of xanthine oxidase and superoxide-scavenging activity in some taxa of the lichen family Graphidaceae. Phytomedicine 10:536–543

Behera BC, Adawadkar B, Makhija U (2004) Capacity of some Graphidaceous lichens to scavenge superoxide and inhibition of tyrosinase and xanthine oxidase activities. Curr Sci 87:83–87

Bézivin C, Tomasi S, Rouaud I, Delcros JG, Boustie J (2004) Cytotoxic activity of compounds from the lichen: *Cladonia convoluta*. Planta Med 70:874–877

Bogo D, Matos MFC, Honda NK, Pontes EC, Oguma PM, Santos ECS, de Carvalho JE, Nomizo A (2010) In vitro antitumor activity of orsellinates. Z Naturforsch 65:43–48

Boustie J, Grube M (2005) Lichens–a promising source of bioactive secondary metabolites. Plant Genet Resour 3:273–287

Brandão LF, Alcantara GB, Matos Mde F, Bogo D, Freitas Ddos S, Oyama NM, Honda NK (2013) Cytotoxic evaluation of phenolic compounds from lichens against melanoma cells. Chem Pharm Bull (Tokyo) 61:176–183

Brisdelli F, Perilli M, Sellitri D, Piovano M, Garbarino JA, Nicoletti M, Bozzi A, Amicosante G, Celenza G (2013) Cytotoxic activity and antioxidant capacity of purified lichen metabolites: an in vitro study. Phytother Res 27:431–437

Brunauer G, Hager A, Krautgartner WD, Turk R, Stocker-Worgotter E (2006) Experimental studies on *Lecanora rupicola* (L.) Zahlbr.: chemical and microscopical investigations of the mycobiont andre synthesis stages. Lichenologist 38:577–585

Brunauer G, Hager A, Grube M, Turk R, Stocker-Worgotter E (2007) Alterations in secondary metabolism of aposymbiotically grown mycobionts of *Xanthoria elegans* and cultured resynthesis stages. Plant Physiol Biochem 45:146–151

Bugrahan E, Ali A, Basak T, Hasan T (2016) In vitro antitumor activities of the lichen compounds olivetoric, physodic and psoromic acid in rat neuron and glioblastoma cells. Pharm Biol 54: 1748–1762

Burlando B, Ranzato E, Volante A, Appendino G, Pollastro F, Verotta L (2009) Antiproliferative effects on tumour cells and promotion of keratinocyte wound healing by different lichen compounds. Planta Med 75:607–613

Burton JF, Cain BF (1959) Antileukæmic activity of polyporic acid. Nature 184:1326–1327

Cain BF (1966) Potential anti-tumour agents. IV. Polyporic acid series. J Chem Soc Perkin 11: 1041–1045

Cardarelli M, Serino G, Campanella L, Ercole P, De Cicco NF, Alesiani O, Rossiello F (1997) Antimitotic effects of usnic acid on different biological systems. Cell Mol Life Sci 53:667–672

Cardile V, Graziano ACE, Avola R, Piovano M, Russo A (2017) Potential anticancer activity of lichen secondary metabolite physodic acid. Chem Biol Interact 263:36–45

Cheng B, Cao S, Vasquez V, Annamalai T, Tamayo-Castillo G, Clardy J, Tse-Dinh YC (2013) Identification of anziaic acid, a lichen depside from *Hypotrachyna* sp., as a new topoisomerase poison inhibitor. PLoS One 8(4):e60770

Cohen PA, Neil Towers GH (1995) Anthraquinones and phenanthroperylenequinones from *Nephroma laevigatum*. J Nat Prod 58:520–526

Correché ER, Carrasco M, Giannini F, Piovano M, Garbarino J, Enriz D (2002) Cytotoxic screening activity of secondary lichen metabolites. Acta Farm Bonaer 21:273–278

Correché ER, Enriz RD, Piovano M, Garbarino J, Gómez-Lechón MJ (2004) Cytotoxic and apoptotic effects on hepatocytes of secondary metabolites obtained from lichens. Altern Lab Anim 32:605–615

Crittenden D, Porter N (1991) Lichen-forming fungi: potential sources of novel metabolites. Trends Biotechnol 9:409–414

Demleitner S, Kraus J, Franz G (1992) Synthesis and antitumour activity of derivatives of curdlan and lichenan branched at C-6. Carbohydr Res 226:239–246

Einarsdóttir E, Groeneweg J, Björnsdóttir GG, Harethardottir G, Omarsdóttir S, Ingólfsdóttir K, Ogmundsdóttir HM (2010) Cellular mechanisms of the anticancer effects of the lichen compound usnic acid. Planta Med 76:969–974

Elix JA (1996) Biochemistry and secondary metabolites. In: Nash TH III (ed) Lichen biology, 1st edn. Cambridge University Press, Cambridge, pp 155–180

Elix JA, Stocker-Worgotter E (2008) Biochemistry and secondary metabolites. In: Nash TH III (ed) Lichen biology, 2nd edn. Cambridge University Press, Cambridge, pp 104–133

Ernst-Russell MA, Chai CLL, Hurne AM, Waring P, Hockless DCR, Elix JA (1999) Revision and cytotoxic activity of the scabrosin esters, epidithiopiperazinediones from the lichen *Xanthoparmelia scabrosa*. Aust J Chem 52:279–283

Fahselt D (1994) Secondary biochemistry of lichens. Symbiosis 16:117–165

Fakim AG (2006) Medicinal plants: traditions of yesterday and drugs of tomorrow. Mol Asp Med 27:1–93

Fujikawa F, Hirai K, Hirayama T, Toyota T, Urasaki M (1972) Influence of cytochalasin B on hyphal morphogenesis of *Aspergillus nidulans*. Yakugaku Zasshi 92:823–827

Fukuoka F, Nakanishi M, Shibata S, Nishikawa Y, Takeda T, Tanaka M (1968) Polysaccharides in lichens and fungi. II. Antitumor activities on sarcoma-180 of the polysaccharide preparations from *Gyrophora esculenta* Miyoshi, *Cetraria islandica* (L.) Ach. var. *orientalis* Asahina, and some other lichens. Gann 59:421–432

Guo J, Li ZL, Wang AL, Liu XQ, Wang J, Guo X, Jing YK, Hua HM (2011) Three new phenolic compounds from the lichen *Thamnolia vermicularis* and their antiproliferative effects in prostate cancer cells. Planta Med 77:2042–2046

Hager A, Brunauer G, Türk R, Stocker-Wörgötter E (2008) Production and bioactivity of common lichen metabolites as exemplified by *Heterodea muelleri* (Hampe) Nyl. J Chem Ecol 34: 113–120

Han D, Matsumaru K, Rettori D, Kaplowitz N (2004) Usnic acid-induced necrosis of cultured mouse hepatocytes: inhibition of mitochondrial function and oxidative stress. Biochem Pharmacol 67:439–451

Haraldsdóttir S, Guðlaugsdóttir E, Ingólfsdóttir K, Ögmundsdóttir HM (2004) Anti-proliferative effects of lichen-derived lipoxygenase inhibitors on twelve human cancer cell lines of different tissue origin in vitro. Planta Med 70:1098–1100

Hirayama T, Fujikawa F, Yosioka I, Kitagawa I (1974) A new depsidone oxyphysodic acid isolated from a lichen *Parmelia enteromorphia* Ach. Chem Pharm Bull (Tokyo) 22:1678–1680

Hirayama T, Fujikawa F, Kasahara T, Otsuka M, Nishida N, Mizuno D (1980) Anti-tumor activities of some lichen products and their degradation products. Yakugaku Zasshi 100:755–759

Huneck S (1999) The significance of lichens and their metabolites. Naturwissenschaften 86: 559–570

Huneck S, Himmelreich U (1995) Arthogalin, a cyclic depsipeptide from the lichen *Arthothelium galapagoense*. Z Naturforsch 50:1101–1103

Ingólfsdóttir K (2002) Usnic acid. Phytochemistry 61:729–736

Ivanova V, Backor M, Dahse HM, Graefe U (2010) Molecular structural studies of lichen substances with antimicrobial, antiproliferative, and cytotoxic effects from *Parmelia subrudecta*. Prep Biochem Biotechnol 40:377–388

Johnson CJ, Bennett JP, Biro SM, Duque-Velasquez JC, Rodriguez CM, Bessen RA, Rockel TE (2011) Degradation of the disease-associated prion protein by a serine protease from lichens. PLoS One 6:19836–19837

Ju-Mi H, Sung-S S, Tai KK, Jung EK, Se JH, Ui JY, Joung HY, Il-C K (2018) Anti-cancer activity of lobaric acid and lobarstin extracted from the Antarctic lichen *Stereocaulon alpnum*. Molecules 23:658

Karikas GA (2010) Anticancer and chemopreventing natural products: some biochemical and therapeutic aspects. J BUON 15:627–638

Koparal AT, Tuylu BA, Turk H (2006) In vitro cytotoxic activities of (?)-usnic acid and (−)-usnic acid on V79, A549, and human lymphocyte cells and their nongenotoxicity on human lymphocytes. Nat Prod Res 20:1300–1307

Koparal AT, Ulus G, Zeytinoglu M, Tay T, Turk AO (2010) Angiogenesis inhibition by a lichen compound olivetoric acid. Phytother Res 24:754–758

Kosanic M, Rankovic B, Stanojkovic T (2012a) Antioxidant, antimicrobial, and anticancer activity of 3 *Umbilicaria* species. J Food Sci 77:20–25

Kosanic MM, Rankovic BR, Stanojkovic TP (2012b) Antioxidant, antimicrobial and anticancer activities of three *Parmelia* species. J Sci Food Agric 92:1909–1916

Kosanic M, Manojlovic N, Jankovic S, Stanojkovic T, Rankovic B (2013) *Evernia prunastri* and *Pseudoevernia furfuraceae* lichens and their major metabolites as antioxidant, antimicrobial and anticancer agents. Food Chem Toxicol 53:112–118

Kristmundsdóttir T, Aradóttir HA, Ingólfsdóttir K, Ogmundsdóttir HM (2002) Solubilization of the lichen metabolite (+)-usnic acid for testing in tissue culture. J Pharm Pharmacol 54:1447–1452

Kumar KCS, Muller K (1999) Lichen metabolites. 2. Antiproliferative and cytotoxic activity of gyrophoric, usnic, and diffractaic acid on human keratinocyte growth. J Nat Prod 62:821–823

Kupchan SM, Kopperman HL (1975) L-Usnic acid: tumor inhibitor isolated from lichens. Experientia (Basel) 31:625

Lahlou M (2013) The success of natural products in drug discovery. Pharmacol Pharm 4:17–31

Lawrey JD (1986) Biological role of lichen substances. Bryologist 89:111–122

Liu H, Liu Y, Liu Y, Xu A, Young CYF, Yuan H, Lou H (2010) A novel anticancer agent, retigeric acid B, displays proliferation inhibition, S phase arrest and apoptosis activation in human prostate cancer cells. Chem Biol Interact 188:598–606

Magaya MC, Yeang DP, Ahkeboyee BGW (2013) Arthogalin inhibits the growth of murine malignant prostate sarcoma cells in vitro. J Pharm Pharmacol Res 3:22–26

Manojlovic N, Rankovic B, Kosanic M, Vasiljevic P, Stanojkovic T (2012) Chemical composition of three *Parmelia* lichens and antioxidant, antimicrobial and cytotoxic activities of some of their major metabolites. Phytomedicine 19:1166–1172

Mayer M, O'Neill MA, Murray KE, Santos-Magalhães NS, Carneiro-Leão AM, Thompson AM, Appleyard VC (2005) Usnic acid: a non-genotoxic compound with anti-cancer properties. Anti-Cancer Drugs 16:805–809

Miao V, Coeffet M, Brown D, Sinnemann S, Donaldson G, Davies J (2001) Genetic approaches to harvesting lichen products. Trends Biotechnol 19:349–355

Millot M, Tomasi S, Articus K, Rouaud I, Bernard A, Boustie J (2007) Metabolites from the lichen *Ochrolechia parella* growing under two different heliotropic conditions. J Nat Prod 70:316–318

Moerman KL, Chai CL, Waring P (2003) Evidence that the lichen-derived scabrosin esters target mitochondrial ATP synthase in P388D1 cells. Toxicol Appl Pharmacol 190:232–240

Molnár K, Farkas E (2010) Current results on biological activities of lichen secondary metabolites: a review. Z Naturforsch 65:157–173

Muggia L, Schmitt I, Grube M (2009) Lichens as treasure chests of natural products. SIM News:59:85–59:97

Muller K (2001) Pharmaceutically relevant metabolites from lichens. Appl Microbiol Biotechnol 56:9–16

Nil K, Mehmet KD, İlker B, Semra SA, Sümer A, Demet CD (2018) Evaluation of in vitro anticancer activity of vulpinic acid and its apoptotic potential using gene expression and protein analysis. Indian J Pharm Educ 52(4):626–634

Nishikawa Y, Ohno H (1981) Studies on the water-soluble constituents of lichens. IV. Effect of antitumor lichen-glucans and related derivatives on the phagocytic activity of the reticuloendothelial system in mice. Chem Pharm Bull (Tokyo) 29:3407–3410

Nishikawa Y, Takeda T, Shibata S, Fukuoka F (1969) Polysaccharides in lichens and fungi. 3. Further investigation on the structures and the antitumor activity of the polysaccharides from *Gyrophora esculenta* Miyoshi and *Lasallia papulosa* Llano. Chem Pharm Bull (Tokyo) 17:1910–1916

Nishikawa Y, Yoshimoto K, Horiuchi R, Michishita K, Okabe M, Fukuoka F (1979) Studies on the water-soluble constituents of lichens. III. Changes in antitumor effect caused by modifications of pustulan- and lichenan-type glucans. Chem Pharm Bull (Tokyo) 27:2065–2072

Ögmundsdóttir HM, Zoëga GM, Gissurarson SR, Ingolfsdóttir K (1998) Natural products: antiproliferative effects of lichen-derived inhibitors of 5-lipoxygenase on malignant cell-lines and mitogen-stimulated lymphocytes. J Pharm Pharmacol 50(1):107–115

O'Neill MA, Mayer M, Murray KE, Rolim-Santos HML, Santos-Magalhães NS, Thompson AM, Appleyard VCL (2010) Does usnic acid affect microtubules in human cancer cells? Braz J Biol 70:659–664

Oksanen I (2006) Ecological and biotechnological aspects of lichens. Appl Microbiol Biotechnol 73:723–734

Pejin B, Iodice C, Bogdanović G, Kojić V, Tešević V (2013) Stictic acid inhibits cell growth of human colon adenocarcinoma HT-29 cells. Arab J Chem 10:S1240. https://doi.org/10.1016/j.arabjc.2013.03.003

Rankovic BR, Kosanic MM, Stanojkovic TP (2011) Antioxidant, antimicrobial and anticancer activity of the lichens *Cladonia furcata, Lecanora atra* and *Lecanora muralis*. BMC Complement Altern Med 11:97

Rankovic B, Kosanic M, Stanojkovic T, Vasiljevic P, Manojlovic N (2012) Biological activities of *Toninia candida* and *Usnea barbata* together with their norstictic acid and usnic acid constituents. Int J Mol Sci 13:14707–14722

Rankovic B, Kosanic M, Manojlovic N, Rančic A, Stanojkovic T (2014) Chemical composition of *Hypogymnia physodes* lichen and biological activities of some its major metabolites. Med Chem Res 23:408–416

Ren MR, Hur JS, Seo KI et al (2009) Anti-proliferative effects of *Lethariella zahlbruckneri* extracts in HT-29 human colon cancer cells. Food Chem 47:2157–2162

Ribeiro-Costa RM, Alves AJ, Santos NP, Nascimento SC, Gonçalves EC, Silva NH, Honda NK, Santos-Magalhães NS (2004) In vitro and in vivo properties of usnic acid encapsulated into PLGA-microspheres. J Microencapsul 21:371–384

Richardson DHS (1991) Lichens as biological indicators – recent developments. In: Jeffrey DW, Madden B (eds) Bioindicators and environmental management. Academic Press, London, San Diego, pp 263–272

Ristic S, Rankovic B, Kosanić M, Stamenkovic S, Stanojković T, Sovrlić M, Manojlović N (2016a) Biopharmaceutical potential of two *Ramalina* lichens and their metabolites. Curr Pharm Biotechnol 17(7):651–658

Ristic S, Rankovic B, Kosanic M, Stanojkovic T, Stamenkovic S, Vasiljevic P, Manojlovic I, Manojlovic N (2016b) Phytochemical study and antioxidant, antimicrobial and anticancer activities of *Melanelia subaurifera* and *Melanelia fuliginosa* lichens. J Food Sci Technol 53(6):2804–282816

Romagni JG, Dayan FE (2002) Structural diversity of lichen metabolites and their potential use. In: Upadhyay RK (ed) Advances in microbial toxin research and its biotechnological exploitation. Kluwer Academic/Plenum, New York, pp 151–169

Rui Z, Yi Y, So-YP TTN, Young WS, Kyung HL, Jae HL, Kyung KK, Jae SH, Hangun K (2017) The lichen secondary metabolite atranorin suppresses lung cancer cell motility and tumorigenesis. Sci Rep 7(1):8136

Russo A, Piovano M, Lombardo L, Vanella L, Cardile V, Garbarino J (2006) Pannarin inhibits cell growth and induces cell death in human prostate carcinoma DU-145 cells. Anti-Cancer Drugs 17:1163–1169

Russo A, Piovano M, Lombardo L, Garbarino J, Cardile V (2008) Lichen metabolites prevent UV light and nitric oxide-mediated plasmid DNA damage and induce apoptosis in human melanoma cells. Life Sci 83:468–474

Russo A, Caggia S, Piovano M, Garbarino J, Cardile V (2012) Effect of vicanicin acid on human prostate cancer cells: role of Hsp70 protein. Chem Biol Interact 195:1–10

Sahu SC, Amankwa-Sakyi M, O'Donnell MW Jr, Sprando RL (2011) Effects of usnic acid exposure on human hepatoblastoma HepG2 cells in culture. J Appl Toxicol. https://doi.org/10.1002/jat.1721

Shibamoto T, Wei CI (1984) Mutagenicity of lichen constituents. Environ Mutagen 6:757–762

Shibata S, Nishikawa Y, Tanaka M, Fukuoka F, Nakanishi M (1968) Antitumour activities of lichen polysaccharides. Z Krebsforsch 71:102–104

Shrestha G, St. Clair LL (2013) Lichens: a promising source of antibiotic and anticancer drugs. Phytochem Rev 12:229–244

Srinivas G, Babykutty S, Sathiadevan PP, Srinivas P (2007) Molecular mechanism of emodin action: transition from laxative ingredient to an antitumor agent. Med Res Rev 27:591–608

Stocker-Worgotter E (2008) Metabolic diversity of lichen-forming ascomycetous fungi: culturing, polyketide and shikimate metabolite production, and PKS genes. Nat Prod Rep 25:188–200

Stocker-Worgotter E, Elix JA (2002) Secondary chemistry of cultured mycobionts: formation of a complete chemosyndrome by the lichen fungus of *Lobaria spathulata*. Lichenologist 34: 351–359

Sung SS, Tai KK, Jung EK, Ju-MH TTN, Se JH, Ui JY, Joung HY, Il-C K (2017) Anticancer activity of Ramalin, a secondary metabolite from the Antarctic lichen *Ramalina terebrata*, against colorectal cancer cells. Molecules 22:1361

Takai M, Uehara Y, Beisler JA (1979) Usnic acid derivatives as potential antineoplastic agents. J Med Chem 22:1380–1384

Tokiwano T, Satoh H, Obara T, Hirota H, Yoshizawa Y, Yamamota Y (2009) A lichen substance as an antiproliferative compound against HL-60 human leukemia cells: 16-O-acetyl-leucotylic acid isolated from *Myelochroa aurulenta*. Biosci Biotechnol Biochem 73:2525–2527

Williams DE, Bombuwala K, Lobkovsky E, De Silva ED, Karunatne V, Allen TM, Clardy J, Andersen RJ (1998) Ambewelamides A and B, antineoplastic epidithiapiperazinediones isolated from the lichen *Usnea* sp. Tetrahedron Lett 39:9579–9582

Yamamoto Y (2000) Screening of biological activities and isolation of biological active compounds from lichens. Chem Regul Plants 35:169–179

Yamamoto Y, Miura Y, Higuchi M, Kinoshita Y (1993) Using lichen tissue cultures in modern biology. Bryologist 96:384–393

Yamamoto Y, Kinoshita Y, Kurokawa T, Yoshimura I, Ahmadjian V, Yamada Y (1995) Cell growth and pigment production in suspension cultures of a mycobiont isolated from the lichen *Cladonia cristatella*. Can J Bot 73:590–594

Yoshimura I, Kurokawa T, Kinoshita Y, Yamamoto Y, Miyawaki H (1994) Lichen substances in cultured lichens. J Hattori Bot Lab 76:249–261

Antigenotoxic Effect of Some Lichen Metabolites

6

Hülya Sivas

Abstract

Naturally occurring compounds can have protective effects towards mutagens and carcinogens, as shown by numerous studies. Several lichen species have been given much attention by researchers because their extracts and compounds have been used in traditional medicine to cure diseases such as ulcer, arthritis, tuberculosis, and cancer throughout the ages. Although a wide variety of scientific investigations on the biological activities of lichen extracts and their constituents have been performed, there is rather less research on their genotoxicity and antigenotoxic activity. To date, most results for the genotoxic/antigenotoxic activities of lichens have been obtained for lichen extracts and their constituents using the Ames/Salmonella and the *E. coli* WP2 microsome, chromosome aberration, micronucleus, sister chromatid exchange, single-cell gel electrophoresis, 8-OH-dG activity, and wing somatic mutation and recombination assays. In the present chapter, findings on antigenotoxic/genotoxic activities and its mechanisms are evaluated. By using the most common bacterial and nonbacterial assays, extracts of various lichen species have been shown to have promising antigenotoxic activity with somewhat less genotoxic activity. Lichen extracts may have a possible therapeutic potential. Therefore, these extracts must be further investigated by other multiple in vitro bioassays for development of therapeutic agents.

Abbreviations

2-AF	2-Aminoflourene
4-NPD	4-Nitrophenylenediamine

H. Sivas (✉)
Faculty of Sciences, Department of Biology, Eskisehir Technical University, Eskisehir, Turkey
e-mail: hzeytino@eskisehir.edu.tr

B. Ranković (ed.), *Lichen Secondary Metabolites*,
https://doi.org/10.1007/978-3-030-16814-8_6

8-oxo-dG	8-Oxo-2′-deoxyguanosine, 8-hydroxy-2′-deoxyguanosine
9-AA	9-Aminoacridine
AFB1	Aflatoxin B_1
BrdU	Bromodeoxyuridine
CA	Chromosome aberration
CBS	Colloidal bismuth subcitrate
COMET	Single-cell gel electrophoresis
DN	Dioxidine
DXR	Doxorubicin
HPL	Human peripheral blood lymphocytes
IMA	Imazalil
MI	Mitotic index
MMC	Mitomycin C
MMS	Methyl methanesulfonate
MN	Micronucleus
MNNG	Methylnitronitrosoguanidine
SCE	Sister chromatid exchange
SCGE	Single-cell gel electrophoresis
SMART	Wing somatic mutation and recombination test

6.1 Introduction

Naturally occurring organic compounds from a variety of organisms, including medicinal plants, are very important. Development of these compounds for new therapeutic agents has become a popular research objective. Investigation of the biological potential of several classes of natural agents, dietary constituents, hormones, and vitamins using in vitro and in vivo bioassays have shown that they can act as genotoxicity inhibitors as well as protectors against cytostasis or environmental carcinogens (Okai et al. 1996; Scarpato et al. 1998; Ipek et al. 2003, 2005; Mersch-Sundermann et al. 2004; Jayaprakasha et al. 2007; Zeytinoglu et al. 2008; Hoshina and Marin-Morales 2014; Delebassée et al. 2017; Makhuvele et al. 2018; Kaura et al. 2018; Kuete et al. 2018). Also, investigation of the possible genotoxicity of such agents captures the attention of researchers because of their use in folk medicine or potential applications. Most of the medicinal plants used traditionally have never been subjected to toxicological tests such as that required for modern pharmaceutical compounds. However, research has shown that quite a few plants that are used in traditional medicine or similar applications may have genotoxic or carcinogenic properties (Santos et al. 2009; Nieminen et al. 2002; Sponchiado et al. 2016; Quadros et al. 2017; Kahaliw et al. 2018; Prinsloo et al. 2018). Recently, development of an in silico method by using different computer models for genotoxicity and carcinogenicity prediction of secondary plant metabolites suggested that these potentials can be a first step for prioritization (Glück et al.

2018). Therefore, it becomes very important to search extracts derived from plants that contain a variety of compounds for their nontoxic, antigenotoxic, or genotoxic properties.

Lichen species have attracted much of the attention of researchers because their extracts and compounds have been used in traditional medicine in Europe, Asia, and North America (Richardson 1988; Cabrera 1996; Tilford 1997). Although lichen extracts and their compounds have been subjected to many scientific investigations for their biological activities such as immunostimulation, analgesic, anti-ulcerogenic, antipyretic, antimicrobial, antioxidative, and antitumor (Kumar and Müller 1999; Ingolfsdottir et al. 2000; Ingolfsdottir 2002; Türk et al. 2003; Tay et al. 2004; Yılmaz et al. 2004; Halici et al. 2005; Karunaratne et al. 2005; Behera et al. 2006; Zeytinoglu et al. 2008; Yeash et al. 2017; Tomović et al. 2017; reviewed in Prashith Kekuda et al. 2018), there is rather less research on their genotoxic or antigenotoxic activity. Scientific investigation of the genotoxic and antigenotoxic properties of lichens includes in vitro and in vivo studies, mostly using the aqueous, methanol, acetone, or *n*-hexane extracts. In the recent past few years, the genotoxicity and antigenotoxicity of isolated compounds from lichen species has attracted attention. In the present chapter, findings on the antigenotoxic and genotoxic activity of lichen extracts or second metabolites and the relevant mechanisms are evaluated.

6.2 Bioassays for the Antigenotoxicity/Genotoxicity of Lichens

The Organization for Economic Co-operation and Development (OECD 2012) and the European Centre for the Validation of Alternative Methods (ECVAM 2012) have widely investigated the validation of mutagenicity tests. A set of assays is recommended to determine the genotoxicity of a test agent. The methods most frequently used for the assessment of genotoxic/antigenotoxic activity of lichen extracts or components based on bacterial short-term assays and mammalian test system are recommended by the OECD and the ECVAM. The Ames/*Salmonella*/ microsome (Ames) and the *Escherichia coli* WP2/microsome (WP2) reverse mutation assays are the most common bacterial systems, and micronucleus (MN), chromosome aberration (CA), sister chromatid exchange (SCE), single-cell gel electrophoresis (COMET), and wing somatic mutation and recombination test (SMART) the most common nonbacterial systems, that have been used to date.

6.2.1 Bacterial Short-Term Assays

The Ames and the WP2 assays are short-term bacterial reverse mutation assays specifically designed to detect a wide range of chemicals or other agents which can produce genetic damage. The combination of *Salmonella typhimurium* and *Eschericha coli* strains in the tests recommended by OECD (1997) are given in Table 6.1, including their characteristics such as mutation types, plasmids, reversion

Table 6.1 Characteristics of the most commonly used strains in the bacterial reverse mutation test, recommended in OECD guideline 471

Original bacteria/ strains	Mutation type	Plasmid	Reversion event	DNA target
S. typhimurium LT2				
TA97/97a	Deletion *hisD6610 hisO1242 rfa Δ(gal chl bio uvrB)*	pKM101	Frameshift	-C-C-C-C-C-C-
TA98	Deletion *hisD3052 rfa Δ(gal chl bio uvrB)*	pKM101	Frameshift	-C-G-C-G-C-G-C-G-
TA100	Deletion *hisG46 rfa Δ(gal chl bio uvrB)*	pKM101	Base-pair substitution	-G-G-G-
TA102	Wild type *hisG428 rfa galE hisΔ(G)8476*	pKM10 pAQ1	Terminating ochre	-T-A-A-
TA1535	Deletion *hisG46 rfa Δ(gal chl bio uvrB)*	None	Base-pair substitution	-G-G-G-
TA1537	Deletion *hisC3076 rfa Δ(gal chl bio uvrB)*	None	Frameshift	+1 near –C-C-C- run of Cs)
E. coli B				
E.coli WP2*uvrA*	*trpE uvrA*	None	Terminating ochre and DNA repair	-A-T-
E.coli WP2*uvrA*	*trpE uvrA*	pKM101		-G-C-

events, and the DNA target of mutations (Maron and Ames 1983; reviewed by Mortelmans and Zeiger 2000). These assays provide a very sensitive study of potentially mutagenic pathways for the metabolism of compounds in both absence and presence of a rat liver microsomal system (S9 mix). The Ames assay employs several histidine-dependent *Salmonella* strains, each carrying different mutations in various genes in the histidine operon, pointing to different mutagenic mechanisms. When the *Salmonella* strains carrying mutations in the *his* gene are grown on a minimal media agar plate with a trace of histidine, only those bacteria that revert to histidine independence are able to form colonies (Fig. 6.1). When a mutagen is added to the plate, the number of revertant colonies per plate is increased. Base-pair substitution (A:T to G:C or G:C to A:T) and frameshift mutations (deletions) in *S. typhimurium* strains are represented to identify both types of mutation caused by a test compound. Therefore, differences in the activity of a test compound acting in these strains may yield some insight into how the compounds interact with the DNA of bacteria. Additionally, some genetic markers have been developed to make the strains more sensitive to certain types of mutagens.

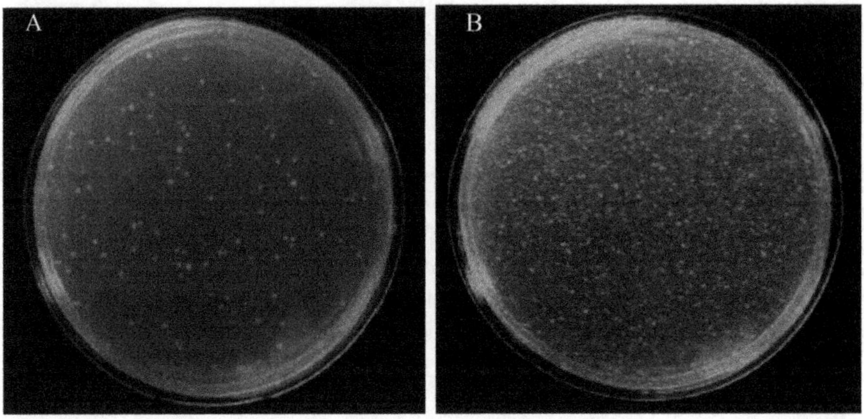

Fig. 6.1 Ames test plates of TA100 strain of *Salmonella typhimurium*. (**a**) Control: spontaneous revertants. (**b**) A mutagenic dose response to sodium azide. (From Mortelmans and Zeiger 2000)

The *Escherichia coli* WP2 assay detects trp (−) to trp (+) reversion at a site blocking a step in the biosynthesis of tryptophan before the formation of anthranilic acid. The different auxotrophic WP2 strains all carry the same A:T base pair at the critical mutation site within the *trpE* gene (see Table 6.1). The most widely used *E. coli* WP2 strains, each carrying the *trpE* mutation, are WP2 (wild type for DNA repair), WP2 (pKM101), WP2 *uvrA*, and WP2 *uvrA* (pKM101) (Mortelmans and Riccio 2000; Sugiyama et al. 2016). The assay is currently used by many researchers in conjunction with the Ames assay for screening chemicals for their mutagenicity. The Ames assay procedures are the same as for the WP2 assay with the exception that limited histidine instead of limited tryptophan is used. International guidelines have been established for performing these mutagenicity assays. These assays are used worldwide as an initial screen to determine the mutagenic/antimutagenic potential of new chemicals, drugs, or natural products from plants or animals (Sponchiado et al. 2016).

Conversely, the anti-mutagenicity of a compound against a selected positive mutagen can be investigated when the two chemicals are co-administered to the bacteria in both test systems. Using known mutagenic compounds as "positive controls," it is possible to study whether tested components can reduce DNA damage.

6.2.2 Nonbacterial Short-Term Assays

At present, several antigenotoxicity/genotoxicity assays, which include the chromosome aberration (CA), micronucleus (MN), somatic mutation and recombination test (SMART), sister chromatid exchange (SCE), single-cell gel electrophoresis (SCGE), or COMET assays, are available, and they are recommended to use as a set for investigations.

According to the literature, the antigenotoxic/genotoxic potential of lichens has been evaluated, commonly by MN, SCE, CA, COMET, 8-oxo-2-deoxyguanosine (8-oxo-dG) activity assays in mammalian cells and the mitotic index (MI) assay in plant cells. The purpose of the MN test is to examine the structural and numerical chromosomal damage which formed small membrane-bound DNA fragments or micronuclei in the cytoplasm of interphase cells caused by a tested agent or by clastogens and aneugens. Micronuclei can be formed by chromosome fragments lacking a centromere or whole chromosomes that are unable to migrate during cell division. The MN test can be conducted in the presence or in the absence of cytochalasin B, which is used to block cell division and generate binucleated cells (Fig. 6.2a). The cytokinesis-block micronucleus assay is a sensitive, comprehensive

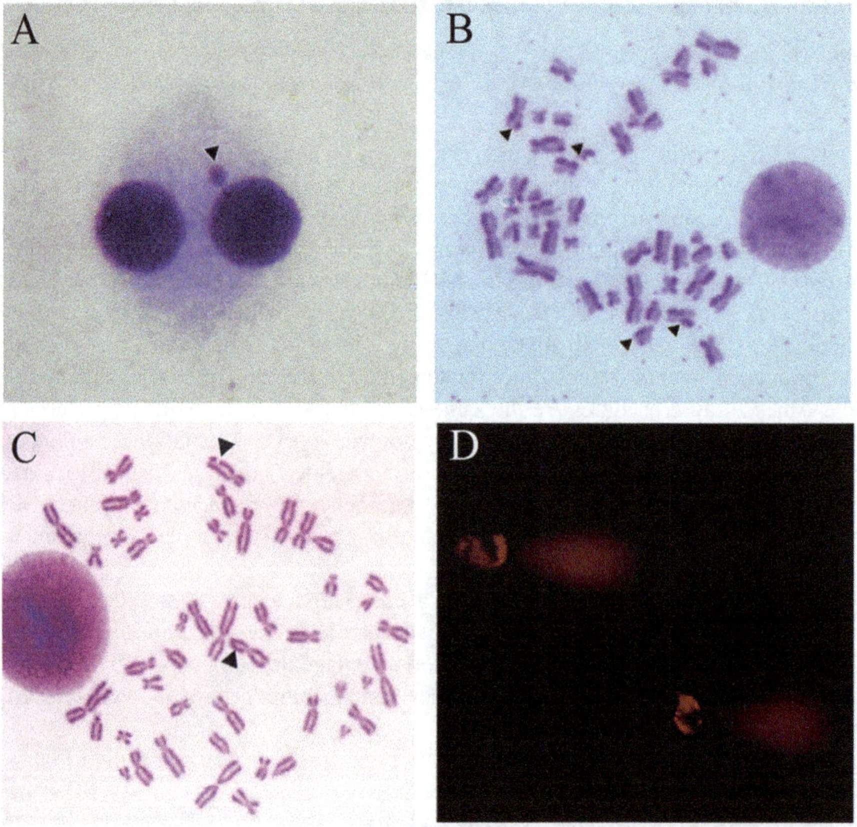

Fig. 6.2 Photomicrographs for nonbacterial genotoxicity assays. (**a**) A mitogen-stimulated cytokinesis-blocked lymphocyte containing one micronucleus (MN). (**b**) Giemsa staining of BrdU-incorporated chromosomes in human lymphocytes for sister chromatic exchange (SCE); arrowheads show chromosome breaks and sister union. (**c**) Chromosome aberration (CA) shows sister chromatids stained at different densities (photograph kindly provided by Dr. B. Ayaz Tuylu). (**d**) COMET tails of chromosomes visualized by an epifluorescence microscope (photograph kindly provided by Dr. A.T. Koparal)

and simple methodology for measuring DNA damage, cytostasis, and cytotoxicity which can be scored easily in a variety of systems 3 and in vivo (Fenech 2007; Kirsch-Volders et al. 2011). The assay is being applied successfully for biomonitoring in vivo genotoxin exposure, for in vitro genotoxicity testing, and in diverse research fields such as nutrigenomics and pharmacogenomics.

The SCE assay is another short-term test useful for the detection of reciprocal exchanges of DNA between two sister chromatids of a duplicating chromosome in mammalian and also nonmammalian cells. Various cytomolecular protocols have been used to perform the SCE assay (Bakkali et al. 2008). SCEs result from the interchange of DNA replication products and involve DNA breakage and reunion (Wilson and Thompson 2007). Detection of SCEs requires the differential staining of sister chromatids, which usually can be achieved by the incorporation of bromodeoxyuridine (BrdU) into chromosomal DNA for two cell cycles (Fig. 6.2b). After labeling, treatment of cells with a spindle inhibitor such as colchicine is required to accumulate cells in a metaphase-like stage of mitosis (Perry and Evans 1975; Ipek et al. 2003).

The short-term in vitro mammalian cell chromosome aberration (CA) test measures the frequency of asymmetrical structural chromosome aberrations after exposure to test chemicals or mutagens. The in vitro CA test may employ cultures of established cell lines or primary cell cultures. Human peripheral blood lymphocytes (HPL) are stimulated to divide by cyclophosphamide in whole blood cultures. Cells in metaphase are analyzed for the presence of chromosomal aberrations (Fig. 6.2c) (Clare 2012).

The COMET assay is used to detect the DNA-strand breaks in eukaryotic cells and named by the shape of the observed DNA distribution, which bears resemblance to a celestial comet. This well-established, highly sensitive, rapid, and simple genotoxicity test is based on the lysing of cells embedded in agarose on a microscope slide to form nucleoids containing supercoiled loops of DNA linked to the nuclear matrix. Then, electrophoresis at high pH results in structures resembling comets, as observed by epifluorescence microscopy (Fig. 6.2d). The intensity of the comet tail relative to the head reflects the number of DNA breaks (Singh et al. 1988; Collins 2004; Speit et al. 2009). Depending on experimental conditions, the migrating DNA reflects the number of single- or double-strand breaks, and alkali-labile sites, including incomplete excision repair sites, but also DNA–DNA and DNA–protein crosslinks (Santos et al. 2009; Verschaeve et al. 2010). A broad spectrum of DNA damage can then be detected either by visual classification of comet morphologies or from morphological parameters obtained by image analysis.

Level of 8-oxo-2′-deoxyguanosine (8-oxo-dG) is a frequently used biomarker of oxidative DNA damage caused by free radicals and other reactive species constantly generated in vivo. 8-Oxo-dG is later removed from DNA by the base excision repair pathway and subsequently transported into body fluids such as saliva, urine, and plasma. Such oxidative damage to DNA is probably the contributor of the age-related development of diseases such as cancer. Agents that decrease oxidative DNA damage should thus decrease the risk of cancer development. Thus, the measurement of 8-oxo-dG is the most common method of assessing DNA damage

(Halliwell 2000; Türkez et al. 2012a). An assay for the measurement of 8-oxo-dG has been developed by using a monoclonal antibody specific to 8-oxo-dG (N45.1), and the enzyme-linked immunosorbent assay (ELISA) has been well established (Toyokuni et al. 1997).

The mitotic index (MI) as a parameter for the evaluation of cytotoxic agents is the ratio of the number of cells undergoing mitosis to the number of cells not undergoing mitosis in a cell population. Mutagens can be detected cytologically by cellular inhibition, disruption in metaphase, induction of chromosomal aberrations and chromosomal fragmentation, and disorganization of the mitotic spindle and consequently of all subsequent dependent mitotic phases. MI is used as an indicator of adequate cell proliferation that can be measured by various plant test systems. Cytotoxicity tests using plant test systems in vivo, such as *Allium cepa* and *Zea mays*, are validated by several researchers who performed these tests jointly with other organism testing for genotoxicity (Agar et al. 2010; Gökbayrak and Sivas 2011; Aslan et al. 2012b).

Somatic mutation and recombination tests (SMART) is a fast and inexpensive mutagenicity detection test using a higher eukaryotic organism in vivo. Genetic damages induced by compounds that generate reactive oxygen species through different action mechanisms such as point mutations, deletions, and somatic recombination can be detected in cells of the wing imaginal disks of *Drosophila melanogaster* in this assay (Fig. 6.3). In particular, the *w/w*$^{+}$ eye and the *mwh/flr* wing variants of *D. melanogaster* have been extensively used to investigate the genotoxicity of a great number of chemicals, including plant extracts and pure compounds (Graf et al. 1984; Sponchiado et al. 2016; Teixeira Zafred et al. 2016).

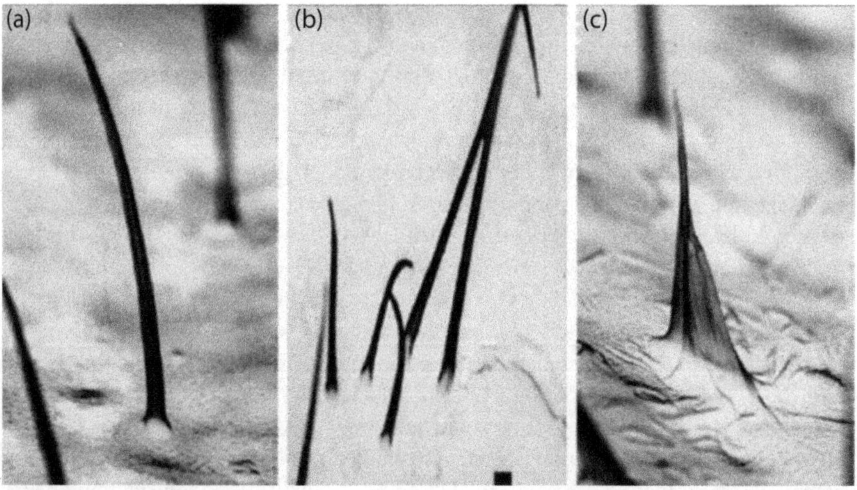

Fig. 6.3 Wings of *Drosophila melanogaster* illustrated by scanning electronic microscopy after a SMART mutagenicity assay: wild (**a**); multiple wing hair (mwh) mutation (**b**); and flare (flr3) mutation (**c**). (From Teixeira Zafred et al. 2016)

Overall these assays are those most used worldwide to determine the genotoxic/antigenotoxic potential of chemicals or natural products (as reviewed by Sponchiado et al. 2016). Thus, it is obvious that it is very important to always include bacterial and mammalian tests for genotoxicity/antigenotoxicity screening of compounds. Antigenotoxicity analysis of plant products is of valuable interest because they are a potential source of therapeutic agents, and thus safety assessments are crucial to validate their use in various areas, not only in phototherapy.

6.3 Antigenotoxic or Genotoxic Potential of Lichen Extracts

The first report on the genotoxicity of lichens using the Ames mutagenicity assay was on a secondary metabolite, physodic acid isolated from *Hypogymnia enteromorpha* (Ach.) Nyl. by Shibamoto and Wei (1984); the second one was on usnic acid by Al-Bekairi et al. (1991). Then, after 2007, biological activities of lichen compounds have been received more attention and several studies have been performed on the genotoxicity/antigenotoxicty of lichens in about the past 10 years. The studies up to date are summarized in two separate tables according to their activities. In Table 6.2, lichen species tested for only their antigenotoxicity or both genotoxicity and antigenotoxicity are listed. The lichen species which were tested for only their genotoxicity are listed in Table 6.3.

The first report describing the therapeutic potential of lichens against drug genotoxicity was from Geyikoglu et al. (2007) (Table 6.2). Aqueous extracts of four common lichen species collected from Giresun Province in Turkey, namely, *Dermatocarpon intestiniforme, Pseudevernia furfuracea, Parmelia pulla, Ramalina capitata, and Rhizoplaca melanophthalma* were tested for their genotoxic and antigenotoxic potentials. *Dermotocarpon intestiniforme, Pseudevernia furfuracea, Parmelia pulla*, and *Ramalina capitata* were found to be antigenotoxic at 5–10 μg/ml against colloidal bismuth subcitrate (CBS)-induced SCE and MN formation in human peripheral lymphocytes (HPL) in vitro. However, one other species, *Rhizoplaca melanophthalma*, was not antigenotoxic. The order of antigenotoxicity efficacy against CBS was *Pseudevernia furfuracea, Dermotocarpon intestiniforme, Ramalina capitata*, and *Parmelia pulla*. On the other hand, all lichen extracts tested were not genotoxic alone (Table 6.2).

After this work, a fresh aqueous extract of *Cetraria aculeatea* (Schreb.) Fr., one of the common species in Turkey, was studied for its genotoxic/antigenotoxic activities in both Ames and mammalian cell systems (Zeytinoglu et al. 2008). The extract (at 0.1–500 μg/ml) exhibited strong antigenotoxic activity against three known mutagenic agents, 4-nitrophenylenediamine (4-NPD), 2-aminoflourene (2-AF), and sodium azide (NaN_3), in TA98 and TA100 strains of *Salmonella typhimurium* in the presence and absence of metabolic activation, without any mutagenic activity (Table 6.2). Preincubation of bacteria with the extract prevented the mutagenic activity of 4-NPD in the higher concentration range in both strains grown without metabolic activation than those grown with metabolic activation. It was suggested that the antigenotoxic potential of the extract was higher in the

Table 6.2 Lichen species tested for only their antigenotoxicity and genotoxicity

Species/type of extract	Against	Assay	Cell types	Genotoxic	Anti-genotoxic	References
Cetraria aculeatea/aqueous	4-NPD	Ames	TA98	No	Yes	Zeytinoglu et al. (2008)
	2-AF		TA100			
	NaN_3					
	MMC	MN	HPL	No	No	
Methanol	MNNG	Ames	TA1535	No	Yes	Ceker et al. (2018)
	NaN_3		TA1537			
	9-AA	WP2	*E. coli*		No	
	AFB1	SCE	HPL			
Cetraria islandica/methanol	AFB1	SCE, MN	HPL	NP	Yes	Kotan et al. (2011)
	9-AA	Ames	TA1535	No	Yes	Aslan et al. (2012b)
	NaN_3		TA1537			
		MI	*Zea mays*			
Cetraria olivetorum/methanol	MNNG	Ames	TA1535	No	Yes	Ceker et al. (2018)
	NaN_3		TA1537			
	9-AA	WP2	*E. coli*			
	AFB1	SCE	HPL			
Cladonia chlorophaea/methanol	MNNG	Ames	TA1535	No	Yes	Ceker et al. (2018)
	NaN_3		TA1537			
	9-AA	WP2	*E. coli*			
	AFB1	SCE	HPL			
Cladonia foliacea/methanol		Ames	TA1535	No	Yes	Anar et al. (2013)
			TA1537			
		WP2	*E. coli*			
	AFB1	SCE	HPL	NP	Yes	

Cladonia rangiformis/methanol	AFB1	SCE, MN	HPL	No	Yes	Kotan et al. (2013)
Dermotocarpon intestiniforme/aqueous	CBS	SCE, MN	HPL	No	Yes	Geyikoglu et al. (2007)
	$CdCl_2$	MN				Guner et al. (2012)
	$HgCl_2$	SCE, MN				Türkez and Dirican (2012)
	Imazalil	CA, MN				Türkez et al. (2012b)
Evernia prunastri/methanol	MNNG	Ames	TA1535	NP	Yes	Alpsoy et al. (2013)
			TA1537			
	Acridin	WP2	*E. coli*			
	AFB1	SCE	HPL			
Lecanora muralis/methanol	AFB1	SCE, MN	HPL	NP	Yes	Alpsoy et al. (2011)
Parmelia pulla/aqueous	CBS	SCE; MN	HPL	No	Yes	Geyikoglu et al. (2007)
Peltigera rufescens (Weis) Humb./aqueous	Imazalil	CA, MN	HPL	No	Yes	Türkez et al. (2012d)
Peltigera canina/methanol		Ames	TA1535	No	Yes	Gormez et al. (2013)
	9-AA		TA1537			
		WP2	*E. coli*			
Pseudevernia furfuracea/aqueous	CBS	SCE, MN	HPL	No	Yes	Geyikoglu et al. (2007)
Methanol, acetone, hexane	AFB1	SCE, MN	HPL	No	Yes	Türkez et al. (2010)
Methanol	9-AA	Ames	TA1535	NP	Yes	Aslan et al. (2012b)
	NaN_3		TA1537			
		MI	*Zea mays*			
Ramalina capitata/aqueous	CBS	SCE, MN	HPL	No	Yes	Geyikoglu et al. (2007)
Rhizoplaca melanophthalma/aqueous	CBS	SCE, MN	HPL	No	No	Geyikoglu et al. (2007)
Methanol	NaN_3	MI	*Zea mays*	NP	Yes	Agar et al. (2010)
	9-AA	Ames	TA1537			
Rhizoplaca chrysoleuca/methanol	AFB1	SCE, MN	HPL	NP	Yes	Alpsoy et al. (2011)
	NaN_3	MI	*Zea mays*			Agar et al. (2010)
	9-AA	Ames	TA1537			

(continued)

Table 6.2 (continued)

Species/type of extract	Against	Assay	Cell types	Genotoxic	Anti-genotoxic	References
Usnea longissima/methanol	AFB1	SCE, MN	HPL	NP	Yes	Agar et al. (2011)
Umbilicaria vellea/methanol	AFB1	SCE, MN	HPL	NP	Yes	Aslan et al. (2012a)
Xanthoria elegans/aqueous	MMC	CA, MN, SCE	HPL	No	Yes	Aydin and Türkez (2011b)
		8-oxo-dG				Türkez et al. (2012a)
Xanthoparmelia somloensis/methanol	AFB1	SCE, MN	HPL	NP	Yes	Aslan et al. (2012a)
Secondary metabolite						
Hypostictic acid (from *Pseudoparmelia sphaerosphora*)						
Protocetraric acid (from *Parmotrema dilatatum*)	DXR	SMART	Drosophila Somatic cells	No	Yes	Guterres et al. (2017)
Psoromic acid (from *Usnea jamaicensis*)						
Salazinic acid (from *Parmotrema cetratum*)						
Usnic acid	MMS	COMET, MN	V79 cells	Yes No	Yes	Leandro et al. (2013)
(+)-Usnic acid (from *Cladonia arbuscula*)	DN	COMET	HPL	No	Yes	Prokopiev et al. (2017) Prokop'ev et al. (2018)
(–)-Usnic acid (from *Cladonia stellaris*)	MMS			Yes		

CA chromosome aberration, *CBS* colloidal bismuth subcitrate, *DN* dioxidine, *DXR* doxorubicin, *HPL* human peripheral blood lymphocytes, *MN* micronucleus, *MMS* methyl methanesulfonate, *MNNG* methylnitronitrosoguanidine, *NP* not performed, *NaN_3* sodium azide, *SMART* Somatic mutation and recombination tests, *8-OH-dG* 8-Oxo-2?-deoxyguanosine

Table 6.3 Lichen species tested for only their genotoxicity

Species/type of extract	Assay	Cell types	Genotoxic	References
Aspicilia calcerea/aqueous	CA, MN	HPL	No	Aydın and Türkez (2011a)
Bryoria capillaris/aqueous			No	Aydın and Türkez (2011b)
Cetraria chlorophylla/aqueous			No	Aydın and Türkez (2011a)
Hypogymnia physodes/methanol			Yes	Ari et al. (2012)
aqueous			No	Türkez et al. (2012c)
Peltigera rufescens/aqueous			No	Aydın and Türkez (2011b)
Physcia aipolia/aqueous			No	Aydın and Türkez (2011a)
Ramalina polymorpha/aqueous			No	Türkez et al. (2012c)
Usnea florida/aqueous			No	
Secondary metabolite				
Phyosodic acid (*Hypogymnia enteromorph*)	Ames	TA100	No	Shibamoto and Wei (1984)
Physodalic acid (*Hypogymnia enteromorph*)			Yes	
Phyosodic acid (from *Pseudevernia furfuracea* (L) Zopf)	8-OH-dG	HPL PRCC U87MG	Yes	Emsen et al. (2016, 2018)
Psoromic acid (from *R. melanophthalma*)		PRCC U87MG		
Olivetoric acid (from *Pseudevernia furfuracea* (L) Zopf)		HPL PRCC U87MG		
Usnic acid	MNPCEs CA, MN	Mouse PCEs HPL	Yes No	Al-Bekairi et al. (1991) Polat et al. (2013)
(+)-Usnic acid (from *Cladonia arbuscula*)	MN	HPL	No	Koparal et al. (2006)
(-)-usnic acid (from *Cladonia stellaris*)				

CA chromosome aberration, *HPL* human peripheral blood lymphocytes, *MN* micronucleus, *PCEs* polychromatic erythrocytes, *PRRC* primary rat cerebral cortex cells, *MNPCEs* micronucleated PCEs, *DN* dioxidine, *DXR* doxorubicin, *8-OH-dG* 8-Oxo-2′-deoxyguanosine, *U87MG* human glioblastoma cells

absence of the metabolic system and in inhibiting frameshift mutations. Result indicates a direct and specific activation of the extracts. However, in a further investigation, the extract of *Cetraria aculeata* (Schreb.) Fr. did not have antigenotoxic activity against mitomycin C (MMC) in term of MN formation in

HPL. The extract was not also genotoxic alone in the mammalian system. As reported very recently by Ceker et al. (2018), non-genotoxicity of methanol extracts of *C. aculeata* has been confirmed by Ames, WP2, and SCE assays. The extract at 5–20 μg/ml also had antigenotoxic activity, interestingly including a mammalian system. Furthermore, methanol extracts of two other species of *Cetraria*, *C. islandica* and *C. olivetorum*, showed activities similar to the activities of *C. aculetea* (Kotan et al. 2011; Aslan et al. 2012b; Ceker et al. 2018). According to the overall results, the extract of *C. aculeata* is significantly antigenotoxic in a bacterial system whereas it is not capable of inhibiting MN formation in MMC-induced human peripheral blood cells, which suggests variation in the effects of different solvent extracts and bioassay systems.

Recently, more investigation has been performed with an aqueous extract of *Dermatocarpon intestiniforme* in cultured HPL (Table 6.2). The extract at 25 and 50 ppm concentration conferred protection against cadmium chloride ($CdCl_2$) (30 ppm)-induced MN formation despite its non-genotoxicity in the cells (Guner et al. 2012). It was also shown that the SCE and MN rates induced by mercury chloride ($HgCl_2$) were alleviated in the cells treated with 50 μg/ml of the extract (Türkez and Dirican 2012). The extract was also antigenotoxic against imazalil (IMA)-induced CA and MN formation in cultured HPL. The lymphocytes were treated in vitro with varying concentrations of the lichen extract (25–100 μg/ml), and tested in combination with IMA (336 μg/ml). The extract alone was not genotoxic, and when combined with IMA treatment, it reduced the frequency of CAs and the rate of MNs (Türkez et al. 2012b). According to overall results of the MN, CA, and SCE assays, the extract of *Dermatocarpon intestiniforme* (5–100 μg/ml) is quite antigenotoxic against different types of clastogens or aneugens, which causes structural and numerical chromosomal damage.

One other aqueous extract of the lichen species *Peltigera rufescens* and *Xanthoria elegans* (25–100 μg/ml) has been assessed by four genotoxicity endpoints including CA, MN, SCE, and 8-oxo-dG assays in HPL (Türkez et al. 2012a, d; Aydin and Türkez 2011b). IMA- and MMC-induced frequencies of four genotoxic indices were diminished by the extract, indicating its inhibitory effect on oxidative DNA damage of reactive agents besides the structural and numerical chromosomal damages. The extract and its secondary metabolites may have potential to decrease the risk of cancer development.

The antigenotoxic effects of methanol extracts of *Rhizoplaca melanophthalma* and *Rhizoplaca chrysoleuca* against known mutagens have been evaluated in two different organisms, a plant and bacteria, using different assays (Agar et al. 2010). The extract (5–40 μg/plate) prevented NaN_3-induced mitotic index partially in *Zea mays* seeds. Furthermore, it was antimutagenic against 9-aminoacridine (9-AA)-induced mutation in the TA1537 strain at all tested concentrations (0.5–5 μg/plate) in the Ames test. The inhibition rates ranged from 70.73% to 85.71%. The extract was also found to have antigenotoxic activity against AFB1-induced genetic damage (Alpsoy et al. 2011).

Several investigators have focused on the possible antigenotoxic potential of lichens against the well-known mutagen aflatoxin B_1 (AFB1). Türkez et al. (2010)

reported the antigenotoxic activity of another lichen species, *Pseudevernia furfuracea*, using its methanol, acetone and *n*-hexane extracts. None of these lichen extracts induced a significant number of SCEs and MN in cytokinesis-blocked HPL. Moreover, the results indicated that AFB1-induced SCEs were inhibited by the application of 50 μM methanol or acetone extracts. The positive effect of methanol, acetone, and ether extracts in decreasing the incidence of MN in comparison with an unprotected level was attained when cultures were treated simultaneously with AFB1 and the extracts. Agar et al. (2011) reported that methanol extracts obtained from *Usnea longissima* suppress the mutagenic effects of AFB1 in HPL examined by the SCE and MN tests. Kotan et al. (2011, 2013) also found that AFB1-induced genotoxicity has been suppressed by the methanol extract of other lichen species, *Cetraria islandica* and *Cladonia rangiformis*. The results showed that the frequencies of SCE and MN level decreased when 5 and 10 mg/ml of extract was added to AFB1-treated cultures. The methanol extracts of *Rhizoplaca chrysoleuca* and *Lecanora muralis*, at 5 and 10 μg/ml (Alpsoy et al. 2011), and *Umbilicaria vellea* and *Xanthoparmelia somloensis* (Aslan et al. 2012a) were antigenotoxic against AFB1-induced SCE and MN formation in HPL in vitro.

The methanol extract of *Evernia prunastri* (Huds.) Willd. was a strong antimutagenic on TA1537 and WP2 strains, with 37.70% and 69.70% inhibition rates against methylnitronitrosoguanidine (MNNG) and acridine-induced mutagenicity, respectively. Co-treatments of HPL with the extract and AFB_1 decreased the frequencies of SCE (Alpsoy et al. 2013).

The genotoxic and antigenotoxic effects of the methanol extract of *Cladonia foliacea* (Huds.) Willd., *Cladonia chlorophaea*, and *Cetraria olivetorum* were studied using WP2, Ames (TA1535 and TA1537), and SCE test systems (Anar et al. 2013; Ceker et al. 2018). According to their results, a 5 μM concentration of AFB1 changed the frequencies of SCE. When 5 and 10 μg/ml concentrations of extract were added to AFB1, the frequencies of SCE were decreased. The extracts of *Cladonia chlorophaea* and *Cetrelia olivetorum* had further antigenotoxic activity against a series of known mutagens, MNNG, NaN3, and 9-AA besides AFB1 (Ceker et al. 2018). On the other hand, the extracts of both species were not mutagenic in tested organisms although each has antimutagenic activity.

Pseudevernia furfuracea and *Cetraria islandica* were tested using their methanol extracts for both genotoxic and antigenotoxic activities. The extracts of these two species were not mutagenic in Ames and *Zea mays* mitotic index test systems. Furthermore, some extracts showed significant antimutagenic activity against 9-AA in the Ames test. Inhibition rates for 9-AA mutagenicity ranged from 25.51% (*Pseudevernia furfuracea*, 0.05 μg/plate) to 66.14% (*Cetraria islandica*, 0.05 μg/plate). In addition, all the extracts were significantly antimutagenic against NaN_3, increasing the MI values of *Zea mays* (Aslan et al. 2012b). Gormez et al. (2013) showed that the methanol extract of *Peltigera canina* posses an antigenotoxic potential in Ames and WP2 tests.

Another seven lichen species were collected from Erzurum and Artvin provinces in Turkey. *Aspicilia calcarea, Bryoria capillaris, Cetraria chlorophylla, Hypnogymnia physodes, Peltigera rufescens, Physcia aipolia, Ramalina polymorpha*, and *Usnea florida* have been tested only for genotoxicity of the water extracts in cultured HPL (Table 6.3) (Aydin and Türkez 2011a, b; Türkez et al. 2012c). All tested lichen extracts up to 500 or 1000 mg/l concentration had no genotoxic effects on the cells by the application of CA and MN assays; however, antioxidant properties were shown. The methanol extract of *Hypogymnia physodes* (L.) Nyl. was also studied for its genotoxicity using CA and MN tests. Relatively higher concentrations are required for its genotoxic activity (Ari et al. 2012).

6.4 Antigenotoxic/Genotoxic Potential of Lichen Secondary Metabolites

Lichen secondary metabolites exert various biological actions such as antitumor, antimicrobial, antiinflammatory, apoptotic, and cytotoxic activities (Ingolfsdottir et al. 1997; Vijayakumar et al. 2000; Huneck 2001; Tay et al. 2004; Yılmaz et al. 2004; Karunaratne et al. 2005; Mayer et al. 2005; Einarsdottir et al. 2010; Mitrovic et al. 2011; Molnar and Farkas 2010). Usnic acid is one of the most abundant lichen secondary metabolites studied for its biological activities. It has been used widely in the pharmaceutical and cosmetic industry because it has high antimicrobial activity (Ingolfsdottir 2002). Furthermore, usnic acid exhibited an antiproliferative effect on human leukemia (K562) and endometrial carcinoma (Ishikawa, HEC-50) cells (Cardarelli et al. 1997; Kristmundsdóttir et al. 2002).

A few findings present the genotoxic/antigenotoxic activities of lichen secondary metabolites. The earliest genotoxicity reports for the secondary metabolites of lichens come from Shibamoto and Wei (1984). They have tested physodic and physodalic acids isolated from *Hypogymnia enteromorpha* (Ach.) Nyl. for their mutagenicity in the Ames assay (Table 6.3). Among them only physodalic acid exhibited significant genotoxicity against *Salmonella typhimurium* strain TA 100 with or without S9 mix in both plate-incorporation and preincubation assays. Later, after several studies, usnic acid was shown to possess genotoxic activity. In a study by Al-Bekairi et al. (1991), mice were treated orally with aqueous suspensions of (+)-usnic acid enantiomers in a single dose of either 100 or 200 mg/kg. A slight increase in the micronucleated polychromatic erythrocytes (MNPCEs) without affecting DNA synthesis was reported and an effect of usnic acid on spindle apparatus was suggested. (+)-Usnic acid and (−)-usnic acid enantiomers isolated from *Ramalina farinacea* and *Cladonia foliacea*, respectively, have been found to be non-genotoxic because of the absence of MN induction in HPL (Koparal et al. 2006). Polat et al. (2013) also showed the nonmutagenicity of usnic acid by CA and MN assays.

The first findings of genotoxic and antigenotoxic potentials of (+)-usnic acid against methylmethanesulfonate (MMS)-induced chromosomal and genome damage in mammalian cells in vitro and in vivo have been shown by Leandro et al. (2013). Usnic acid alone induced DNA damage at concentrations of 60 and 120 g/ml determined by COMET assay. However, it did not induce MN formation in V79 cells at the concentrations tested, and no genotoxic effects were observed in vivo. The combined administration of usnic acid and MMS significantly reduced the frequencies of MN and DNA damage in vitro and in vivo when compared to treatment with MMS alone (Table 6.2). Recently, more results about genotoxic/antigenotoxic activities of usnic acid enantiomers isolated from two different lichen species, *Cladonia arbuscula* and *Cladonia stellaris*, have been reported by Prokopiev et al. (2017), Prokop'ev et al. (2018). Both (+)- and (−)-usnic acids at 0.01–1.00 μM were inhibited by genotoxicity around 37–70%, induced by DN and MMS in HPL cells, as indicated by COMET assay. On the other hand, both enantiomers exhibited genotoxic effects at concentrations of 40–300 μM, but (−)-usnic acid induced 2.5–3.5 times more genotoxicity in COMET than (+)-usnic acid. Usnic acid triggered oxidative stress and disruption of the normal metabolic processes of breast cancer cell line MCF7 and lung cancer cell line H1299 (null for p53); however, it was not involved in DNA damage. It was suggested that the property of usnic acid as a non-genotoxic anticancer agent that works in a p53-independent manner makes it a potential candidate for novel cancer therapy (Mayer et al. 2005).

Increasing amounts information on the genotoxicity and antigenotoxicity of other secondary metabolites of lichens are appearing. Lately, different species of lichen isolated by hypostictic acid, protocetraric acid, salazinic acid, and psoromic acid have been found to act as inhibitors of DXR-induced mutagenicity in the somatic cells of *Drosophila melanogaster* (Guterres et al. 2017). Tested concentrations of the compounds at 6 mM were not genotoxic. However, psoromic acid isolated from a different species, *Rhizoplaca melanophthalma*, and other two compounds as phyosodic acid and olivetoric acid were genotoxic by inducing DNA damage at different concentrations from each other in HPL, PRCC, and U87MG cells (Emsen et al. 2016, 2018). There were notable differences among the 8-OH-dG levels caused by olivetoric acid and phyosodic acid (5–40 mg/l) showing the highest genotoxic effects. Therefore, it was suggested that olivetoric acid and then phyosodic acid may have high potential in the treatment of glioblastoma multiform. On the other hand, phyosodic acid can be used as a natural antioxidant because of its low genotoxic but high antioxidant capacity.

6.5 Conclusion

The methods most frequently used for the assessment of genotoxic and antigenotoxic activities of lichen extracts and products in vitro and in vivo are described here. These methods are not meant to be comprehensive of all existing methods, but more must be in consideration for further investigation of the genotoxicity for their safety assessment or antigenotoxicity, especially of secondary metabolites alone or in combination for their synergetic activities. Positive results of in vitro/in vivo testing indicate that the tested substance is genotoxic or antigenotoxic, and negative results indicate that the test substance is not genotoxic under the conditions of the assay performed. Genotoxicity and antigenotoxicity of lichens have appeared to be evaluated using several type of assays by detecting direct or indirect base substitution and frameshift mutagenicity (Ames and WP2), clastogenicity (chromosome breakage), and aneugenicity (chromosome lagging from dysfunction of mitotic apparatus) (MN), numerical and structural DNA damage (CA), DNA-strand breaks (COMET), oxidative DNA damage (8-oxo-dG), and somatic mutation and recombination (SMART).

Accumulated data from short-term in vitro and in vivo studies have shown that lichen extracts could possess antigenotoxic effects. There are a small number of results for extracts which do not have antigenotoxic effects and some for those having genotoxic effects. The tests generally used for this purpose were common bacterial tests such as Ames and WP2 and human lymphocytes tests such as MN and SCE. However, there are gaps in the lichen genotoxicity/antigenotoxicity data because some groups studied only mutagenicity and others antigenotoxicity without genotoxicity. Most findings are extremely promising in that lichens may have therapeutic potential, at least for cancer, because of their antigenotoxic activities without genotoxic activity. The extracts of 13 species of the 20 lichen species tested, namely, *Cetraria aculeata, Cetraria islandica, Cetralia olivetorum, Cladonia chlorophaea, Cladonia foliacea, Cladonia rangiformis, Dermatocarpon intestiniforme, Parmelia pulla, Peltigera rufescens, Peltigera canina, Pseudevernia furfuracea, Ramalina capitata*, and *Xanthoria elegans*, have antigenotoxic activities but they are not genotoxic (Table 6.2). The extracts of 6 species, that is, *Evernia prunastri, Lecanora muralis, Rhizoplaca chrysoleuca, Usnea longissima, Umbilicaria vellea*, and *Xanthoparmelia somloensis* are antigenotoxic, but not tested for their genotoxic activities. On the other side, the aqueous extracts of *Cetraria aculeatea* and *Rhizoplaca melanophthalma* are neither genotoxic nor antigenotoxic for *E. coli* WP2 and human peripheral blood lymphocytes, respectively. The aqueous extracts of the other eight lichen species tested are not also genotoxic, except for the methanol extract of *Hypogymnia physodes* (Table 6.2).

There is minor evidence about the genotoxic and antigenotoxic activities of the secondary metabolites of lichens. Hypostictic acid, protocetraric acid, psoromic acid, and salazinic acid are antigenotoxic without genotoxic activity in *Drosophila* somatic cells. However, psoromic acid induced oxidative DNA damage in mammalian cells. Although phyosodic acid is nonmutagenic in the Salmonella assay, it induces oxidative DNA damage in mammalian cells.

Interestingly, secondary metabolites of the lichens tested show variation in their effects because each is either genotoxic or antigenotoxic according to the type of assays and laboratories where the work was done. Also, variation in the effective doses of the extract on different cells or test systems suggests the necessity of more in vitro and in vivo antigenotoxicity studies to know the exact potential of the extract. It may then find application for treatment. Further investigation to complete the gap and more data for other lichen species will be very useful for possible therapeutic applications.

The mechanism of antigenotoxic action of all these lichen extracts is not completely known but appears to be the antioxidative potential of their secondary metabolites, as described in Chap. 1. Because most of the extracts have been investigated for their antigenotoxicity and antioxidant activities, quite strong antioxidative activity is also indicated (Aydin and Türkez 2011a, b; Kotan et al.

2011; Polat et al. 2013; Kekuda et al. 2018). The chemoprevention potential of several lichen extracts or secondary metabolites against DNA damage induced by the known compounds such as AFB1, MNNG, MMS, IMA, and CBS strongly indicates that lichens can be a resource of new therapeutics.

References

Agar G, Gulluce M, Aslan A et al (2010) Mutation preventive and antigenotoxic potential of methanol extracts of two natural lichen. J Med Plants Res 4(20):2132–2137

Agar G, Aslan A, Sarioglu EK et al (2011) Protective activity of the methanol extract of *Usnea longissima* against oxidative damage and genotoxicity caused by aflatoxin B(1) *in vitro*. Turk J Med Sci 41(6):1043–1049

Al-Bekairi AM, Qureshi S, Chaudhry MA et al (1991) Mitodepressive, clastogenic and biochemical effects of (+)-usnic acid in mice. J Ethnopharmacol 33(3):217–220

Alpsoy L, Aslan A, Kotan E et al (2011) Protective role of two lichens in human lymphocytes in vitro. Fresenius Environ Bull 20(7):1661–1666

Alpsoy L, Orhan F, Nardemir G et al (2013) Antigenotoxic potencies of a lichen species, *Evernia prunastri*. Toxicol Ind Health 31:153. https://doi.org/10.1177/0748233712469655

Anar M, Orhan F, Alpsoy L et al (2013) The antioxidant and antigenotoxic potential of methanol extract of *Cladonia foliacea* (Huds.) Willd. Toxicol Ind health 32:721–729. https://doi.org/10.1177/0748233713504805

Ari F, Celikler S, Oran S et al (2012) Genotoxic, cytotoxic, and apoptotic effects of *Hypogymnia physodes* (L.) Nyl. On breast cancer cells. Environ Toxicol 29(7):804–813. https://doi.org/10.1002/tox.21809

Aslan A, Agar G, Alpsoy L et al (2012a) Protective role of methanol extracts of two lichens on oxidative and genotoxic damage caused by AFB(1) in human lymphocytes in vitro. Toxicol Ind Health 28(6):505–512

Aslan A, Gulluce M, Agar G et al (2012b) Mutagenic and antimutagenic properties of some lichen species grown in the eastern Anatolia region of Turkey. Cytol Genet 46(5):291–296

Aydin E, Türkez H (2011a) Antioxidant and genotoxicity screening of aqueous extracts of four lichens collected from North East Anatolia. Fresenius Environ Bull 20(8A):2085–2091

Aydin E, Türkez H (2011b) Effects of lichenic extracts (*Bryoria capillaris, Peltigera rufescens* and *Xanthoria elegans*) on human blood cells: a cytogenetic and biochemical study. Fresenius Environ Bull 20(11A):2992–2998

Bakkali F, Averbeck S, Averbeck D et al (2008) Biological effects of essential oils-a review. Food Chem Toxicol 46:446–475

Behera BC, Verma N, Sonone A, Makhija U (2006) Determination of antioxidative potential of lichen *Usnea ghattensis in vitro*. LWT 39:80–85

Cabrera C (1996) Materia medica: *Usnea* spp. Eur J Herbal Med 2:11–13

Cardarelli M, Serino G, Campanella L et al (1997) Antimitotic effects of usnic acid on different biological systems. Cell Mol Life Sci 53(8):667–672

Ceker S, Orhan F, Sezen S et al (2018) Anti-mutagenic and anti-oxidant potencies of *Cetraria aculeata (Schreb.) Fr., Cladonia chlorophaea (Flörke ex Sommerf.) Spreng.* and *Cetrelia olivetorum (Nyl.) W.L. Culb. & C.F. Culb.)* Iran J Pharm Res 17(1):326–335

Clare G (2012) The *in vitro* mammalian chromosome aberration test. Methods Mol Biol 817:69–91

Collins AR (2004) The comet assay for DNA damage and repair: principles, applications, and limitations. Mol Biotechnol 26(3):249–261

Delebassée S, Mambu L, Pinault E et al (2017) Cytochalasin E in the lichen *Pleurosticta acetabulum*. Anti-proliferative activity against human HT-29 colorectal cancer cells and quantitative variability. Fitoterapia 121:146–151

ECVAM (2012, Jan). http://ecvam.jrc.it/consulted

Einarsdottir E, Groeneweg J, Björnsdottir GG et al (2010) Cellular mechanisms of the anticancer effects of the lichen compound usnic acid. Planta Med 76:969–974

Emsen B, Aslan A, Togar B et al (2016) In vitro antitumor activities of the lichen compounds olivetoric, physodic and psoromic acid in rat neuron and glioblastoma cells. Pharm Biol 54 (9):1748–1762

Emsen B, Togar B, Türkez H et al (2018) Effects of two lichen acids isolated from *Pseudevernia furfuracea* (L.) Zopf in cultured human lymphocytes. Z Naturforsch 73(7–8C):303–312

Fenech M (2007) Cytokinesis-block micronucleus cytome assay. Nat Protoc 2:1084–1104

Geyikoglu F, Türkez H, Aslan A (2007) The protective roles of some lichen species on colloidal bismuth subcitrate genotoxicity. Toxicol Ind Health 23(8):487–492

Glück J, Buhrke T, Frenzel F et al (2018) In silico genotoxicity and carcinogenicity prediction for food-relevant secondary plant metabolites. Food Chem Toxicol 116:298–306

Gökbayrak S, Sivas H (2011) Investigation of cytotoxic effects of pyridine in root meristem cells of *Allium cepa*. BioDiCon 4(2):92–98

Gormez A, Karadayi M, Güllüce M et al (2013) Determination of genotoxic and antigenotoxic effects of *Peltigera canica* by the bacterial reverse mutation assays. Curr Opin Biotechnol 24S: S112

Graf U, Würgler FE, Katz AJ et al (1984) Somatic mutation and recombination test in *Drosophila melanogaster*. Environ Mol Mutagen 6:153–188

Guner A, Türkez H, Aslan A (2012) The in vitro effects of *Dermotocarpon intestiniforme* (a lichen) extracts against cadmium induced genetic and oxidative damage. Ekoloji 21(84):38–46

Guterres ZR, Honda NK, Coelho RG et al (2017) Antigenotoxicity of depsidones isolated from Brazilian lichens. Orbital: Electron J Chem 9(1):50–54

Halici M, Odabasoglu F, Suleyman H et al (2005) Effects of water extract of *Usnea longissima* on antioxidant enzyme activity and mucosal damage caused by indomethacin in rats. Phytomedicine 12:656–662

Halliwell B (2000) Why and how should we measure oxidative DNA damage in nutritional studies? How far have we come? Am J Clin Nutr 72(5):1082–1087

Hoshina MM, Marin-Morales MA (2014) Anti-genotoxicity and anti-mutagenicity of *Apis mellifera* venom. Mutat Res Genet Toxicol Environ Mutagen 762:43–48

Huneck S (2001) New results on the chemistry of lichen substances. In: Falk H, Kirby GW, Moore RE (eds) Progress in the chemistry and organic natural products, vol 81. Springer, New York, pp 1–276

Ingolfsdottir K (2002) Molecules of interest usnic asid. Phytochemistry 61:729–736

Ingolfsdottir K, Hjalmarsdottir MA, Guojonsdottir GA et al (1997) *In vitro* susceptibility of *Helicobacter pylori* to protolichesterinic acid from *Cetraria islandica*. Antimicrob Agents Chemother 41:215–217

Ingolfsdottir K, Lee SK, Bhat KPL et al (2000) Evaluation of selected lichens from Iceland for cancer chemopreventive and cytotoxic activity. Pharm Biol 38:313–317

Ipek E, Tuylu BA, Zeytinoglu H (2003) Effects of carvacrol on sister chromatid exchanges in human lymphocyte culture. Cytotechnology 43:145–148

Ipek E, Zeytinoglu H, Okay S et al (2005) Genotoxicity and antigenotoxicity of *Origanum* oil and carvacrol evaluated by Ames Salmonella/microsomal test. Food Chem 93:551–556

Jayaprakasha GK, Negi PS, Jena BS et al (2007) Antioxidant and antimutajenic activities of *Cinnamomum zeylanicum* fruit extracts. J Food Compos Anal 20:330–336

Kahaliw W, Hellman B, Engidawork E (2018) Genotoxicity study of Ethiopian medicinal plant extracts on HepG2 cells. BMC Comp Altern Med 18:45. https://doi.org/10.1186/s12906-017-2056-x

Karunaratne V, Bombuwela K, Kathirgamanathar S et al (2005) Review lichens: a chemically important biota. J Natl Sci Found 33(3):169–186

Kaura V, Kumarb M, Kumara A et al (2018) Pharmacotherapeutic potential of phytochemicals: implications in cancer chemoprevention and future perspectives. Biomed Pharmacother 97:564–586

Kekuda TRP, Vinayaka KS, Sachin MB (2018) Chemistry, ethnobotanical uses and biological activities of the lichen genus Heterodermia Trevis. (Physciaceae; Lecanorales; Ascomycota): a comprehensive review. J Appl Pharm Sci 8(05):148–155

Kirsch-Volders M, Decordier I, Elhajouji A et al (2011) *In vitro* genotoxicity testing using the micronucleus assay in cell lines, human lymphocytes and 3D human skin models. Mutagenesis 26(1):177–184

Koparal AT, Tüylü BA, Türk H (2006) In vitro cytotoxic activities of (+)-usnic acid and (−)-usnic acid on V79, A549, and human lymphocyte cells and their non-genotoxicity on human lymphocytes. Nat Prod Res 20(14):1300–1307

Kotan E, Alpsoy L, Anar M et al (2011) Protective role of methanol extract of *Cetraria islandica* (L.) against oxidative stress and genotoxic effects of AFB(1) in human lymphocytes in vitro. Toxicol Ind Health 27(7):599–605

Kotan E, Agar G, Alpsoy L et al (2013) Anti-genotoxic and anti-oxidative effects of *Cladonia* extracts against aflatoxin B-1 *in vitro*. Fresenius Environ Bull 22(4A):1139–1143

Kristmundsdóttir T, Aradóttir HA, Ingólfsdóttir K et al (2002) Solubilization of the lichen metabolite (+)-usnic acid for testing in tissue culture. J Pharm Pharmacol 54(11):1447–1452

Kuete V, Ngnintedo D, Fotso GW et al (2018) Cytotoxicity of seputhecarpan D, thonningiol and 12 other phytochemicals from African flora towards human carcinoma cells. BMC Complement Altern Med 18:36. https://doi.org/10.1186/s12906-018-2109-9

Kumar SKC, Müller K (1999) Lichen metabolites. 1. Inhibitory action against leukotriene B_4 biosynthesis by a non-redox mechanism. J Nat Prod 62:817–820

Leandro LF, Munari CC, Ferreira LS et al (2013) Assessment of the genotoxicity and antigenotoxicity of (+)-usnic acid in V79 cells and Swiss mice by the micronucleus and comet assays. Mutat Res Genet Toxicol Environ Mut 753(2):101–106

Makhuvele R, Matshoga RG, Antonissen R et al (2018) Genotoxicity and antigenotoxicity of selected South African indigenous plants. S Afr J Bot 114:89–99

Maron DM, Ames BN (1983) Revised methods for the Salmonella mutagenicity test. Mutat Res 113:173–215

Mayer M, O'Neill MA, Murray KE et al (2005) Usnic acid: a non-genotoxic compound with anti-cancer properties. Anti-Cancer Drug 16(8):805–809

Mersch-Sundermann V, Kassie F, Böhmer S et al (2004) Extract of *Toxicodendron quercifolium* caused genotoxicity and antigenotoxicity in bone marrow cells of CD1 mice. Food Chem Toxicol 42:1611–1617

Mitrovic T, Stamenkoviç S, Cvetkovic V et al (2011) Lichens as source of versatile bioactive compounds. Biol Nyssana 2(1):1–6

Molnar K, Farkas E (2010) Current results on biological activities of lichen secondary metabolites: a review. Z Naturforsch C 65:157–173

Mortelmans K, Riccio ES (2000) The bacterial tryptophan reverse mutation assay with *Escherichia coli* WP2. Mutat Res 455(1–2):61–69

Mortelmans K, Zeiger E (2000) The Ames Salmonella/microsome mutagenicity assay. Mutat Res 455(1–2):29–60

Nieminen SM, Maki-paakkanen J, Hirvonen MR et al (2002) Genotoxicity of gliotoxin, a secondary metabolite of *Aspergillus fumigatus*, in a battery of short-term test systems. Mutat Res 520:161–170

OECD (1997) Test No. 471: bacterial reverse mutation test, OECD guidelines for the testing of chemicals, section 4. OECD Publishing, Paris. https://doi.org/10.1787/9789264071247-en

OECD (2012) Consulted, 4, Jan 2012. https://www.oecd-ilibrary.org/environment/oecd-guidelines-for-the-testing-of-chemicals-section-4-health-effects_20745788

Okai Y, Higashi-Okai K, Nakamura S et al (1996) Suppressive effects of retinoids, carotenoids and antioxidant vitamins on heterocyclic amine-induced Umu C gene expression in *Salmonella typhimurium* (TA 1535/Psk 1002). Mutat Res 368:133–140

Perry P, Evans HJ (1975) Cytological detection of mutagen-carcinogen exposure by sister chromatid exchange. Nature 258(5531):121–125

Polat Z, Aydın E, Türkez H et al (2013) In vitro risk assessment of usnic acid compound. Toxicol Ind Health 32:1–8. https://doi.org/10.1177/0748233713504811

Prashith Kekuda TR, Vinayaka KS, Sachin MB (2018) Chemistry, ethnobotanical uses and biological activities of the lichen genus *Heterodermia* Trevis. (Physciaceae; Lecanorales; Ascomycota): a comprehensive review. J Appl Pharm Sci 8(05):148–155

Prinsloo G, Nogemane N, Street R (2018) The use of plants containing genotoxic carcinogens as foods and medicine. Food Chem Toxicol 116(part B):27–39

Prokop'ev IA, Filippov EV, Filippova GV et al (2018) Pro/antigenotoxic activity of usnic acid enantiomers in vitro. Bull Exp Biol Med Pharm Toxicol 164(9):289–293

Prokopiev A, Filippov EV, Filippova GV et al (2017) Genotoxicity of usnic-acid enantiomers in vitro in human peripheral-blood lymphocytes. Cell Tissue Biol 11(2):141–146

Quadros APO, Mazzeo DEC, Marin-Morales MA et al (2017) Fruit extract of the medicinal plant *Crataegus oxyacantha* exerts genotoxic and mutagenic effects in cultured cells. J Toxicol Environ Health Part A 80(3):161–170

Richardson DHS (1988) Medicinal and other economic aspects of lichens. In: Galun M (ed) CRC handbook of lichenology, vol 3. CRC Press, Boca Raton, pp 93–108

Santos DB, Schiar VP, Ribeiro MC et al (2009) Genotoxicity of organo selenium compounds in human leukocytes *in vitro*. Mutat Res 676:21–26

Scarpato R, Bertoli A, Naccarati A et al (1998) Different effects of newly isolated saponins on the mutagenicity and cytotoxicity of the anticancer drugs mitomycin C and bleomycin in human lymphocytes. Mutat Res 420:49–54

Shibamoto T, Wei CI (1984) Mutagenicity of lichen constituents. Environ Mutagen 6(5):757–762

Singh NP, McCoy MT, Tice RR et al (1988) A simple technique for quantitation of low levels of DNA damage in individual cells. Exp Cell Res 175:184–191

Speit G, Vasquez M, Hartmann A (2009) Thc comet assay as an indicator test for germ cell genotoxicity. Mutat Res 681:3–12

Sponchiado G, Adam ML, Silva CD et al (2016) Quantitative genotoxicity assays for analysis of medicinal plants: a systematic review. J Ethnopharmacol 178(3):289–296

Sugiyama K, Yamada M, Awogi T et al (2016) The strains recommended for use in the bacterial reverse mutation test (OECD guideline 471) can be certified as non-genetically modified organisms. Genes Environ 38:2

Tay T, Türk AÖ, Yılmaz M et al (2004) Evaluation of the antimicrobial activity of the acetone extract of the lichen *Ramalina farinacea* and its (+)-usnic acid, norstictic acid and protocetraric acid constituents. Z Naturforsch 59C:384–388

Teixeira Zafred RR, Spano MA, Martins GR et al (2016) Pro-oxidant activity and genotoxicity of the *Astronium fraxinifolium* using wing SMART and *Allium cepa* test. Res J Med Plants 10:276–285

Tilford GL (1997) Edible and medicinal plants of the west. Mountain Press, Missoula

Tomović J, Kosanić M, Ristić S et al (2017) Chemical composition and bioactive properties of the lichen, *Pleurosticta acetabulum*. Trop J Pharm Res 16(12):2977–2984

Toyokuni S, Tanaka T, Hattori Y et al (1997) Quantitative immunohistochemical determination of 8-hydroxy-2′-deoxyguanosine by a monoclonal antibody N45.1: its application to ferric nitrilotriacetate-induced renal carcinogenesis model. Lab Investig 76:365–374

Türk AÖ, Yılmaz M, Kıvanç M et al (2003) The antimicrobial activity of extracts of the lichen *Cetraria aculeata* and its protolichesterinic acid constituent. Z Naturforsch 58C:850–854

Türkez H, Dirican E (2012) A modulator against mercury chloride-induced genotoxic damage: *Dermatocarpon intestiniforme* (L.). Toxicol Ind Health 28(1):58–63

Türkez H, Geyikoglu F, Aslan A et al (2010) Antimutagenic effects of lichen *Pseudovernia furfuracea* (L.) Zoph. extracts against the mutagenicity of aflatoxin B-1 in vitro. Toxicol Ind Health 26(9):625–631

Türkez H, Aydin E, Aslan A (2012a) *Xanthoria elegans* (Link) (lichen) extract counteracts DNA damage and oxidative stress of mitomycin C in human lymphocytes. Cytotechnology 64 (6):679–686

Türkez H, Aydin E, Aslan A (2012b) An antidote for imazalil-induced genotoxicity in vitro: the lichen *Dermatocarpon intestiniforme* (Korber) Hasse. Acta Biol Hung 63(3):354–361

Türkez H, Aydin E, Aslan A (2012c) Effects of lichenic extracts (*Hypogymnia physodes*, *Ramalina polymorpha* and *Usnea florida*) on human blood cells: cytogenetic and biochemical study. Iran J Pharm Res 11(3):889–896

Türkez H, Aydin E, Sisman T et al (2012d) Role of *Peltigera rufescens* (Weis) Humb. (a lichen) on imazalil-induced genotoxicity analysis of micronucleus and chromosome aberrations in vitro. Toxicol Ind Health 28(6):492–498

Verschaeve L, Juutilainen J, Lagroye I et al (2010) *In vitro* and *in vivo* genotoxicity of radiofrequency fields. Mutat Res 705:252–268

Vijayakumar CS, Viswanathan S, Reddy MK et al (2000) Anti-inflammatory activity of (+)-usnic acid. Fitoterapia 71:564–566

Wilson DM III, Thompson LH (2007) Molecular mechanisms of sister-chromatid exchange. Mutat Res-Fund Mol M 616(1–2):11–23

Yeash EA, Letwin L, Malek L et al (2017) Biological activities of undescribed north American lichen species. J Sci Food Agric 97:4721–4726

Yılmaz M, Türk AÖ, Tay T et al (2004) The antimicrobial activity of extracts of the lichen *Cladonia foliacea* and its (−)-usnic acid, atranorin, and fumarprotocetraric acid constituents. Z Naturforsch 59C:249–254

Zeytinoglu H, Incesu Z, Tuylu Ayaz B et al (2008) Determination of genotoxic, antigenotoxic and cytotoxic potential of the extract from lichen *Cetraria aculeata* (Schreb.) Fr. *in vitro*. Phytother Res 22(1):118–123

Lichen Secondary Metabolites as Possible Antiviral Agents

7

Damian C. Odimegwu, Kenneth Ngwoke, Chika Ejikeugwu, and Charles O. Esimone

Abstract

The revised edition of this chapter has been carefully and entirely rewritten. As a general rule, all sections have been revised and updated with current knowledge, more recent references included, and some sections expanded. Of particular note is the incorporation of significant additional materials in both sections dealing with some biological activities of lichen-derived secondary metabolites with particular emphasis on antitumor and anti-cellular agents and antiviral activities of lichen secondary metabolites. Every other section has been thoroughly revised and enriched with more current references.

Section 7.1 deals with broad and specific definitions of lichens and antiviral agents. Here, we will focus briefly on viruses and antiviral agents: their nature, source, general characteristics, and effects. Attempt is made to briefly classify antiviral agents based on their mode of action and virus targets. Lichen as a broad group of useful source of phytochemical agents is briefly described with emphasis on their unique niche in natural product research.

Section 7.2 deals with issues related to secondary metabolites (phytochemicals) from lichen. Lichen compounds because they possess some biological effects will form also the focus of this section. Some classes of lichen compounds are presented as well as their molecular structures. How they influence biological host and agents would be described briefly in the latter part of this section with presentation of some that display antiviral activities. The exact sources of these compounds are fully elucidated. Attempt is also made to mention lichen-derived compounds yet-to-be-fully-validated antiviral property.

D. C. Odimegwu
Faculty of Pharmaceutical Sciences, University of Nigeria, Nsukka, Enugu State, Nigeria

K. Ngwoke · C. Ejikeugwu · C. O. Esimone (✉)
Faculty of Pharmaceutical Sciences, Nnamdi Azikiwe University, Awka, Anambra State, Nigeria
e-mail: co.esimone@unizik.edu.ng

B. Ranković (ed.), *Lichen Secondary Metabolites*,
https://doi.org/10.1007/978-3-030-16814-8_7

Section 7.3 will focus on conclusion beginning with a summary of narrated up-to-date available data of antiviral lichen compounds currently undergoing preliminary and extensive research and development. How these agents have been applied or would be applied to biomedical and pharmaceutical utility would be addressed. The pharmaceutical industry involvement in further product development would be mentioned to display existing link between basic research setting and industry development endpoint. Finally, statement is made speculating on the future of lichen research and the positive expected outcomes of improved research interest in this unique group of medicinal plants.

7.1 Antiviral Agents and Lichens

7.1.1 Antiviral Agents

Antiviral agents refer to chemicals, moiety, natural products, and medical and therapeutic substances employed for the prevention and treatment of virus infections and diseases. Therefore, viruses which cause a number of infections and diseases in plants, animals, and humans are targeted by these agents. Viruses are very tiny infectious agents that cannot carry out metabolism or replicate exclusively on their own. Following infection of a host (e.g., a cell), it can direct the cell machinery to produce viral proteins and eventually more viruses. Viruses have either RNA or DNA as their genetic material. This genetic material or nucleic acid may be either single or double stranded (designated ss or ds). The whole intact virus (virion) consists of the nucleic acid and an outer shell of protein. Antiviral agents could target the different proteins (substructure) of the respective virus (virion), its nucleic acid, and other virus or host cell factors to terminate or control the putative viral infection/disease. Viruses in contrast to bacteria and fungi are notoriously intracellular parasites living inside the cell of their host and harness the metabolic and replicative ability of their host cell to drive their own multiplication and proliferation within the host, thereby making intervention against them difficult. Consequently, antiviral agents have been shown to act by one or more of the following processes:

1. Inactivating extracellular virus particles
2. Preventing viral attachment and/or entry
3. Preventing functioning or replication of viral genome
4. Preventing synthesis or functioning of specific viral protein
5. Preventing assembly or release of new infectious virions

Extensive screening efforts have been made to find antiviral agents from synthetic and natural sources. Natural products from plants traditionally have provided the pharmaceutical industry with one of its most important sources of lead compounds, and over 50% of modern drugs are derived from natural sources (Jassim and Naji

2003; Gautam et al. 2007; Harvey 2008; Newman and Cragg 2012; Atanasov et al. 2015). The bioactive molecules occur in plants as secondary metabolites and as defense mechanisms against predation, herbivores, fungal attack, microbial invasion, and viral infection. During the past three decades, potent agents have become available against viral infections, and extensive studies have shown that medicinal plants of several parts of the world contain compounds active against viruses that cause human diseases (Semple et al. 1998; Kott et al. 1999; Sindambiwe et al. 1999; Akram et al. 2018). Not to be left out of these are lichens which are also known as promising sources of diverse secondary metabolites, the majority of which exhibit various biological properties including extensive antiviral properties (Pengsuparp et al. 1995; Cohen et al. 1996; Neamati et al. 1997; Esimone et al. 2007; Stocker-Worgotter 2008; Molnar and Farkas 2010; Lai et al. 2013; Basile et al. 2015; Paluszczak et al. 2018).

7.1.2 Lichens

A lichen is a stable, ecologically obligate, self-supporting mutualism between an exhabitant fungus and one or more inhabitant, extracellularly located unicellular or filamentous photoautotrophic partners (Farrar 1976; Hawksworth and Honegger 1994). Thus, lichens are symbiotic organisms of fungi (mycobiont) and algae or cyanobacteria (photobiont) (Voss et al. 1983). Lichen thalli are complex ecosystems rather than organisms (Farrar 1976; Lumbsch 1998). Lichens form the largest mutualistic group among fungi especially given that about 21% of all fungi are able to act as a mycobiont (Honegger 1991). It is estimated that about 40 fungi genera are involved as photobiont in lichen formation (Kirk et al. 2008). Lichens can be found in a wide range of habitats ranging from aquatic to terrestrial terrain (Brightman and Seaward 1977; Seaward 2008; Lisicka 2008; Nash 2008). They are well distributed all over the world and are able to survive in extreme environmental conditions (Nash 2008).

Lichens are known to synthesize a great variety of secondary metabolites, many of which are unique.

7.2 Lichen Secondary Metabolites

Lichens are known to produce tremendous amounts of chemically diverse group of relatively low molecular weight secondary metabolites, ranging from aliphatic compounds like caperatic acid to aromatic like lobaric acid (Turk et al. 2003). Many examples of lichen secondary metabolites (LSM) abound in literature; a few are shown in Fig. 7.1. For instance, ramalin (γ-glutamyl-N′-(2-hydroxyphenyl) hydrazide) is a secondary metabolite isolated from *Ramalina terebrata* (Suh et al. 2017; Paudel et al. 2011). Other examples include norstictic acid, a depsidone derivative which has been isolated from different lichen genera including *Usnea* and *Ramalina* (Ebrahim et al. 2016). Atranorin is another LSM, a derivative of

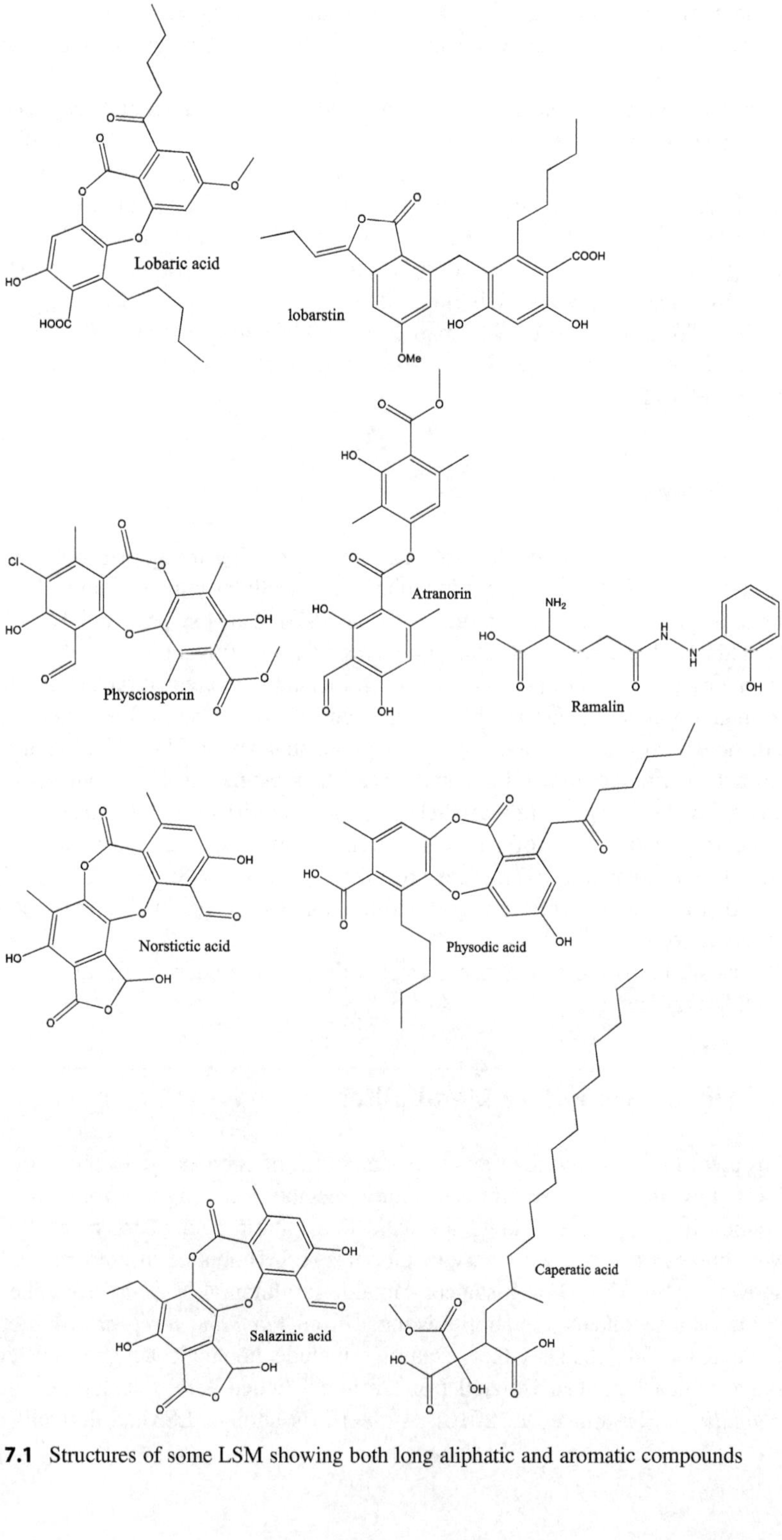

Fig. 7.1 Structures of some LSM showing both long aliphatic and aromatic compounds

β-orcinol which is widely distributed in many lichen families (Studzinska-Sroka et al. 2017). Physodic acid from *Hypogymnia physodes*, squamatic acid from *Cladonia uncialis*, and salazinic acid and caperatic acid from *Platismatia glauca* (Paluszczak et al. 2018) are all examples of LSM. Furthermore, physciosporin from *Pseudocyphellaria coriacea* (Yang et al. 2015), lobaric acid and lobarstin from *Stereocaulon alpnum* (Hong et al. 2018), depsides from *Stereocaulon evolutum* (Vu et al. 2015), and usimines A–C, usnic acid derivatives, and stereocalpin A (Seo et al. 2008a, b) from *Stereocaulon alpnum* are other examples of secondary metabolites isolated from lichens.

Lichen has three major pathways that yield secondary metabolites. They include the shikimic pathway, the mevalonic acid pathway, and acetate-polymalonate pathway. The shikimic pathway yields pulvinic acid derivatives, while the mevalonic pathway yields the terpenoids and sterols. The acetate-polymevalonate yields the depside, depsidones, depsipeptides, depsone, etc. Most of the compounds mentioned in this subsection are products of acetate-polymalonate pathway.

7.2.1 Some Biological Activities of Lichen-Derived Secondary Metabolites

LSM have been used in lichen chemotaxonomy (Xu et al. 2016) and systematics (Molnar and Farkas 2010). They have also been studied for bioactivity. Lichen secondary metabolites have been reported to have several bioactive properties that have been demonstrated in the laboratories, including antitumor, antibacterial, anti-herbivore, antioxidant, and antiviral actions. Usnic acid which is wildly available from the *Usnea* genera has been reported to decrease proliferation of human breast cancer cells and human lung cancer cells without any DNA damage (Mayer et al. 2005). Many of the secondary metabolites from the acetate-polymalonate pathway have been known for different antitumor properties. For instance, ramalin from *Ramalina terebrata* has strong multiple point inhibition of colorectal cancer replication, migration, invasion, and proliferation (Suh et al. 2017). Norstictic acid inhibited the proliferation, migration, and invasion of breast cancer cells in vitro as well as in vivo invasion growth (Ebrahim et al. 2016). Caperatic acid and physodic acid have also been reported to potently inhibit Wnt signaling in colorectal cancer (Paluszczak et al. 2018). Wnt signaling is a major aberration associated with colorectal cancer cell survival, proliferation, and migration. Kwon et al. (2016) recently reported the inhibition of VCAM-1 expression in TNF-α-stimulated vascular smooth muscle cells via modulation of NF-kB and MAPK signaling pathways by lobaric acid. VCAM-1 expression is important in the development of atherosclerosis making lobaric a potential agent in the treatment inflammatory diseases. Physciosporin, a lichen secondary metabolite, has been reported to inhibit cell motility in lung cancer models A549, H1650, and H1975 (Yang et al. 2015). Atranorin also inhibits cell motility and tumorigenesis in cancer cell model (Zhou et al. 2017; Tian et al. 2017). The depside sphaerophorin isolated from *Sphaerophorus globosus* and the depsidone pannarin isolated from *Psoroma pholidotoides*, *P. pulchrum*, and *P. pallidum* also displayed anticancer activity against the M14 human melanoma cells (Russo et al. 2008).

It was reported by Bezivin et al. (2004) that (−)-usnic acid, the enantiomer of usnic acid isolated from *Cladonia convoluta*, showed obvious cytotoxic activities against cancer cell lines, such as murine Lewis lung carcinoma, human chronic myelogenous leukemia, human brain metastasis of a prostate carcinoma, human breast adenocarcinoma, and human glioblastoma, while its enantiomer (+)-usnic acid was reported to inhibit non-small cancer cell motility (Yang et al. 2015). Strong anti-larvicidal activity against the third and fourth instar larvae of the house mosquito (*Culex pipiens*) and against the polyphagous larvae of the herbivorous insect *Spodoptera littoralis* were observed with the two enantiomers of usnic acid and vipiric acids (Cetin et al. 2008; Emmerich et al. 1993).

Lichen have also been found to contain a variety of secondary lichen substances with strong antioxidant activity scavenging toxic free radicals (Jha et al. 2017; Thadhani and Karunaratne 2017; Kumar et al. 2014). Vulpinic acid, usnic acid, and pinastric acids from lichen species were reported to have UV blocking and thus photo-protecting properties (Legouin et al. 2017). The common existence of phenolic groups among lichens could account for this. The depsides atranorin isolated from *Placopsis* sp. and divaricatic acid isolated from *Protousnea malacea* and the depsidones pannarin isolated from *Psoroma pallidum* and 10-chloropannarin isolated from *Erioderma chilense* displayed some antioxidant properties (Hidalgo et al. 1994). Similarly, both sphaerophorin and pannarin (a depsidone) inhibited superoxide anion formation in vitro (Russo et al. 2008).

Lichen is also known to provide sources of metabolites with notable antimicrobial properties. Good antimicrobial properties against a wide range of microbes by parietin isolated from *Xanthoria parietina* have been reported (Basile et al. 2015). Also, isodivaricatic acid, divaricatinic acid, and usnic acid isolated from *Protousnea poeppigii* were also reported to have displayed antifungal action against *Microsporum gypseum*, *Trichophyton mentagrophytes*, and *T. rubrum* (Schmeda-Hirschmann et al. 2008). Moderate to strong antifungal action against several fungal species was displayed by extracts from *Alectoria sarmentosa* and *Cladonia rangiferina* (Ranković and Mišić 2007) and *Evernia prunastri* and *Hypogymnia physodes* (Halama and Van Haluwin 2004).

Additionally, it was reported (Ranković and Mišić 2008; Ranković et al. 2008) that physodic acid from *Hypogymnia physodes*, stictic acid from *Xanthoparmelia conspersa*, protocetraric acid from *Flavoparmelia caperata*, atranorin sourced from *Physcia aipolia*, fumarprotocetraric acid isolated from *Cladonia furcata*, 140 gyrophoric acid from *Umbilicaria polyphylla*, lecanoric acid from *Ochrolechia androgyna*, and usnic acid from *Flavoparmelia caperata* all showed relatively strong antimicrobial effects against some bacteria and fungi pathogens.

7.2.2 Antiviral Activities of Lichen Secondary Metabolites

Lichens have proven a useful source of antiviral agents. Investigation of the antiviral properties of atranorin showed that it exhibited strong inhibitory properties against hepatitis C virus (Vu et al. 2015). Usnic acid enantiomers and derivatives showed significant antiviral properties against influenza virus in in vitro studies (Sokolov et al. 2012). Established antiviral activities have been attributed to some other lichen secondary metabolites including parietin, anthraquinones, hypericin, lichenan, etc. For instance, lichenan, which is widely distributed in lichens, is demonstrated to inhibit the tobacco mosaic virus (Stubler and Buchenauer 1996). Similarly, some workers (Cohen and Towers 1995a, b; Cohen et al. 1996) reported the antiviral activity of some naturally occurring anthraquinones, bianthrones, and hypericin derivatives isolated from the lichens *Nephroma laevigatum* and *Heterodermia obscurata*. They particularly reported their activities against the virus herpes simplex virus type 1 (HSV-1) where 6 out of 13 compounds at not more than 2 μg/ml could completely inactivate HSV-1 in an endpoint cytopathic effect (CPE) assay. The compounds include emodin, 7-chloroemodin, and 7-chloro-1-O-methylemodin, and 5,7-dichloroemodin being the most active anthraquinone (0.25 μg/ml). The hypericin derivatives 7,70 dichlorohypericin and hypericin equally showed comparable inhibitory activity against HSV-1 with complete inactivation at less than 0.06 μg/ml. Thus, the potential of these compounds as possible tools for further development against HSV-1 is not in question. Moreover, these metabolites if tested against a wider array of viruses may be quite effective as is reflected in the study by Andersen et al. (1991) who showed that hypericin and emodin isolated from *Hypericum perforatum* were inhibitory against vesicular stomatitis virus (VSV), HSV-2, parainfluenza virus, and vaccinia virus. It is therefore potentially useful to promote a more broad-based extensive screening of potential antiviral metabolites against a wide array of viruses.

In a screening exercise involving a total of 69 lichen species from New Zealand by Perry et al. (1999), they identified the three species *Cladia retipora*, *Pseudocyphellaria glabra*, and *P. homoeophylla* with antiviral activity. These antiviral activities were later traced to the usnic acid constituent following a bioactivity-guided isolation process. Viruses strongly inhibited include the herpes simplex virus type 1 (HSV-1) and the polio type 1 virus. It is also well known that usnic acid has multifaceted biological activities. So its antiviral activity equally does not come as a surprise.

In another related study, investigation of some Icelandic lichens resulted in the identification of two antiviral compounds active against three viruses: respiratory syncytial virus RSV, HSV-1, and HSV-2. Activity against HSV 1 was less potent than activity against RSV. The compounds were the depsidone salazinic acid (IC_{50} 11.9 μg/ml) from *Parmelia saxatilis* (L.) Ach. and the benzyl depside alectorialic acid (IC_{50} 17.0 μg/ml) from *Alectoria nigricans* (Ach.) (Omarsdottir et al. 2006). In the year following, Fazio et al. (2007) reported their findings where they described the antiviral activity of parietin isolated from *Teloschistes chrysophthalmus* against some Junin and Tacaribe arenaviruses.

Extracts of the lichen species *Xanthoria parietina* and *Xanthoparmelia tinctina* were evaluated and shown to be active against human parainfluenza virus type 2 (HPIV-2) (Karagöz and Aslan 2005). Other investigations have reported antiviral activities of lichen extracts against the human cytomegalovirus (Wood et al. 1990), HIV (Hirabayashi et al. 1989), and Epstein-Barr virus (Yamamoto et al. 1995).

The search for lichen metabolites with antiviral properties continued with some investigations from African lichens. In some studies involving lichen species from West Africa, Esimone et al. (2007) tested some extracts of the lichen *Parmelia perlata* against some RNA viruses: yellow fever, infectious bursal disease virus, and polio virus. While the water and acetone extracts showed no activity against the viruses when tested at concentrations below the cytotoxic level, the crude polysaccharide fraction on the other hand showed activity against yellow fever virus with an IC_{50} of 15 μg/ml. Following the outcome of further investigation for mechanism of action profile, they proposed that the target of the inhibitory activity is possibly mediated against the envelope of the yellow fever virus. Furthermore, Esimone and coworkers screened derivatives from the lichen *Ramalina farinacea* against lentiviruses and adenoviruses. Their reported work (Esimone et al. 2005) initially showed that the ethyl acetate-soluble fraction (ET4) from *R. farinacea* inhibited the infectivity of lentiviral and adenoviral vectors, as well as HIV-1. Recorded antiviral activity was about 20 μg/ml. Preliminary mechanistic studies based on the addition of the extracts at different time points in the viral infection cycle (kinetic studies) led to the suggestion that early steps in the lentiviral or adenoviral replication cycle could be the major target of ET4.

Later, in another related study (Esimone et al. 2009), they determined the antiviral activity of ET4 against other wild-type viruses, including the herpes simplex virus type1 (HSV-1) and the respiratory syncytial virus (RSV). They found ET4 to inhibit HSV-1 and RSV potently with IC_{50} of 6.09 and 3.65 μg/ml, respectively. Attempt to elucidate the possible mechanism of action of the ET4 revealed through the time-of-addition studies suggests that both entry and post entry steps of the HIV-1 replication cycle and the entry step of the RSV replication cycle are targeted, respectively. Thus, ET4 is expected to inhibit HIV-1 at some point around the viral envelope, while RSV is inhibited downstream beyond the viral envelope. Additionally, they reported that ET4 inhibited HIV-1 reverse transcriptase with an IC_{50} of 0.022 μg/ml. Further activity-guided fractionation of ET4 yielded two subfractions rfO and rfM with varied activities. While rfM demonstrated activity against HSV-1 (DNA virus) but not against the lentiviral vector, rfO was active against lentiviral vector and HIV-1 (RNA viruses) but not against HSV-1.

In a follow-up continuity study by Lai et al. (2013), further investigation involving the ethyl acetate-soluble fraction (ET4) of *R. farinacea* led to the isolation of 13 phenolic compounds including 1 new hydroquinone depside (5-hydroxysekikaic acid) and 1 new orsellinic acid derivative (2,3-dihydroxy-4-methoxy-6-pentylbenzoic acid). All 13 compounds displayed varying degrees of antiviral activity against the respiratory syncytial virus (RSV) with one of them sekikaic acid showing the most potent inhibition toward a recombinant strain of the respiratory syncytial virus rgRSV (IC_{50} 5.69 μg/ml) and the clinical RSV A2 strain (IC_{50}

7.73 μg/ml). The time-of-addition assay for mechanistic elucidation revealed that sekikaic acid clearly interferes with viral replication at a viral post-entry step.

7.2.3 Experimental Methods in the Discovery of Antiviral Lichen Compounds

Viral infection could be evaluated by direct cytological and histological examination under the light microscope for signs of viral-induced cytopathic effect (CPE) (Esimone et al. 2008; Lai et al. 2013), through antigen detection by immunological and serological assays (Odimegwu 2018), immunohistochemistry (Lai et al. 2013), immunofluorescence (Sloan et al. 2015), viral proteomics (Tavalai et al. 2008; Schreiner et al. 2010), and viral genome detection by hybridization with specific nucleic acid probe polymerase chain reaction (PCR) (Ternette et al. 2007). They could also be further indirectly detected by hemabsorption (Drake et al. 2016), hemagglutination (Motohashi et al. 2013; Li et al. 2015), and, in the case of egg viral cultivation pocks on chorioallantoic membrane (CAM) (Wang et al. 2011), inclusion bodies (Sloan et al. 2015). Enzyme-linked immunosorbent assay (ELISA) (Ternette et al. 2007), complement fixation tests (CFT) (Ronalds et al. 1983), protein immunoblot (Western blot) (Kohlmann et al. 2009), and other serological assays are equally useful toward viral detection. Moreover, disease or death could also confirm viral infection especially in animal models. Additionally, there are various tools and vehicles for antiviral screening.

Cell Lines and Medium Antiviral screening is largely done by in vitro methods requiring the use of cell lines of mammalian origin. Therefore, cell lines utilized are greatly widespread and commonly include Vero (a monkey kidney-derived primary cell line), Hep2 (of human airway epithelial lineage), Hela (human cervical carcinoma cell line), human tsA201 cell (a derivative of human embryonic kidney cell line 293), 293A cell line (a permanent line of primary human embryonal kidney transformed by sheared human adenovirus type 5 (Ad 5) DNA), and TZM-bl cell (a transformed cell line of cervical cancer origin expressing the classical HIV-1 entry receptors). Madin-Darby canine kidney (MDCK) cells and several other cell lines may be utilized in various antiviral screenings. These cells may for instance usually be grown in Dulbecco's Modified Eagle Medium (DMEM) consisting of high glucose and 2 mM L-glutamine and supplemented with heat-inactivated fetal calf serum and a mixture of penicillin and streptomycin (Pen-Strep) (Ternette et al. 2007; Esimone et al. 2008; Odimegwu et al. 2018). There are other media which are enriched to accommodate additional requirements of certain specialized cells utilized in the antiviral screening.

Viral Particles and Vectors Viruses are usually utilized in these screenings since the effects of antiviral compounds of lichen origin must be assayed in the presence of the target virus. In this case, the virus could be utilized live and unaltered (e.g., RSV), or the virus is slightly genetically modified without necessarily altering its infectivity while imparting maybe some sort of regulatory or reporter protein property (an example is the rgRSV expressing the green fluorescent protein) (Ternette et al. 2007; Lai et al. 2013). In some cases, viruses can be wholly redesigned to produce a viral particle with clearly transformed property referred to as viral vectors. Viral vectors are designed to act as the virus and to effectively deliver the viral genome into susceptible host cell. They can be genetically engineered to either delete unwanted genes/antigen/structures or to introduce new foreign genes/antigen/structures from another virus to create pseudo-typed viruses (vectors) (Esimone et al. 2005, 2009). They find great utility in the evaluation of antiviral lichen compounds.

Antiviral Assays A variety of assay types exist to enable the effective evaluation of antiviral agents. The preliminary stage of screening of novel antiviral compound from lichen may begin with an initial simple screening of crude or broad-based fraction of the lichen containing the supposed antiviral substance. Screening is completed through utilization of various bioassays presented sometimes in the format of a bioassay-guided fractionation process.

Bioassay-Guided Fractionation Bioassay-guided isolation involves the use of some well-defined protocol for the isolation of pure compounds with defined biological activity from a crude extract of a natural product which could be plant or animal parts or a cocktail of mixture (Esimone et al. 2008). It involves step-by-step separation of fractions/constituents contained within the crude extracts or mixture based on their physicochemical activities and through assaying for biological activity. Several rounds of separation of fractions/constituents and bioassay may be required to finally identify pure molecular hit compound(s).

Antiviral Cytopathic and Plaque Reduction Assays Screening of novel antiviral constituent requires the presence of the screened substance in flask or system containing appropriate mammalian cells either adhered or in suspension within cell culture medium including the investigated lichen antiviral substance (Cohen et al. 1996; Kott et al. 1999; Esimone et al. 2009; Lai et al. 2013). Antibiotics are normally included in these media to prevent contamination. Normally, routine screening could take between 24 h and several days following which plaques representing regions of virus activity/replication are counted and reported relative to control setup excluding the antiviral substance. Viral-induced cytopathy or plaque formation may be observed directly under the light microscope or may be done following specific/immune staining of responsible viral protein present in the infected cell (Kott et al. 1999; Esimone et al. 2008). Where a fluorescent protein has been introduced into virus or vector, then observation could also be analyzed using fluorescent microscopy (Lai et al. 2013).

Additional antiviral assays to further define certain specific properties of the identified compound could still be analyzed employing the viral plaque reduction assay, for instance, assays investigating the mechanism of antiviral activity in a broader sense (e.g., time point of optimal antiviral effect during viral infectivity cycle), interference of compound with some measurable viral or host cell enzymes, and protein systems.

Viral Cytopathic Effect Measurement As with viral plaque reduction assays above, a generalized viral cytopathic effect reduction by investigated antiviral lichen compound could be carried out to characterize the antiviral activity of the substance. The experimental setup remains essentially the same as above, but measurement of antiviral activity may be done by analyzing for the presence of cell pathology-indicating enzymes in cell culture supernatant or the measurement for any reduction in substance-treated cell oxidative activity for sign of cell pathology (Semple et al. 1998; Odimegwu 2018).

Further Antiviral Assays to Discover Antiviral Lichen Compounds Additional antiviral assays to further define certain specific properties of the identified compound could still be analyzed employing the viral plaque reduction assay, for instance, assays investigating the mechanism of antiviral activity in a broader sense (e.g., time point of optimal antiviral effect during viral infectivity cycle) (Stubler and Buchenauer 1996; Lai et al. 2013) and interference of compound with some measurable viral or host cell enzymes and protein systems (Odimegwu 2018). In existence are other virological and immune-based assays that have been reportedly employed for the antiviral screening of lichen-derived compounds, which include virucidal effect (Andersen et al. 1991; Lai et al. 2013), protein immunoblot assay/Western blot analysis (Neamati et al. 1997), enzyme-linked immunosorbent assay (ELISA) (Semple et al. 1998), fluorescence-activated cell sorting assays (Odimegwu 2018), DNA and RNA polymerase inhibition assays (Pengsuparp et al. 1995), and some other antiviral assays. Clearly, antiviral assays could always be developed to fill the gap for any specific analysis of the candidate lichen-derived constituents/compounds.

7.3 Conclusion

Secondary metabolites of lichens with virus-inhibiting effects as explained so far (summarized in Table 7.1) constitute potent lead compounds for the development of novel antiviral agents. Thus, secondary metabolites of lichens represent a promising source of novel bioactive lead compounds for the medical and pharmaceutical industry in particular. Though many lichen compounds have been well characterized and proved to be efficacious against certain viruses, the potential of discovering more hitherto unknown antiviral compounds of lichen origin remains very promising.

Indeed, lichens have proven to be useful source of antiviral agents. There seems to have been a recent upsurge in the recognition of the potential of lichens as useful source of secondary metabolites with varied biological activities. And this is partly due to the increasingly fairly well-documented biological and antiviral properties of lichen secondary metabolites. And the fact that new compounds continue to be

Table 7.1 Antiviral lichen secondary metabolites

Lichen	Promising secondary metabolites	Antiviral activity (viruses inhibited)	References
Umbilicaria esculenta	Lichen polysaccharide sulfate	HIV	Hirabayashi et al. (1989)
Cladonia cristatella	Acetone extract	Epstein-Barr virus	Yamamoto et al. (1995)
Several lichens	Lichenan	Tobacco mosaic virus	Stubler and Buchenauer (1996)
Nephroma laevigatum, *Heterodermia obscurata*	Emodin, 7-chloroemodin, 7-chloro-1- O-methylemodin, 5,7- dichloroemodin, 7,7′ dichlorohypericin, and hypericin	Herpes simplex virus type 1 (HSV-1)	Cohen et al. (1996)
Cladia retipora, *Pseudocyphellaria glabra*, *P. homoeophylla*	Usnic acid	Herpes simplex virus type 1 (HSV-1), Polio type 1 virus	Perry et al. (1999)
Xanthoria parietina and *Xanthoparmelia tinctina*	Aqueous and ethanol extracts	Human Parainfluenza virus 2	Karagöz and Aslan (2005)
Ramalina farinacea	Ethyl acetate-soluble fraction (*ET4*)	Lentiviral and adenoviral vectors, HIV-1	Esimone et al. (2005)
Parmelia saxatilis, *Alectoria nigricans*	Salazinic acid, alectorialic acid	Respiratory syncytial virus (RSV), HSV-1, HSV-2	Omarsdottir et al. (2006)
Teloschistes chrysophthalmus	Parietin	Junin and Tacaribe arenaviruses	Fazio et al. (2007)
Parmelia perlata	Crude polysaccharide fraction (extract)	Yellow fever virus	Esimone et al. (2007)
Ramalina farinacea	Ethyl acetate-soluble fraction (*ET4*)	HSV-1, RSV	Esimone et al. (2009)
Ramalina farinacea	Sekikaic acid	RSV	Lai et al. (2013), Odimegwu (2018)

isolated from lichens makes them an attractive potential source of novel biologicals and antivirals of the future. Another obvious advantage is the relative ease of their growth and survival given the right environment; therefore, special biotechnological platform can be designed for their optimal cultivation and exploitation.

Although the rate of reportage of antiviral agents of lichen origin is low, it is equally important to understand perhaps why this is so. Perhaps, antiviral evaluation is not being pursued by many investigations on lichens, and where is it is carried out the virus focus is usually too narrowed on one a few viruses. Nevertheless, as currently available reports and data show, lichens widely exist waiting to be discovered and investigated for possible antiviral potential. So what is required is the need to speed up extensive screening of potential lichen candidates/metabolites. Screenings should equally be made broad-based to discover even the remotest virus that could be covered by any one single metabolite. Given the varied opportunities offered by lichens, it is expected that they would thus remain an interesting focus of our searchlight in the pursuit of effective antiviral remedies against the hordes of viral infections and diseases facing humanity.

References

Akram M, Tahir IM, Shah SMA, Mahmood Z, Altaf A, Ahmad K, Munir N, Daniyal M, Nasir S, Mehboob H (2018) Antiviral potential of medicinal plants against HIV, HSV, influenza, hepatitis, and coxsackievirus: a systematic review. Phytother Res 32(5):811–822

Andersen DO, Weber ND, Wood SG, Hughes BG, Murray BK, North JA (1991) In vitro virucidal activity of selected anthraquinones and anthraquinone derivatives. Antivir Res 16:185–196

Atanasov AG, Waltenberger B, Pferschy Wenzig EM, Linder T, Wawrosch C, Uhrin P, Temml V, Wang L, Schwaiger S, Heiss EH (2015) Discovery and resupply of pharmacologically active plant-derived natural products: a review. Biotechnol Adv 33:1582–1614

Basile A, Rigano D, Loppi S et al (2015) Antiproliferative, antibacterial and antifungal activity of the lichen Xanthoria parietina and its secondary metabolite parietin. Int J Mol Sci 16:7861–7875

Bezivin C, Tomasi S, Rouaud I, Delcros J, Boustie J (2004) Cytotoxic activity of compounds from the lichen: *Cladonia convoluta*. Planta Med 70:874–877

Brightman FH, Seaward MRD (1977) Lichens of man-made substrates. In: Seaward MRD (ed) Lichen ecology. Academic Press, London, pp 253–293

Cetin H, Tufan-Cetin O, Turk AO, Tay T, Candan M, Yanikoglu A, Sumbul H (2008) Insecticidal activity of major lichen compounds, (−)- and (+)-usnic acid, against the larvae of house mosquito, *Culex pipiens* L. Parasitol Res 102:1277–1279

Cohen PA, Towers GHN (1995a) Anthraquinones and phenanthroperylenequinones from Nephroma laevigatum. J Nat Prod 58:520

Cohen PA, Towers GHN (1995b) The anthraquinones of *Heterodermia obscurata*. Phytochemistry 40:911

Cohen PA, Hudson JB, Towers GHN (1996) Antiviral activities of anthraquinones, bianthrones and hypericin derivatives from lichens. Experientia 52:180–183

Drake MG, Bivins-Smith ER, Proskocil BJ et al (2016) Human and mouse eosinophils have antiviral activity against parainfluenza virus. Am J Respir Cell Mol Biol 55(3):387–394

Ebrahim HY, Elsayed HE, Mohyeldin MM et al (2016) Norstictic acid inhibits breast cancer cell proliferation, migration, invasion, and in vivo invasive growth through targeting C-met. Phytother Res 30:557–566

Emmerich R, Giez I, Lange OL, Proksch P (1993) Toxicity and antifeedant activity of lichen compounds against the polyphagous herbivorous insect *Spodoptera littoralis*. Phytochemistry 33:1389–1394

Esimone CO, Grunwald T, Wildner O, Nchinda G, Tippler B, Proksch P, Uberla K (2005) In vitro pharmacodynamic evaluation of antiviral medicinal plants using a vector-based assay technique. J Appl Microbiol 99(6):1346–1355

Esimone CO, Ofokansi KC, Adikwu MU, Ibezim EC, Abonyi DO, Odaibo GN, Olaleye DO (2007) In vitro evaluation of the antiviral activity of extracts from the lichen *Parmelia perlata* (L.) Ach. against three RNA viruses. J Infect Dev Ctries 1(3):315–320

Esimone CO, Eck G, Duong TN, Uberla K, Proksch P, Grunwald T (2008) Potential anti-respiratory syncytial virus lead compounds from *Aglaia* species. Pharmazie 63:1–6

Esimone CO, Grunwald T, Nworu CS, Kuate S, Proksch P, Uberla K (2009) Broad spectrum antiviral fractions from the lichen *Ramalina farinacea* (L.) Ach. Chemotherapy 55:119–126

Farrar JF (1976) The lichen as an ecosystem: observation and experiment. In: Brown DH, Hawksworth DL, Bailey RH (eds) Lichenology: progress and problems, Special volume – The Systematics Association, vol 8. Academic Press, London, pp 385–406

Fazio AT, Adler MT, Bertoni MD, Sepulveda CS, Damonte EB, Maier MS (2007) Lichen secondary metabolites from the cultured lichen mycobionts of *Teloschistes chrysophthalmus* and *Ramalina celastri* and their antiviral activities. Z Naturforsch 62:543–549

Gautam R, Saklani A, Jachak SM (2007) Indian medicinal plants as a source of antimycobacterial agents. J Ethnopharmacol 110:200–234

Halama P, Van Haluwin C (2004) Antifungal activity of lichen extracts and lichenic acids. BioControl 49:95–107

Harvey AL (2008) Natural products in drug discovery. Drug Discov Today 13:894–901

Hawksworth DL, Honegger R (1994) The lichen thallus: a symbiotic phenotype of nutritionally specialized fungi and its response to gall producers. In: Williams MAJ (ed) Plant galls, Special volume – The Systematics Association, vol 49. Clarendon Press, Oxford, pp 77–98

Hidalgo ME, Fernandez E, Quilhot W, Lissi E (1994) Antioxidant activity of depsides and depsidones. Phytochemistry 37:1585–1587

Hirabayashi K, Iwata S, Ito M, Shigeta S, Narui T, Mori T, Shibata S (1989) Inhibitory effect of a lichen polysaccharide sulfate, GE-3-S, on the replication of human immunodeficiency virus (HIV) in vitro. Chem Pharm Bull 37:2410–2412

Honegger R (1991) Functional aspects of the lichen symbioses. Annu Rev Plant Physiol Plant Mol Biol 42:553–578

Hong JM, Suh SS, Kim TK et al (2018) Anti-cancer activity of lobaric acid and lobarstin extracted from the Antarctic lichen *Stereocaulon alpnum*. Molecules 23:658. https://doi.org/10.3390/molecules23030658. eCollection 2018 Mar

Jassim SAA, Naji MA (2003) Novel anti-viral agents: a medicinal plant perspective. J Appl Microbiol 95:412–427

Jha BN, Shrestha M, Pandey DP et al (2017) Investigation of antioxidant, antimicrobial and toxicity activities of lichens from high altitude regions of Nepal. BMC Complement Altern Med 17:282. eCollection 2017

Karagöz A, Aslan A (2005) Antiviral and cytotoxic activity of some lichen extracts. Biol Bratis 60:281–286

Kirk PM, Cannon PF, Minter DW, Stalpers JA (eds) (2008) Dictionary of the fungi, 10th edn. CAB International, Wallingford

Kohlmann R, Schwannecke S, Tippler B, Ternette N, Temchura V, Tenbusch M, Überla K, Grunwald T (2009) Protective efficacy and immunogenicity of an adenoviral vector vaccine encoding the codon-optimized F protein of respiratory syncytial virus. J Virol 83 (23):12601–12610

Kott V, Barbini L, Cruanes M et al (1999) Anti-viral activity in Argentine medicinal plants. J Ethnopharmacol 64:79–84

Kumar J, Dhar P, Tayade AB et al (2014) Antioxidant capacities, phenolic profile and cytotoxic effects of saxicolous lichens from trans-Himalayan cold desert of Ladakh. PLoS One 9:e98696

Kwon IS, Yim JH, Lee HK et al (2016) Lobaric acid inhibits VCAM-1 expression in TNF-Î-±-stimulated vascular smooth muscle cells via modulation of NF-Î°B and MAPK signaling pathways. Biomol Ther (Seoul) 24:25–32

Lai D, Odimegwu DC, Esimone C, Grunwald T, Proksch P (2013) Phenolic compounds with in vitro activity against respiratory syncytial virus from the Nigerian lichen *Ramalina farinacea*. Planta Med 79(15):1440–1446

Legouin B, Lohézic-Le Dévéhat F, Ferron S et al (2017) Specialized metabolites of the lichen *Vulpicida pinastri* act as photoprotective agents. Molecules 22:1162. https://doi.org/10.3390/molecules22071162. eCollection 2017 Jul

Li F, Ma C, Wang J (2015) Inhibitors targeting the influenza virus hemagglutinin. Curr Med Chem 22:1361–1382

Lisicka E (2008) Lichens on an acrylic-coated aluminium roof. Graph Scr 20:9–12

Lumbsch HT (1998) Taxonomic use of metabolic data in lichen-forming fungi. In: Frisvad JC, Bridge PD, Arora DK (eds) Chemical fungal taxonomy. Marcel Dekker, New York, pp 345–387
Mayer M, O'Neill MA, Murry KE, Santos-Magalhaes NS, Carneiro-Leao AMA, Thompson AM, Appleyard VCL (2005) Usnic acid: a non-genotoxic compound with anti-cancer properties. Anti-Cancer Drugs 16:805–809
Molnar K, Farkas E (2010) Current results on biological activities of lichen secondary metabolites: a review. Z Naturforsch 65c:157–173
Motohashi Y, Igarashi M, Okamatsu M et al (2013) Antiviral activity of stachyflin on influenza A viruses of different hemagglutinin subtypes. Virol J 10:118. https://doi.org/10.1186/1743-422X-10-118
Nash TH III (ed) (2008) Lichen biology, 2nd edn. Cambridge University Press, Cambridge
Neamati N, Hong H, Mazumder A, Wang S, Sunder S, Nicklaus MC, Milne GWA, Proksa B, Pommier Y (1997) Depsides and depsidones as inhibitors of HIV-1 integrase: discovery of novel inhibitors through 3D database searching. J Med Chem 40:942–951
Newman DJ, Cragg GM (2012) Natural products as sources of new drugs over the 30 years from 1981 to 2010. J Nat Prod 75:311–335
Odimegwu DC (2018) Low-dose Sekikaic acid modulates host immunity and protects cells from Respiratory Syncytial Virus infection. Biotechnol J Int 21(2):1–10
Odimegwu DC, Okoye FBC, Nworu CS, Esimone CO (2018) Anti-respiratory syncytial virus activities of leaf extracts of *Alchornea cordifolia* and *Alchornea floribunda*. Afr J Pharm Pharmacol 12(8):97–105
Omarsdottir S, Óladóttir AK, Árnadóttir T, Ingólfsdóttir K (2006) Antiviral compounds from Icelandic lichens. Planta Med 72. https://doi.org/10.1055/s-2006-949742
Paluszczak J, Kleszcz R, Studzińska-Sroka E et al (2018) Lichen-derived caperatic acid and physodic acid inhibit Wnt signaling in colorectal cancer cells. Mol Cell Biochem 441:109–124
Paudel B, Bhattarai HD, Koh HY et al (2011) Ramalin, a novel nontoxic antioxidant compound from the Antarctic lichen *Ramalina terebrata*. Phytomedicine 18:1285–1290
Pengsuparp T, Cai L, Constant H, Fong HHS, Lin LZ, Kinghorn AD, Pezzuto JM, Cordell GA, Ingolfsdöttir K, Wagner H, Hughes SH (1995) Mechanistic evaluation of new plant-derived compounds that inhibit HIV-1 reverse transcriptase. J Nat Prod 58:1024–1031
Perry NB, Benn MH, Brennan NJ, Burgess EJ, Ellis G, Galloway DJ, Lorimer SD, Tangney RS (1999) Antimicrobial, antiviral and cytotoxic activity of New Zealand lichens. Lichenologist 31 (6):627–636
Ranković B, Mišić M (2007) Antifungal activity of extracts of the lichens *Alectoria sarmentosa* and *Cladonia rangiferina*. Mikol Fitopatol 41:276–281
Ranković B, Mišić M (2008) The antimicrobial activity of the lichen substances of the lichens *Cladonia furcata*, *Ochrolechia androgyna*, *Parmelia caperata* and *Parmelia conspersa*. Biotechnol Biotechnol Equip 22:1013–1016
Ranković B, Mišić M, Sukdolak S (2008) The antimicrobial activity of substances derived from the lichens *Physcia aipolia, Umbilicaria polyphylla, Parmelia caperata* and *Hypogymnia physodes*. World J Microbiol Biotechnol 24:1239–1242
Ronalds CJ, Hardiman AE, Griffiths PD (1983) Hotting up the complement fixation test. J Hyg (Lond) 90:127–134
Russo A, Piovano M, Lombardo L, Garbarino J, Cardile V (2008) Lichen metabolites prevent UV light and nitric oxide-mediated plasmid DNA damage and induce apoptosis in human melanoma cells. Life Sci 83:468–474
Schmeda-Hirschmann G, Tapia A, Lima B et al (2008) A new antifungal and antiprotozoal depside from the Andean lichen *Protousnea poeppigii*. Phytother Res 22:349–355
Schreiner S, Wimmer P, Sirma H, Everett RD, Blanchette P, Groitl P, Dobner T (2010) Proteosome-dependent degradation of Daxx by the Viral E1B-55K protein in human adenovirus-infected cells. J Virol 84(14):7029–7038
Seaward MRD (2008) Environmental role of lichens. In: Nash TH III (ed) Lichen biology, 2nd edn. Cambridge University Press, Cambridge, pp 274–298

Semple SJ, Reynolds GD, O'Leary GD, Flower RLP (1998) Screening of Australian medicinal plants for anti-viral activity. J Ethnopharmacol 60:163–172

Seo C, Sohn JH, Park SM et al (2008a) Usimines A-C, bioactive usnic acid derivatives from the Antarctic lichen Stereocaulon alpinum. J Nat Prod 71:710–712

Seo C, Yim JH, Lee HK et al (2008b) Stereocalpin A, a bioactive cyclic depsipeptide from the Antarctic lichen Stereocaulon alpinum. ChemInform 49:29–31

Sindambiwe JB, Calomme M, Cos P et al (1999) Screening of seven selected Rwandan medicinale plants for antimicrobial and anti-viral activities. J Ethnopharmacol 65:71–77

Sloan E, Tatham MH, Groslambert M, Glass M, Orr A, Hay RT, Everett RD (2015) Analysis of the SUMO2 Proteome during HSV-1 Infection. PLoS Pathog 11(7):e1005059. https://doi.org/10.1371/journal.ppat.1005059

Sokolov DN, Zarubaev VV, Shtro AA et al (2012) Anti-viral activity of (−)- and (+)-usnic acids and their derivatives against influenza virus A(H1N1)2009. Bioorg Med Chem Lett 22:7060–7064

Stocker-Worgotter E (2008) Metabolic diversity of lichen-forming ascomycetous fungi: culturing, polyketide and shikimate metabolite production, and PKS genes. Nat Prod Rep 25:188–200

Stubler D, Buchenauer H (1996) Antiviral activity of the glucan lichenan (poly-β{1 → 3, 1 →4} Danhydroglucose) 1. Biological activity in tobacco plants. J Phytopathol 144:37–43

Studzinska-Sroka E, Galanty A, Bylka W (2017) Atranorin - an interesting lichen secondary metabolite. Mini Rev Med Chem 17:1633–1645

Suh SS, Kim TK, Kim JE et al (2017) Anticancer activity of ramalin, a secondary metabolite from the antarctic lichen Ramalina terebrata, against colorectal cancer cells. Molecules 22:1361. https://doi.org/10.3390/molecules22081361. eCollection 2017 Aug

Tavalai N, Papior P, Rechter S, Stamminger T (2008) Nuclear domain 10 components promyelocytic leukemia protein and hDaxx independently contribute to an intrinsic antiviral defense against human cytomegalovirus infection. J Virol 82:126–137

Ternette N, Tippler B, Uberla K, Grunwald T (2007) Immunogenicity and efficacy of codon optimized DNA vaccines encoding the F-protein of respiratory syncytial virus. Vaccine 25:7271–7279

Thadhani VM, Karunaratne V (2017) Potential of lichen compounds as antidiabetic agents with antioxidative properties: a review. Oxid Med Cell Longev 2017:2079697

Tian Y, Li YL, Zhao FC (2017) Secondary metabolites from polar organisms. Mar Drugs 15:28. https://doi.org/10.3390/md15030028. eCollection 2017 Mar

Turk AO, Yilmaz M, Kivanc M, Turk H (2003) The antimicrobial activity of extracts of the lichen *Cetraria aculeata* and its protolichesterinic acid constituent. Z Naturforsch 58c:850–854

Voss EG, Burdet HM, Chaloner WG, Demoulin V, Hiepko P, Mcneill J, Meikle RD, Nicolson DH, Rollins RC, Silva PC, Greuter W (1983) International code of botanical nomenclature (Sydney Code). Regnum Veg 111:1–472

Vu TH, Le Lamer AC, Lalli C et al (2015) Depsides: lichen metabolites active against hepatitis C virus. PLoS One 10:e0120405

Wang X, Chi X, Wang M (2011) Structural characteristics and antiviral activity of multiple peptides derived from MDV glycoproteins B and H. Virol J 8:190

Wood S, Huffman J, Weber N, Andersen D, North J, Murray B, Sidwell R, Hughes B (1990) Antiviral activity of naturally occurring anthraquinones and anthraquinone derivatives. Planta Med 56:65–52

Xu M, Heidmarsson S, Olafsdottir ES et al (2016) Secondary metabolites from cetrarioid lichens: chemotaxonomy, biological activities and pharmaceutical potential. Phytomedicine 23:441–459

Yamamoto Y, Kinoshita Y, Kurokawa T, Yoshimura I, Ahmadjian V, Yamada Y (1995) Cell growth and pigment production in suspension cultures of a mycobiont isolated from the lichen *Cladonia cristatella*. Can J Bot 73:590–594

Yang Y, Park SY, Nguyen TT et al (2015) Lichen secondary metabolite, physciosporin, inhibits lung cancer cell motility. PLoS One 10:e0137889

Zhou R, Yang Y, Park SY et al (2017) The lichen secondary metabolite atranorin suppresses lung cancer cell motility and tumorigenesis. Sci Rep 7:8136. E Collection 2017

Antineurodegenerative and Antidiabetic Activity of Lichens

8

Marijana Kosanić and Branislav Ranković

Abstract

Neurodegenerative diseases are phenomena that occur in the central nervous system through the hallmarks associating the loss of neuronal structure and function. These diseases affect many of our body's activities, such as balance, movement, talking, breathing, and heart function. These disorders are very common and their frequency increases with age, in mid-to-late adult life. Except neurodegeneration, a very common disease is diabetes mellitus. Diabetes mellitus is a chronic disease caused by inherited and/or acquired deficiency in the production of insulin by the pancreas or by the ineffectiveness of the insulin produced. Such a deficiency results in increased concentrations of glucose in the blood, which in turn damage many of the body's systems. Since the past few decades, diabetes has become a global health problem. Apart from conventional therapy, several studies have shown that some natural products have beneficial effects in neurological and diabetic patients. Various natural products emerged as interesting molecules with antineurodegenerative and antidiabetic potential. However, little is known in relation to lichens as potential therapeutics in these disorders. Therefore, this chapter focuses on the antineurodegenerative and antidiabetic potential of lichens that have received considerable attention in the recent past.

8.1 Antineurodegenerative Activity of Lichens

Neurodegenerative diseases are a progressive loss of structure and/or function of neurons, which leads to deficits in specific brain functions (e.g., movement, memory, balance, talking, cognition, swallowing, and breathing). Neurodegenerative diseases

M. Kosanić (✉) · B. Ranković
Department of Biology, Faculty of Science, University of Kragujevac, Kragujevac, Serbia

B. Ranković (ed.), *Lichen Secondary Metabolites*,
https://doi.org/10.1007/978-3-030-16814-8_8

of the brain often develop slowly over many years before they are diagnosed, and this can make it difficult to study factors which may be linked to their development (Kim et al. 2017). Some of the commonly studied factors that cause neurodegenerative diseases are protein degradation, oxidative stress, inflammation, environmental factor, mitochondrial defects, familial history, abnormal protein accumulation in neuron, smoking, alcohol, obesity, etc. Also, aging plays a very important role in neurodegenerative diseases (Fig. 8.1).

Of these hundreds of different disorders, so far attention has been mainly focused on only a handful, with the most notable being Alzheimer's disease and Parkinson's disease, while less common diseases include Huntington disease, Amyotrophic lateral sclerosis, and Spinocerebellar ataxia.

Alzheimer's Disease Alzheimer's disease is an irreversible, progressive brain disorder that slowly destroys memory and thinking skills and eventually the ability to carry out the simplest tasks. In most people with Alzheimer's, symptoms first appear in their mid-60s. In people with Alzheimer's, the brain is changed, it has fewer healthy cells, and it gets smaller over time. Most of the time, the brain cells also form two types of defects, neurofibrillary tangles and beta-amyloid plaques. Neurofibrillary tangles are twisted fibers inside brain cells that keep nutrients and other important things from moving from one part of the cell to another. Beta-amyloid plaques are sticky clumps of proteins that build up between nerve cells

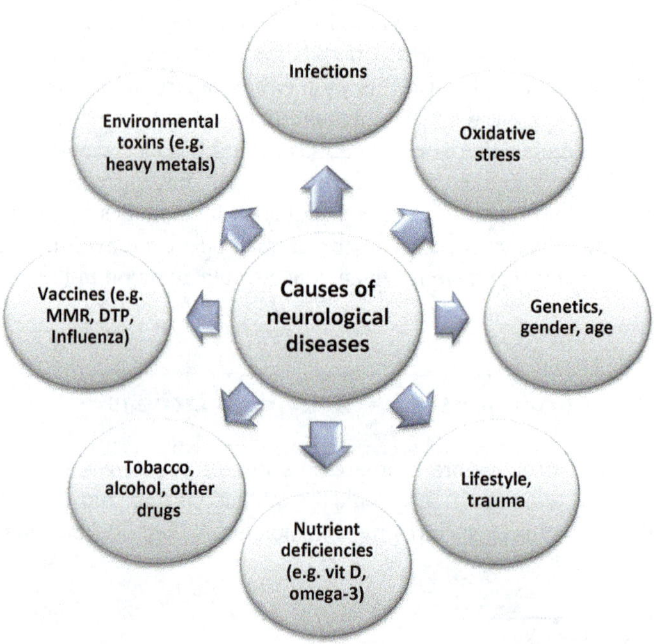

Fig. 8.1 Causes of neurological diseases

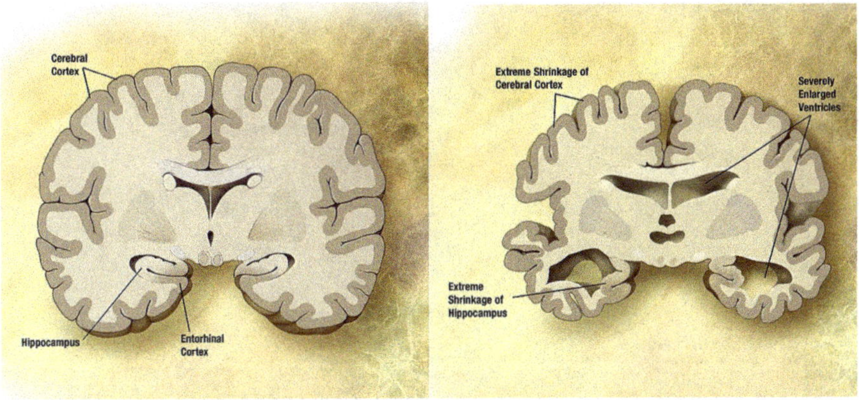

Fig. 8.2 Comparison between normal brain and Alzheimer's brain (figure © ADEAR: "Alzheimer's Disease Education and Referral Center, a service of the National Institute on Aging," via Wikimedia Commons)

instead of breaking down like they do in healthy brains. Plaques and tangles damage the healthy brain cells around them (Hippius and Neundörfer 2003; Jack et al. 2011). The damaged cells die, and the brain shrinks (Fig. 8.2).

While there is no known cause for Alzheimer's disease, some research studies have indicated that the following factors may play an important role in the development of the condition:

- Genetic factors, such as the presence of, or changes to, certain genes
- Environmental factors, such as long-term exposure to some environmental solvent (e.g., pesticides, glues, and paints) or infection with certain viruses or bacteria
- Lifestyle factors, such as a lack of exercise, poor-quality sleep, and a diet lacking fruit and vegetables

Scientists now think that a combination of these genetic, environmental, and lifestyle risk factors triggers an abnormal biological process in the brain that, over decades, results in Alzheimer-type dementia. Identified risk factors for developing the condition include increasing age, history of a head injury, risk factors for blood vessel disease such as smoking, Down syndrome, family history of Alzheimer's disease, high blood pressure, high cholesterol, obesity, and insulin resistance (Ikeda and Yamada 2010; Crous-Bou et al. 2017).

There is some evidence of a slightly higher prevalence of Alzheimer's disease in females than males, but this may reflect their longer life expectancy. The symptoms of Alzheimer's disease typically develop quite slowly. The time between the onset of the disease and death can range from 5 to 20 years. Symptoms of Alzheimer's disease occur gradually over a period of several years and vary from person to person. The first symptoms that appear are usually oblivious and small confusion.

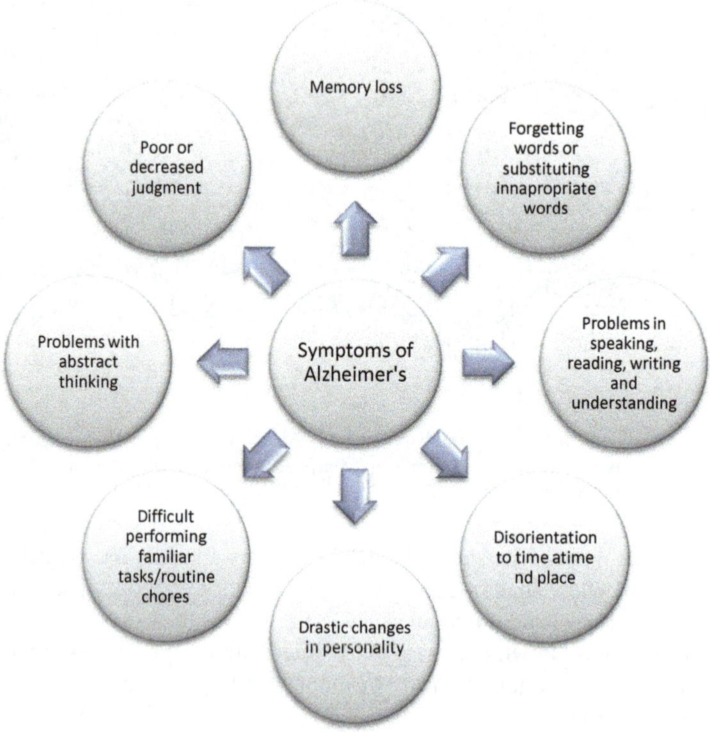

Fig. 8.3 Symptoms of Alzheimer's disease

Symptoms of Alzheimer's disease (Fig. 8.3) can spread to categorize into cognitive, functional, and behavioral/psychological changes:

Cognitive changes—weakened short-term memory (damaged memory of recent events, repetition, poor organization, loss of things around the house), decision-making difficulties, reduced understanding of space and time concepts, reduced learning ability, and the problem of recognizing family members and friends

Functional changes—reduced ability to carry out everyday activities, as well as more complex tasks: difficulties in maintaining personal hygiene, incontinence, difficulties in handling money, difficulty in dressing, and imbalance and unstable movement

Behavioral/psychological changes—patients can develop behavioral disorders as well as mental disorders, which increase the burden of care for them. They can develop and manifest socially inadequate behaviors such as delusions, depression/dysphoria irritability/aggression, and changes in appetite/diet

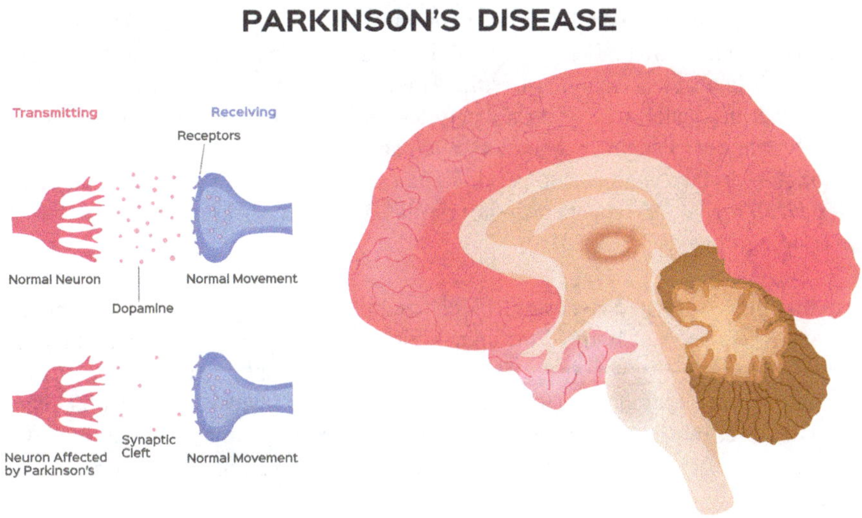

Fig. 8.4 Comparison between normal neuron and Parkinson's affected neuron (figure © gritsalak/stock.adobe.com)

Parkinson's Disease Parkinson's disease is a disease of the nervous system, which implies a gradual loss of functions of certain nerve cells in the brain, which leads to disorders of mobility and motor activity, but with further development and other mental and psychological disorders.

By definition, Parkinson's disease is a chronic neurodegenerative disease that arises as a result of a marked lack of dopamine neurotransmitter in the part of the brain responsible for controlling the willingly coordinated movement and the beginning of a motor action (Fig. 8.4).

It is thought that the cause of the disease is a combination of genetic factors and environmental impact. It is known that there is a greater possibility to get a disease in those in whom one of the lowest family suffer from Parkinson's disease, but this probability is estimated to be about 5%. The main symptoms of Parkinson's disease are tremor; stiffness of the muscles of the arms, legs, and neck; and slowing movement. With each case of Parkinson's disease, the symptoms or severity can vary from person to person, so it is important to take note of each individual's strengths, weaknesses, and pain levels. Although Parkinson's disease cannot be cured, with adequate therapy, symptoms of the disease can be significantly alleviated (Kremens et al. 2014; Chonpathompikunlert et al. 2018).

Less Common Neurodegenerative Diseases Less common neurodegenerative diseases are Huntington disease, amyotrophic lateral sclerosis, and spinocerebellar ataxia.

Huntington Disease It is a rare neurodegenerative disorder of the central nervous system characterized by unwanted choreatic movements, behavioral and psychiatric disturbances, and dementia. Mean age at onset of symptoms is 30–50 years. In some cases symptoms start before the age of 20 years with behavior disturbances and learning difficulties at school (juvenile Huntington disease). The classic sign is chorea that gradually spreads to all muscles. All psychomotor processes become severely retarded. Patients experience psychiatric symptoms and cognitive decline. Huntington disease is an autosomal dominant inherited disease caused by an elongated CAG repeat (36 repeats or more) on the short arm of chromosome 4p16.3 in the Huntingtin gene (Roos 2010).

Amyotrophic Lateral Sclerosis It is characterized by progressive loss of motor neurons in the anterior horn of the spinal cord. It is classified as either familial or sporadic depending on whether there is a clearly defined, inherited genetic element. Sporadic Amyotrophic lateral sclerosis typically emerges between 50 and 60 years old. The onset of sporadic amyotrophic lateral sclerosis is unknown, and thus the identification of causal genes and environmental factors remains elusive (Pratt et al. 2012).

Spinocerebellar Ataxia It is an autosomal dominant disease characterized by progressive neurodegeneration. Common symptoms associated with spinocerebellar ataxia are ataxic gait, oculomotor disorders, dysarthria, and cognitive impairment, which can ultimately cause death. Over 20 types of spinocerebellar ataxia have been identified based on genetic descriptions. An expansion of repeated CAG trinucleotides accounts for the major pathogenic mutation in spinocerebellar ataxia. This mutation leads to the overexpression of mutant ataxin 1 protein that has an expanded polyglutamine. Mutant ataxin 1 can affect the stability of RAR-related orphan receptor alpha, which plays an important role in Purkinje cell functions. Decreased gene expression of RAR-related orphan receptor alpha has been associated with ataxia and cerebellar hypoplasia. While most types of Spinocerebellar ataxia are suggested as genetic diseases correlated with ataxin mutation, other pathogenic mechanisms that involve the dysfunction of mitochondria have been proposed (Paulson 2009).

8.2 Methods to Determine Neuroprotective Potential

The most used methods to evaluate neuroprotective capacity of natural products are acetylcholinesterase (AChE) inhibitory activity assay and tyrosinase inhibitory activity assay.

The most remarkable biochemical change in Alzheimer's disease patients is a reduction of acetylcholine (ACh) levels in the hippocampus and cortex of the brain. Acetylcholine (ACh) is the most abundant neurotransmitter in the body and the primary neurotransmitter in the brain which is responsible for cholinergic transmission. Therefore, inhibition of AChE, the enzyme responsible for hydrolysis of ACh at the cholinergic synapse, is currently the most established approach to treating Alzheimer's disease.

Because most of AChE inhibitors are alkaloids that often possess several side effects, such as galantamine (GAL) used clinically for the disease treatment, it is important to search for new natural inhibitors without adverse effects.

The most common test for AChE activity assay is based on Ellman's method (1961) using an alternative substrate acetylthiocholine. Upon hydrolysis, these substrate analogs produce acetate and thiocholine. Thiocholine in the presence of the highly reactive dithiobisnitro-benzoate (DTNB) ion reacts to generate the yellow color of 5-thio-2-nitrobenzoate anion. The yellow color can be quantified by its absorbance at 412 nm. As the substrate is hydrolyzed, the increase of absorbance at 412 nm is recorded and the activity of the enzyme calculated as the reaction rate from the slope of the linear part of the time-dependence of the absorbance. The method was developed by Ellman and coworkers in early 1961, and it is still used up to now, generally with significant modifications.

Parkinson's disease is another neurodegenerative disease, resulting from dopamine deficiency in the brain. Tyrosinase is an enzyme that plays a role in neuromelanin production and in oxidizing excess dopamine to produce dopamine quinones, highly reactive species which induce neural damage and cell death associated with Parkinson's disease, and, therefore, inhibition of tyrosinase has become a popular target in drug research for Parkinson's disease.

Many potential tyrosinase inhibitors are examined in the presence of L-tyrosine or 3,4-dihydroxy-l-phenylalanine (L-DOPA) as the enzyme substrate, and tyrosinase activity is assayed in terms of dopachrome formation. The amount of dopachrome production in the reaction mixture can be determined spectrophotometrically at 450 nm.

Oxidative stress also plays a crucial role in neuronal damage and is responsible for the development of several neurodegenerative diseases. In the brain and neuronal tissues, reactive oxygen species are produced as by-products of excitatory amino acids and neurotransmitter metabolism. They can attack glial cells and neurons which are highly sensitive to free radicals, leading to neuronal damage. It has been demonstrated that cells become more sensitive to oxidative stress by aging, which makes human more vulnerable to the neurodegenerative diseases. Several studies indicated that an antioxidant-rich diet can reduce the deleterious effects of oxidative stress and improve the signs of neurodegeneration (Kim et al. 2015).

Antioxidants appear to have neuroprotective and antineurodegenerative roles, reducing or slowing down neuronal cell death, and have an important role to play in diminishing the cumulative effects of oxidative damage. Numerous studies have been performed to investigate the roles of reactive oxygen species (ROS) in neurodegenerative progression, and some positive results are yielded. Of note is that neuron cells are particularly vulnerable to oxidative damage because of their high polyunsaturated fatty acid content in membranes, high oxygen consumption, and weak antioxidant defense. Under physiological conditions, ROS generated from mitochondria, NADPH oxidase (Nox), and xanthine oxidase (XO) are maintained at relatively low levels by endogenous antioxidants. However, the redox balance can be disturbed by neural inflammation or abnormal mitochondrial function. The pathogenesis of several neurodegenerative disorders such as Alzheimer's disease and Parkinson's disease is associated with the accumulation of misfolded proteins. The aggregation of these modified proteins can in turn trigger inflammatory response in the brain, which induces marked ROS release and subsequent oxidative stress. Mitochondrial dysfunction, which often accompanies aberrant ROS production, is tightly linked with neurodegenerative disorders. For example, in Huntington disease, mutant huntingtin (mHTT) may directly interact with mitochondria, leading to compromised energy supply and increased ROS levels (Liu et al. 2017).

Considering the pivotal roles of oxidative stress in neurodegenerative diseases, the manipulation of ROS levels may represent a promising treatment option to slow down neurodegeneration and alleviate associated symptoms (Ataie et al. 2016). Therefore, many antioxidative tests can be used for investigation of neuroprotective effects, such as method for identification of free radical scavenger, nitric oxide scavenger, and oxygen radical absorbance capacity (ORAC).

Method for identification of free radical scavenger: 2,2-diphenyl-1-picrylhydrazyl (DPPH) is a stable free-radical molecule. The DPPH radical with an odd electron shows a deep violet color in solution, since a strong absorption band is centered at about 517 or 520 nm. It becomes colorless or pale yellow when neutralized with antioxidants. In the cell-free system, this method is readily used to distinguish whether an agent that suppresses ROS in an in vitro or an in vivo system is a free radical scavenger or indirectly affects ROS level in the cells. To understand how a test agent affects cell redox status, a functional study in a living system should be carried out.

Method for identification of nitric oxide (NO) scavenger: This assay relies on a diazotization reaction that was originally described by Griess in 1879. The half-life of NO in biological systems ranges from ≤ 1 to around 30 s. However, NO in biological systems could be converted into nitrite ion (NO_2^-) and nitrate ion (NO_2^+). NO_2^+ can be reduced to NO_2^- by nitrated reductase. The level of NO_2^- can be measured with Griess reagents (1% sulfanilamide, 2% H_3PO_4, and 0.1% naphthyl ethylenediamine dihydrochloride). In the cell-free assay system under a physiological pH, NO is spontaneously generated from sodium nitroprusside in aqueous solution. NO interacts with O_2 to produce NO_2^-. Scavenger of NO, and when incubated with sodium nitroprusside at 250 °C for 150 min, competes with O_2 leading to a decrease in NO_2^- production. The samples from the above react with Griess reagents for 15–30 min. The absorbance of the red-pink chromophore formed during the diazotization of NO_2^- is measured at 546 nm. The assay has been used for determination of NO scavenging activity of plant, biomolecules, and small molecules. Although the Griess reaction is reliable, reproducible, convenient, and easily performed in most laboratories, the limit level of NO_2^- for detection could be 2.5 μM, and the sensitivity is dependent on the sample used.

The (ORAC) assay is a method that measures the antioxidant capacity of a substance. The ORAC assay measures a fluorescent signal from a probe that is quenched in the presence of ROS. Addition of an antioxidant absorbs the generated

ROS, allowing the fluorescent signal to persist. Modifications of the ORAC assay include the use of fluorescein as the fluorescent probe (ORACFL), the separation of hydrophilic and lipophilic antioxidants to obtain total antioxidant capacity, and an adaptation to a high-throughput platform. The ORAC assay is unique in that its ROS generator, AAPH (2,2′-azobis(2-methylpropionamidine) dihydrochloride), produces a peroxyl free radical upon thermal decomposition. This free radical is commonly found in the body, making this reaction biologically relevant. Furthermore, AAPH is reactive with water- and lipid-soluble substances, so it can measure total antioxidant potential.

8.3 Antineurodegenerative Activity of Lichen Extracts and Their Metabolites

Exploration of the antineurodegenerative properties of lichens is still poor compared to that of other natural products. In some previous investigations, several lichen species have demonstrated antineurodegenerative properties (Table 8.1). For example, the results of the screening of 31 species belonging to the lichen family *Graphidaceae* for their ability to inhibit tyrosinase activity have been reported by Behera et al. (2004). Among tested species, six species, namely, *Graphina glaucorufa*, *G. multistriata*, *G. salacinilabiata*, *Graphis assamensis*, *G. nakanishiana*, and *Phaeographopsis indica*, showed tyrosinase inhibition from moderate to high range 48–73%. All other species inhibited tyrosinase below 40%.

Methanolic extracts of lichens *Arthothelium awasthii*, *Heterodermia podocarpa*, and *Parmotrema tinctorum* were taken for evaluating their potential to inhibit tyrosinase enzyme activity by Verma et al. (2008). They found IC_{50} of 8.71 μg (*A. awasthii*), 14.55 μg (*H. podocarpa*), and 12.44 μg (*P. tinctorum*). According to their IC_{50}, the tested extracts turned out to be more effective for the tested parameters than the standard kojic acid tyrosinase inhibitor (IC_{50} for tyrosinase inhibition 17.63 μg). Similarly, Honda et al. (2016) evaluated nine lichen species for their tyrosinase inhibition and found that *Cladia aggregata*, *Cladonia dimorphoclada*, *Stereocaulon ramulosum*, and *S. microcarpum* extracts were active.

Zrnzević et al. (2017) studied the inhibition of *Ramalina capitata* acetone extract on cholinesterase activity. Obtained activity was in a dose-dependent manner. Extract at concentration of 1.0 mg/mL has manifested insignificant activation effect on cholinesterase to extent of 2.8%, while more concentrated extract (10 mg/mL) exhibited the slight inhibition. Zlatanović et al. (2017) found similar activity for *Umbilicaria crustulosa* acetone extract. This extract manifested weak activation effect on cholinesterase to an extent of 1.6%.

Fernandez-Moriano et al. (2015) investigated the in vitro neuroprotective effects of methanol extracts of two lichens *Cetraria islandica* and *Vulpicida canadensis*. They studied neuroprotective activity with respect to the antioxidant properties of the extracts. Tested lichen extracts were able to reverse the oxidative damage caused

Table 8.1 List of lichens and their secondary metabolites used to evaluate antineurodegenerative activity

Tested lichen/compounds	References
Graphina glaucorufa *G. multistriata* *G. salacinilabiata* *Graphis assamensis* *G. nakanishiana* *Phaeographopsis indica*	Behera et al. (2004)
Arthothelium awasthii *Heterodermia podocarpa* *Parmotrema tinctorum*	Verma et al. (2008)
Cladia aggregata *Cladonia dimorphoclada* *Stereocaulon ramulosum* *S. microcarpum*	Honda et al. (2016)
Cetraria islandica *Vulpicida canadensis*	Fernandez-Moriano et al. (2015)
Ramalina capitata	Zrnzević et al. (2017)
Umbilicaria crustulosa	Zlatanović et al. (2017)
Fumarprotocetraric acid	Fernández-Moriano et al. (2017)
Biruloquinone	Luo et al. (2013)
Depsidone $C_{21}H_{18}O_9$	Pejin et al. (2013)
Lobaric acid	Thadhani et al. (2014)
Atranorin Perlatolic acid Usnic acid	Reddy et al. (2016)
Physodic acid	Reddy et al. (2016), Emsen et al. (2016)
Psoromic acid	Emsen et al. (2016)
Ramalin	Paudel et al. (2011)
Stictic acid Salazinic acid	de Paz et al. (2010)

by hydrogen peroxide, thus promoting astrocyte survival. The same authors reported the neuroprotective properties of the major secondary metabolite of the *Cetraria islandica*, i.e., the depsidone fumarprotocetraric acid (Fernández-Moriano et al. 2017). They performed experiments on two models of CNS-like cells, namely, the human neuroblastoma SH-SY5Y and human astrocytoma U373-MG cell lines; the exposure to exogenous H_2O_2 has been used to mimic cellular OS, a key feature in neurodegenerative diseases. The cytoprotective properties of fumarprotocetraric acid against oxidative neurotoxicity seem to be mediated by the protection against mitochondrial dysfunction and the enhancement of the enzymatic and nonenzymatic (glutathione system) antioxidant potential. Such effects in turn resulted in a decrease in ROS production, lipid peroxidation, and levels of pro-apoptotic factors.

Moreover, ramalin isolated from Antarctic *Ramalina terebrata* presented scavenging activity, and it inhibits tyrosine enzyme activity (Paudel et al. 2011).

One AChE inhibitor identified as biruloquinone (IC_{50} = 27.1 μg/mL) was isolated from *Cladonia macilenta*, which showed the most potent AChE inhibitory activity in the previous screening experiment. Inhibitory kinetic assay suggested that biruloquinone is a mixed-II inhibitor on AChE. Meanwhile, biruloquinone improved the viability of the H_2O_2- and β-amyloid-injured PC12 cells at 1–25 μg/mL. The protective effects are proposed to be related to the potent antioxidant activities of biruloquinone. These results suggest that biruloquinone has the potential to be developed as a multifunctional anti-AD agent (Luo et al. 2013).

Pejin et al. (2013) isolated a new depsidone from *Lobaria pulmonaria* with molecular formula $C_{21}H_{18}O_9$ which showed a moderate acetylcholinesterase inhibition activity (1 μg) in vitro. Similarly, it has been reported that lobaric acid extracted from the lichen *Heterodermia* sp. showed moderate inhibition against acetylcholinesterase with an IC_{50} value of 26.86 ± 0.9 μM and butyrylcholinesterase with an IC_{50} value of 36.76 ± 0.8 μM (Thadhani et al. 2014).

Reddy et al. (2016) considered 12 lichen-derived metabolites as possible pharmacological agents for diverse neurodegenerative and neuropsychiatric disorders. Their in vitro neurotrophic activity was assessed through a preliminary investigation using Neuro2A cells. They found four lichen compounds (atranorin, perlatolic acid, physodic acid, and usnic acid) that exhibited potent neurotrophic activity similar to that of a known neurotrophin, NGF. In addition, perlatolic acid exhibited more neurotrophicity than the well-studied neuroactive natural compound, resveratrol. These neurotrophically active lichen compounds were then evaluated at optimum concentrations for cytotoxic side effects in Neuro2A cells. Except for usnic acid, all the other neurotrophic lichen compounds (atranorin and physodic acid at 5 μM, while perlatolic acid at 0.5 μM) proved to be non-cytotoxic in Neuro2A cells. They found that perlatolic acid also exhibited good proneurogenic activity in ex vivo neurosphere cultures at 0.5 μM, while the 1 μM dose result was similar to that of the control. Thus, perlatolic acid appears to be neurogenic at this optimum neurotrophic concentration.

A β-orcinol depsidone, stictic acid, and salazinic acid from *Xanthoparmelia* spp. showed neuroprotective effect on U373MG cell line by diminishing ROS production. These compounds would be useful as antioxidant agents in neurodegenerative diseases. Also, cytotoxic potential of three lichen secondary metabolites including olivetoric acid and physodic acid isolated from *Pseudevernia furfuracea* and psoromic acid isolated from *Rhizoplaca melanophthalma* was investigated on human U87MG-GBM cell lines and primary rat cerebral cortex (PRCC) cells by Emsen et al. (2016). The IC_{50} values of olivetoric acid, physodic acid, and psoromic acid were 125.71, 698.19, and 79.40 mg/L for PRCC cells and 17.55, 410.72, and 56.22 mg/L for U87MG cells, respectively. This finding revealed that primarily olivetoric acid and then psoromic acid had high potential for use in the treatment of glioblastoma multiforme.

8.4 Antidiabetic Activity of Lichens

Diabetes mellitus is one of the common metabolic disorders with micro- and macrovascular complications that result in significant morbidity and mortality. It is considered as one of the five leading causes of death in the world. Diabetes mellitus is characterized by elevated plasma glucose concentrations resulting from insufficient insulin and insulin resistance, or both, along with disturbances in metabolism of lipids, carbohydrates, and proteins (Baynes 2015).

Diabetes mellitus can be categorized into type 1 or insulin-dependent diabetes mellitus also referred to as juvenile-onset diabetes resulting from a cellular-mediated autoimmune destruction of β-cells in the pancreas. It accounts for 5–10% globally of individuals with diabetes. The other category is type 2 or non-insulin-dependent diabetes mellitus whose onset is usually after 40 years of age and accounts for approximately 90–95% of the diabetes mellitus cases worldwide. It is also called adult-onset diabetes which affects individuals who have insulin resistance and usually have relative insulin deficiency (Baynes 2015; Khan et al. 2018).

Type 1 diabetes occurs when the immune system, the body's system for fighting infection, attacks and destroys the insulin-producing β-cells of the pancreas. Scientists think that type 1 diabetes is caused by genes and environmental factors, such as viruses, which could trigger the disease. Type 2 diabetes is caused by several factors, including lifestyle factors and genetics (Fig. 8.5).

The clinical manifestation of diabetes mellitus includes excessive excretion of urine (polyuria), thirst (polydipsia), constant hunger (polyphagia), weight loss,

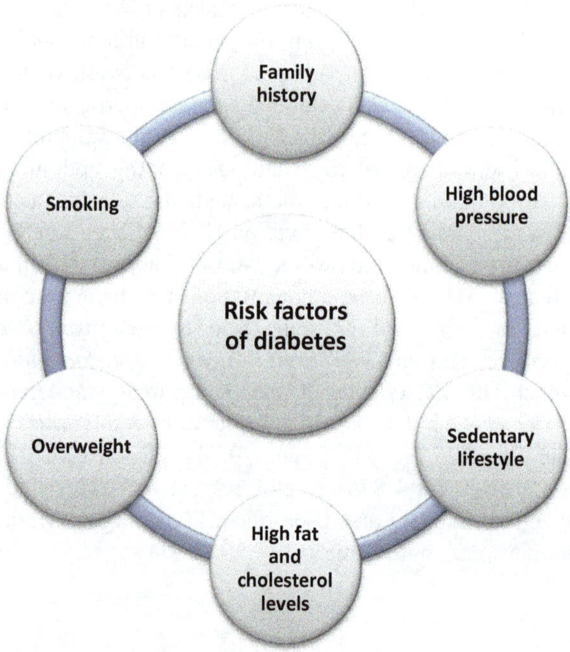

Fig. 8.5 Causes of diabetes mellitus

blurred vision, and fatigue. Acute, life-threatening consequences of uncontrolled diabetes are hyperglycemia with ketoacidosis or the nonketotic hyperosmolar syndrome. Long-term complications of diabetes include retinopathy with potential loss of vision; nephropathy leading to renal failure; peripheral neuropathy with risk of foot ulcers, amputations, and Charcot joints; and autonomic neuropathy causing gastrointestinal, genitourinary, and cardiovascular symptoms and sexual dysfunction (Barquilla García 2017).

Diabetes mellitus can be managed by diet and physical exercise, as well as modern drugs (insulin and/or oral hypoglycemic drugs such as sulfonylureas and biguanides). However, all hypoglycemic drugs are expensive and have many side effects. Different natural products have also been used traditionally to manage diabetes globally, and these are considered as relatively inexpensive and less toxic and with relatively little or no side effects. Hence, the search for safer and more effective hypoglycemic agents continues (Dsouza and Lakshmidevi 2015; Gupta et al. 2017).

8.5 In Vitro Methods Used in Diabetic Research

Currently, several methods have been applied to measure antidiabetic activity of lichens, and some of the most commonly used are listed below.

Alpha amylase is a prominent enzyme found in the pancreatic juice and saliva which hydrolyzes starch into mixture of smaller oligosaccharides comprising of maltose, maltotriose, and oligoglucans which are further degraded by glucosidase into glucose that enters the bloodstream upon absorption. Alpha glucosidase is confined to the mucosal brush border of the small intestine and catalyzes the final step of digestion of starch and sucrose. Inhibitors of alpha amylase and alpha glucosidase delay the breaking down of carbohydrates and diminish the postprandial blood glucose excursion. One of the strategies and methods adopted to cure diabetes mellitus involves the inhibition of carbohydrate-digesting enzymes such as alpha amylase and alpha glucosidase in the gastrointestinal glucose absorption thereby lowering postprandial glucose level (Abusufyan et al. 2018). This is an attempt to search for alternative drugs from natural products with increased potency and lesser adverse effects than existing drugs (Dsouza and Lakshmidevi 2015).

Alpha Amylase Activity can be measured in vitro by hydrolysis of starch in the presence of alpha amylase enzyme. This process was quantified by using iodine, which gives a blue color with starch. The reduced intensity of blue color indicates the enzyme-induced hydrolysis of starch into monosaccharides. If the substance/extract possesses alpha amylase inhibitory activity, the intensity of blue color will be more. In other words, the intensity of blue color in test sample is directly proportional to alpha amylase inhibitory activity (Parimelazhagan 2016).

Alpha Glucosidase The most common method for evaluation of alpha glucosidase inhibitory activity uses *p*-nitrophenyl-α-D-glucopyranoside (PNP-G) as a substrate.

In this method, alpha glucosidase enzyme will hydrolyze *p*-nitrophenyl α-D-glucopyranoside to become *p*-nitrophenol with yellow color and glucose.

Dipeptidyl Peptidase IV (DPP IV) Sequentially, it removes dipeptides from the N-terminus of polypeptides, leaving alanine and proline as the ultimate residue. Dipeptidyl peptidase IV is a multifunctional protein, located on the surface of epithelial, endothelial, and lymphoid cells. The soluble form of the enzyme is present in plasma. Substrates of DPP IV are a variety of proline-containing peptides such as growth factors, chemokines, neuropeptides, and vasoactive peptides. A significant therapeutic approach for type 2 diabetes is the inhibition of DPP IV; thus the activity of incretin hormones, responsible for glucose homeostasis, is unhindered. Besides that, DPP IV has a role in immune regulation, signal transduction, and apoptosis, and furthermore it has a major role in tumor progression (Villhauer et al. 2003; Dsouza and Lakshmidevi 2015).

Protein Tyrosine Phosphatase 1B (PTP 1B) Tyrosine phosphorylation of proteins is a basic mechanism to control cell growth and differentiation. Protein tyrosine phosphatase 1B has been implicated in modulating signaling pathways initiated by the activation of the tyrosine kinase receptor super family. It regulates insulin signaling by dephosphorylating and inactivating insulin receptor. Protein tyrosine phosphatasc 1B inhibition could provide a new therapeutic option in the treatment of type 2 diabetes. It is directed by the opposing activities of protein tyrosine kinases which catalyze phosphorylation and protein tyrosine phosphatases that are responsible for dephosphorylation. An inappropriate operation leads to aberrant tyrosine phosphorylation, contributing to the development of many diseases, viz., cancer and diabetes (Haj et al. 2003).

Cell Lines as Models to Study Antidiabetic Activity In Vitro Cell lines simplify examination of physiological and pathophysiological processes in cells, making the use of animal models unnecessary. Cell lines are used as models in studying the physiology of cells and in testing the effects of different compounds on specific cell types. A great advantage of using cell lines is attaining reliable and reproducible results. Additionally, there are some disadvantages. Main disadvantages are modifications of characteristics of the cell line, due to the continuous growth of cells, and limitless growth of cell cultures. Those disadvantages are associated with their tumor origin, which is why they usually have modified metabolism, abnormal chromosomal content, or abnormal protein expression, along with other genetic mutations. Bioengineering technologies have contributed to the establishment of a more adequate cultured cell line, which helped in studying the mechanism involved in insulin secretion and glucose uptake by target organs and in studying transcriptional factors such as peroxisome proliferator-activated receptor-γ (PPAR-γ), which are the targets of nowadays modern therapy. Malfunctions of insulin secretion lead to a number of diseases, including diabetes, hyperinsulinemic hypoglycemia, and insulinoma. By understanding the underlying mechanisms, responsible for insulin secretion, new diabetes treatments can be developed, such as engineering insulin-producing cells, which can be used as a replacement. Glucose phosphorylation in β-cells is the key regulatory event in joining insulin secretion and extracellular glucose concentrations. Glucose is a major physiological insulin secretagogue and a potent regulator of the β-cell activity, which produces insulin, in response to a high blood glucose concentration and with an increase in cytoplasmic calcium concentration (Ca^{2+}). In the absence of glucose, the ATP-sensitive K-channels (KATP) remain closed due to low cytoplasmic ATP concentration. High extracellular glucose concentration eases glucose entry into cells through GLUT2 transporters, and glycolysis increases production of ATP, thereby closing KATP in the plasma membrane; depolarization takes place resulting in opening of voltage-gated Ca^{2+} channels and increasing the influx of extracellular Ca^{2+}. Electrical activity together with Ca^{2+} influx occurs in bursts and results in insulin release (Dsouza and Lakshmidevi 2015).

Insulin Secretion Assay Using Pancreatic RIN m5F Cells RIN m5F (Rat insulinoma) cells are one of the most widely used insulin-secreting cell lines, containing insulin, a small amount of somatostatin and glucagon. Different aberrant features of glucose transport have been noted for RINm5F and RINr cells, and both cells exhibit inadequate sensitivity to glucose. A solid-phase two-site sandwich immunoassay is used in Mercodia Rat insulin ELISA (enzyme-linked immunosorbent assay) based on a technique described by Persaud (Dsouza and Lakshmidevi 2015). Two monoclonal antibodies react with different antigenic determinants on a molecule of insulin. During incubation, insulin in the sample reacts with anti-insulin antibodies and peroxidase-conjugated anti-insulin antibodies which are bound for a better microtitration. The unbound enzyme-labeled antibodies are removed by washing, and the bound conjugate is detected using 3,3′,5,5′-tetramethylbenzidine with the addition of acid, for achieving a colorimetric end point, which is read spectrophotometrically.

Insulin Secretion Assay Using Pancreatic β-Cell Line The release of insulin from the β-cell line depends on glucose concentration, in a dose-dependent manner. Insulin released into the medium, as a response to increasing concentrations of glucose, is used in detecting changes in secretory properties of the cell, like GSIS, basal secretion, and shifts in glucose sensitivity.

β-Cell Proliferation Assay β-Cells, located in islets of Langerhans, produce, store, and secrete insulin. Type 1 and type 2 diabetes develop due to the loss of β-cell mass and function. These cells get destroyed by the immune system (in case of type 1 diabetes) or do not produce the enough amount of insulin to maintain normal blood glucose concentration (in case of type 2 diabetes). β-Cell mass is regulated by various processes, including apoptosis, neogenesis, proliferation, and differentiation. The compounds that can restore β-cells or promote their proliferation are of the utmost importance.

Glucose Uptake Assay Using 3T3 L1 Cells Insulin promotes glucose uptake, metabolism, and storage in adipose tissue and skeletal muscle. Insulin stimulates phosphorylation of insulin receptor substrates (IRS) by kinase, which leads to activation of PI3 kinase, PKB, and protein kinase C isoforms. Activated PKB translocates Glut4 to the cell surface and stimulates glucose transport in muscle and fat cells, whereas it phosphorylates and inhibits GSK-3. Inhibitor of GSK-3 enhances insulin signaling, thereby favoring glucose entry into the muscle cells and adipose tissue. Inactivation of GSK-3 in 3T3 L1-differentiated cells [adipocytes] stimulates glucose uptake which can be measured by incorporation of 2-deoxy-14C-glucose and quantified by using scintillation counter.

Peroxisome Proliferator-Activated Receptor-γ Luciferase Assay Peroxisome proliferator-activated receptor-γ (PPAR-γ) is a nuclear receptor that has an important role in lipid and glucose metabolism. PPAR-γ is located mainly in the white adipose tissue. Upon activation, it stimulates the expression of genes involved in insulin sensitivity and glucose homeostasis. Conformational changes occur upon the binding of ligand, and that stabilizes its interaction with retinoid X receptor (RXR), which leads to formation of heterodimers. The heterodimer (PPAR-γ/RXR) then takes on a set of coactivators to the promoter regions of target genes and brings about transcription (Berger and Wagner 2002; Evans et al. 2004). A ligand-stimulated trans-repression of genes is mediated through a peroxisome proliferator response in which the PPAR-γ ligand binds to the SMRT/NCoR corepressor complex 46. This interaction prevents proteasomal degradation of the corepressor and maintains the gene in a repressed mode. Activation of PPAR-γ is an important step in glucose uptake.

Glucose Uptake Assay For a long time, the gold standard for measuring glucose uptake has been the radiolabeled analog 3H-2DG and the measurement of accumulated 3H-2DG6P (Yamamoto et al. 2015). Although the method is sensitive and works well, it requires multiple wash steps. Many researchers do not use this method because they want to avoid the handling and disposal of radioactive material. Commercial nonradioactive methods use non-radiolabeled 2DG and measure the accumulation of 2DG6P through the action of G6PDH (glucose-6-phosphate dehydrogenase) (Saito et al. 2011). This enzyme oxidizes 2DG6P to form NADPH, which can then be used to generate probes for absorbance or fluorescence assays. Both assays are amenable to multiwell plate formats, but they require multiple processing steps. Due to a cycling detection reaction, the absorbance assay is most sensitive and can detect very low amounts of 2DG6P. However, both assays have small signal windows (i.e., signal above no 2DG6P background) that limit their use.

Recently is used a luminescent glucose uptake assay that is based on the same principle as the aforementioned assays: The accumulation of 2DG6P and detection through the enzymatic action of G6PDH (3). The luminescent assay has a number of advantages:

- Similar sensitivity to the radioactive assay but without any wash steps or handling and disposal of radioactive material
- Larger signal window than the absorbance and fluorescence assays, making it easier to detect small changes and requiring less assay optimization
- Amenable to high-throughput screening due to fewer false hits and steady glow-type signals

The luminescent assay can detect glucose uptake from as few as 5000 cells with a greater than threefold signal-to-background ratio, and the signal remains linear up to at least 50,000 cells. The luminescent assay is the best combination of simplicity and sensitivity available for a nonradioactive multiwell plate method (Valley et al. 2016).

8.6 Antidiabetic Activity of Lichen Extracts and Their Metabolites

A few studies have discovered the positive potential of lichens extracts as potent antidiabetic agents (Table 8.2). Their hypoglycemic effect has been assessed in numerous methods, including through their inhibitory potential of FC;-amylase and FC;-glucosidase (carbohydrate hydrolyzing enzymes) or protein tyrosine phosphatase 1B (PTP1B), which is recognized as the major negative regulator in insulin signaling, or through antioxidative effect, involved in restoring of insulin secreting pancreatic cells.

For numerous lichen extracts was found relatively strong FC;-amylase inhibitory effect. For example, Vinayaka et al. (2013) determined amylase inhibitory activity of methanol extract of six macrolichens including *Everniastrum cirrhatum*, *Usnea sinensis*, *Ramalina conduplicans*, *R. hossei*, *Parmotrema pseudotinctorum*, and *P. tinctorum* collected from Western Ghats, Karnataka. They found that tested extracts caused a dose-dependent inhibition of amylase activity. Among lichens, *R. conduplicans* caused higher inhibition of enzyme activity followed by *R. hossei*, *P. tinctorum*, *U. sinensis*, *E. cirrhatum*, and *P. pseudotinctorum*. *P. tinctorum*. In that study, *P. tinctorum* acetate extract exhibited alpha amylase activity with an estimated IC_{50} value of 587.74 μg/mL.

Shivanna et al. (2015) examined alpha amylase inhibitory effect of methanol and ethyl acetate extracts of *Flavoparmelia caperata* and *Physcia aipolia*. The alpha amylase inhibition assay showed that the methanolic extract of *F. caperata* had highest inhibition of 49% at 15 mg/mL, followed by *P. aipolia* with 46% inhibition. Four different lichen extracts obtained from *Ramalina sinensis*, *Heterodermia leucomelos*, *Herpothallon* sp., and *Parmotrema reticulatum* were evaluated by Hengameh et al. (2016). Investigation was carried out on three different concentrations (5, 10, and 15 mg/mL) to get an idea about inhibitory concentration of each species. In all tested samples, a dose-dependent response was observed. Highest percent of inhibition was found in ethyl acetate extract of *H. leucomelos* with 95% of inhibition followed by *Herpothallon* sp. 85% and *P. reticulatum* 82% in 15 mg/mL concentration. *R. sinensis* showed its highest activity in ethanol extract

Table 8.2 Reported antidiabetic activity of lichens and lichen compounds

Lichens/Compounds	Activity	Reference
Caloplaca biatorina	Alpha glucosidase	Valadbeigi (2016)
Cladonia rei	Alpha amylase	Valadbeigi and Shaddel (2016)
Everniastrum cirrhatum	Alpha amylase	Vinayaka et al. (2013)
Flavoparmelia caperata	Alpha amylase	Valadbeigi and Shaddel (2016)
Herpothallon sp.	Alpha amylase	Hengameh et al. (2016)
Heterodermia leucomela	Alpha amylase	Shivanna et al. (2015), Hengameh et al.
Hyperphyscia adglutinata	Alpha amylase	2016
Parmotrema	Alpha amylase	Valadbeigi and Shaddel (2016)
pseudotinctorum	Alpha amylase	Vinayaka et al. (2013)
P. tinctorum	Alpha amylase	Vinayaka et al. (2013)
P. reticulatum	Alpha amylase	Hengameh et al. (2016)
P. chinense	Alpha amylase	Valadbeigi and Shaddel (2016)
Peltigera praetextata	Alpha amylase	Valadbeigi and Shaddel (2016)
Physcia aipolia	Alpha amylase	Shivanna et al. (2015)
Punctelia borreri	Alpha amylase	Valadbeigi and Shaddel (2016)
P. subrudecta	Alpha amylase	Valadbeigi and Shaddel (2016)
Ramalina sinensis	Alpha glucosidase	Hengameh et al. (2016)
R. nervulosa	Alpha glucosidase	Verma et al. (2012)
R. pacifica	Alpha glucosidase	Verma et al. (2012)
R. celastri	Alpha amylase	Verma et al. (2012)
R. conduplicans	Alpha amylase	Vinayaka et al. (2013)
R. hossei	Alpha amylase	Vinayaka et al. (2013)
R. hyrcana	Alpha amylase	Valadbeigi and Shaddel (2016)
R. pollinaria	Alpha amylase	Valadbeigi and Shaddel (2016)
Usnea articulate	Alpha amylase	Valadbeigi and Shaddel (2016)
U. sinensis		Vinayaka et al. (2013)
Atranorin	Antiglycation, PTP	Thadhani (2013), Seo et al. (2011)
Brialmontin 1	PTP	Seo et al. (2011)
Diterpene furanoids	PTP	Cui et al. (2012)
Divaricatic acid	Antiglycation	Thadhani (2013)
Ethyl hematomate	Antiglycation	Choudhary et al. (2011)
Ethyl orsellinate	Antiglycation	Choudhary et al. (2011)
Gyrophoric acid	Antiglycation, PTP	Choudhary et al. (2011), Seo et al.
Lecanoric acid	Antiglycation, PTP	(2009a)
Lobaric acid	PTP	Choudhary et al. (2011), Seo et al.
Methyl hematomate	PTP	(2009a)
Methylorsellinate	Alpha glucosidase,	Seo et al. (2009b)
Methyl-β-orcinol	PTP	Seo et al. (2011)
carboxylate	PTP	Karunaratne et al. (2014), Seo et al.
Orsellinic acid	Antiglycation	(2009a)
Salazinic acid	Alpha glucosidase	Karunaratne et al. (2014)
Sekikaic acid	Alpha glucosidase	Choudhary et al. (2011)
Zeorin	Alpha glucosidase,	Verma et al. (2012)
Usnic acid	PTP	Verma et al. (2012)
	Alpha glucosidase,	Thadhani (2013), Seo et al. (2011)
	Antiglycation	Verma et al. (2012), Thadhani (2013)

with 80% of inhibition to be last in inhibition order. In case of IC_{50}, there was a positive relationship between its values with inhibition percentage. For inhibition of 50% of enzyme activity by *H. leucomelos*, as the most potent species, 27.6 mg/mL of this lichen extract required, while the same was at 3.31 mg/mL in case of *R. sinensis*.

In study reported by Valadbeigi and Shaddel (2016), the effect of ten Iranian macrolichens on alpha amylase was tested. The extracts showed a dose-dependent inhibitory effect on amylase: *Usnea articulate* > *Ramalina pollinaria* > *R. hyrcana* > *Cladonia rei* > *Flavoparmelia caperata* > *Parmotrema chinense* > *Punctelia subrudecta* > *P. borreri* > *Hyperphyscia adglutinata* > *Peltigera praetextata*. The highest inhibition of amylase was 60% at extract concentration 75 mg/mL in *U. articulate*.

With regard to pure lichen compounds, no reports exist for the evaluation against the alpha amylase. However, it could be resolved that the alpha amylase inhibitory activity shown by lichens could be attributed to their phenolics. The structure activity relationship of polyphenols isolated from plant sources has shown that alpha amylase inhibitory effect is influenced by a number of hydroxyl groups and their positions, methylation, methoxylation, glycosylation, and so on. Broadly, it is considered that hydroxylation of phenols increases the alpha amylase inhibitory activity and methoxylation, blocks the free hydroxyl groups, and reduces the inhibitory activity. However, lichen polyphenols are structurally distinct from other phenols such as flavonoids, catechins, and tannins found in higher plants. Lichen phenolics are mainly monocyclic phenols, depsides, depsidones, and dibenzofurans, derived through the acetyl-polymalonly pathway, with mainly orsellinic acid as the basic unit in the biosynthesis (Thadhani and Karunaratne 2017). Thus, it would be interesting to estimate the alpha amylase inhibitory activity of these different classes of polyphenols and to analyze its structure activity relationship, to describe its mechanism of action.

Compared to alpha amylase inhibitory activity where several different lichens have been examined, an extensive literature survey showed only four lichens extracts were evaluated for their alpha glucosidase inhibitory activity. Valadbeigi (2016) reported that methanol extract of *Caloplaca biatorina* showed strong alpha glucosidase inhibitory activities with IC_{50} value of 17.12 μg/mL. Verma et al. (2012) found that methanol extract of *Ramalina nervulosa*, *R. pacifica*, and *R. celastri* showed high alpha glucosidase inhibition activity of 87.88%, 76.84%, and 69.97% in comparison with other solvent extracts acetone, ethanol, and DMSO (31.93–68.25%). The two species *R. nervulosa* and *R. pacifica* inhibited alpha glucosidase more than 70% and were found to be higher than the standard alpha glucosidase inhibitor acarbose (67.80%), while *R. celastri* had shown alpha glucosidase inhibitor activity almost equivalent to the acarbose.

On the other hand, six common secondary metabolites have been evaluated in two different studies and have shown promising antihyperglycemic effect. Karunaratne et al. (2014) found three potent alpha glucosidase inhibitors from the lichen *Cladonia* sp. Triterpenoid zeorin showed the highest alpha glucosidase inhibitory activity with an IC_{50} value of 100.0 ± 0.3 μM, which is seven times lower than that of acarbose. The two phenolic compounds, methyl β-orcinolcarboxylate ($IC_{50} = 140.0 \pm 0.6$ μM) and methylorsellinate ($IC_{50} = 165.0 \pm 1.2$ μM), also exhibited four- to fivefold higher activity than acarbose. Atranorin and lobaric acid, the other two metabolites isolated from the lichen, did not show any alpha glucosidase inhibitory potential. Similarly, salazinic acid from *Ramalina celastri*, sekikaic acid from *R. nervulosa*, and usnic acid from *R. pacifica* showed relatively strong alpha glucosidase inhibition with 55.1%, 68.1%, and 56.3% (Verma et al. 2012).

A group of researchers evaluated different lichens (*Umbilicaria antarctica*, *Stereocaulon alpinum*, *Lecidella carpathica*, and *Huea* sp.) against protein tyrosine phosphatase inhibitory activity. The compounds responsible for their inhibitory activity were identified as zeorin methyl-β-orcinol carboxylate, methylorsellinate, methyl hematomate, lecanoric acid, gyrophoric acid, atranorin, brialmontin 1, and lobaric acid along with four new diterpene furanoids. Gyrophoric acid consisting of three orsellinic acid units had shown almost nine times more potent protein tyrosine phosphatase inhibitory activity than lecanoric acid which consisted of two orsellinic rings. Similarly, lecanoric acid had shown almost nine times more potent inhibitory activity than monocyclic methylorsellinate. On the other hand, brialmontin 1, with more hydrophobicity than lecanoric acid, has shown higher inhibitory potential. These results reveal that inhibitory potency appears to get stronger with increase in lipophilicity. Zeorin, the most lipophilic compound out of the tested compounds, had shown the highest potential as protein tyrosine phosphatase inhibitor (Seo et al. 2009a, b, 2011; Cui et al. 2012).

With regard to antiglucation activity, in literature no data on crude lichen extracts, however, antiglycation activity of several secondary metabolites has been described. Among the compounds examined, ethyl hematomate, ethyl orsellinate, lecanoric acid, and gyrophoric acid had shown substantial activity as an antiglycation agent (Choudhary et al. 2011). In another study, atranorin, divaricatic acid, and usnic acid had shown antiglycation activity (Thadhani 2013).

References

Abusufyan S, Ibrahim M, Mohib K (2018) Comparative in vitro antidiabetic and antioxidant activity of various extracts of *Ficus* species. Pharm J 10:349–354

Ataie A, Shadifar M, Ataee R (2016) Polyphenolic antioxidants and neuronal regeneration. Basic Clin Neurosci 7:81–90

Barquilla García A (2017) Brief update on diabetes for general practitioners. Rev Esp Sanid Penit 19:57–65

Baynes HW (2015) Classification, pathophysiology, diagnosis and management of diabetes mellitus. J Diabetes Metab 6:541

Behera BC, Adawadkar B, Makhija U (2004) Capacity of some Graphidaceous lichens to scavenge superoxide and inhibition of tyrosinase and xanthine oxidase activities. Curr Sci 87:83–87

Berger J, Wagner JA (2002) Physiological and therapeutic roles of peroxisome proliferatoractivated receptors. Diabetes Technol Ther 4:163–174

Chonpathompikunlert P, Boonruamkaew P, Sukketsiri W et al (2018) The antioxidant and neurochemical activity of *Apium graveolens* L. and its ameliorative effect on MPTP-induced Parkinson-like symptoms in mice. BMC Complement Altern Med 18:103

Choudhary MI, Ali M, Wahab AT et al (2011) New antiglycation and enzyme inhibitors from *Parmotrema cooperi*. Sci China Chem 54:1926–1931

Crous-Bou M, Minguillón C, Gramunt N et al (2017) Alzheimer's disease prevention: from risk factors to early intervention. Alzheimers Res Ther 9:71

Cui Y, Yim JH, Lee DS et al (2012) New diterpene furanoids from the Antarctic lichen *Huea* sp. Bioorg Med Chem Lett 22:7393–7396

de Paz GA, Raggio J, Gómez-Serranillos MP et al (2010) HPLC isolation of antioxidant constituents from *Xanthoparmelia* spp. J Pharm Biomed Anal 53:165–171

Dsouza D, Lakshmidevi N (2015) Models to study in vitro antidiabetic activity of plants: a review. Int J Pharm Biol Sci 6:732–741

Ellman GL, Courtney DK, Andreas V et al (1961) A new and rapid colorimetric determination of acetylcholinesterase activity. Biochem Pharmacol 7:88–95

Emsen B, Aslan A, Togar B et al (2016) In vitro antitumor activities of the lichen compounds olivetoric, physodic and psoromic acid in rat neuron and glioblastoma cells. Pharm Biol 54:1748–1762

Evans RM, Barish GD, Wang YX (2004) PPARs and the complex journey to obesity. Nat Med 10:355–361

Fernandez-Moriano C, Divakar PK, Crespo A et al (2015) Neuroprotective activity and cytotoxic potential of two Parmeliaceae lichens: identification of active compounds. Phytomedicine 22:847–855

Fernández-Moriano C, Divakar PK, Crespo A et al (2017) In vitro neuroprotective potential of lichen metabolite fumarprotocetraric acid via intracellular redox modulation. Toxicol Appl Pharmacol 316:83–94

Griess P (1879) Bemerkungen zu der abhandlung der H.H. Weselsky und Benedikt "Ueber einige azoverbindungen". Chem Ber 12:426–428

Gupta RC, Chang D, Nammi S et al (2017) Interactions between antidiabetic drugs and herbs: an overview of mechanisms of action and clinical implications. Diabetol Metab Syndr 9:59

Haj FG, Markova B, Klaman LD et al (2003) Regulation of receptor tyrosine kinase signaling by protein tyrosine phosphatase-1B. J Biol Chem 278:739–744

Hengameh P, Shivanna R, Rajkumar HG (2016) In-vitro inhibitory activity of some lichen extracts against FC;-amylase enzyme. Eur J Biomed Pharm Sci 3:315–318

Hippius H, Neundörfer G (2003) The discovery of Alzheimer's disease. Dialogues Clin Neurosci 5:101–108

Honda NK, Gonçalves K, Brandão LZG et al (2016) Screening of lichen extracts using tyrosinase inhibition and toxicity against *Artemia salina*. Orbital Electron J Chem 8:181–188

Ikeda T, Yamada M (2010) Risk factors for Alzheimer's disease. Brain Nerve 62:679–690

Jack CR Jr, Albert MS, Knopman DS et al (2011) Introduction to the recommendations from the National Institute on Aging-Alzheimer's Association workgroups on diagnostic guidelines for Alzheimer's disease. Alzheimers Dement 7:257–262

Karunaratne V, Thadhani VM, Khan SN et al (2014) Potent FC;-glucosidase inhibitors from the lichen Cladonia species from Sri Lanka. J Natl Sci Found Sri Lanka 42:95–98

Khan MF, Rawat AK, Khatoon S et al (2018) In vitro and in vivo antidiabetic effect of extracts of *Melia azedarach, Zanthoxylum alatum*, and *Tanacetum nubigenum*. Integr Med Res 7:176–183

Kim GH, Kim JE, Rhie SJ et al (2015) The role of oxidative stress in neurodegenerative diseases. Exp Neurobiol 24:325–340

Kim M, Ho A, Lee JH (2017) Autophagy and human neurodegenerative diseases—a fly's perspective. Int J Mol Sci 18:1596

Kremens D, Hauser RA, Dorsey ER (2014) An update on Parkinson's disease: improving patient outcomes. Am J Med 127:S3

Liu Z, Zhou T, Ziegler AC et al (2017) Oxidative stress in neurodegenerative diseases: from molecular mechanisms to clinical applications. Oxid Med Cell Longev 2017:2525967

Luo H, Li C, Kim JC et al (2013) Biruloquinone, an acetylcholinesterase inhibitor produced by lichen-forming fungus *Cladonia macilenta*. J Microbiol Biotechnol 23:161–166

Parimelazhagan T (2016) Anti-diabetic activity. In: Parimelazhagan T (ed) Pharmacological assays of plant-based natural products. Springer, Cham, pp 139–143

Paudel B, Bhattarai HD, Koh HY et al (2011) Ramalin, a novel nontoxic antioxidant compound from the Antarctic lichen Ramalina terebrata. Phytomedicine 184:1285–1290
Paulson HL (2009) The spinocerebellar ataxias. J Neuroophthalmol 29:227–237
Pejin B, Tommonaro G, Iodice C et al (2013) A new depsidone of *Lobaria pulmonaria* with acetylcholinesterase inhibition activity. J Enzyme Inhib Med Chem 28:876–878
Pratt AJ, Getzoff ED, Perry JJ (2012) Amyotrophic lateral sclerosis: update and new developments. Degener Neurol Neuromuscul Dis 2:1–14
Reddy RG, Veeraval L, Maitra S et al (2016) Lichen-derived compounds show potential for central nervous system therapeutics. Phytomedicine 23:1527–1534
Roos RA (2010) Huntington's disease: a clinical review. Orphanet J Rare Dis 5:40
Saito K, Lee S, Shiuchi T et al (2011) An enzymatic photometric assay for 2-deoxyglucose uptake in insulin-responsive tissues and 3T3-L1 adipocytes. Anal Biochem 412:9–17
Seo C, Choi YH, Ahn JS et al (2009a) PTP1B inhibitory effects of tridepside and related metabolites isolated from the Antarctic lichen *Umbilicaria antarctica*. J Enzyme Inhib Med Chem 24:1133–1137
Seo C, Sohn JH, Ahn JS et al (2009b) Protein tyrosine phosphatase 1B inhibitory effects of depsidone and pseudodepsidone metabolites from the Antarctic lichen *Stereocaulon alpinum*. Bioorg Med Chem Lett 19:2801–2803
Seo C, Yim JH, Lee HK et al (2011) PTP1B inhibitory secondary metabolites from the Antarctic lichen *Lecidella carpathica*. Mycology 2:18–23
Shivanna R, Hengameh P, Rajkumar HG (2015) Screening of lichen extracts for in-vitro antidiabetic activity using alpha amylase inhibitory assay. Int J Biol Pharm Res 6:364–367
Thadhani VM (2013) Semisynthesis and bioactivities of lichen metabolites. Scholars' Press, Saarbrücken
Thadhani VM, Karunaratne V (2017) Potential of lichen compounds as antidiabctic agcnts with antioxidative properties: a review. Oxid Med Cell Longev 2017:2079697
Thadhani VM, Naaz Q, Choudhag MI et al (2014) Enzyme inhibitory and immunomodulatory activities of the depsidone lobaric acid extracted from the lichen *Heterodermia* sp. J Natl Sci Found Sri Lanka 42:193–196
Valadbeigi T (2016) Chemical composition and enzymes inhibitory, brine shrimp larvae toxicity, antimicrobial and antioxidant activities of *Caloplaca biatorina*. Zahedan J Res Med Sci 18: e4267
Valadbeigi T, Shaddel M (2016) Amylase inhibitory activity of some macro lichens in Mazandaran province, Iran. Physiol Pharmacol 20:215–219
Valley MP, Karassina N, Aoyama N et al (2016) A bioluminescent assay for measuring glucose uptake. Anal Biochem 505:43–50
Verma N, Behera BC, Sonone A et al (2008) Lipid peroxidation and tyrosinase inhibition by lichen symbionts grown in vitro. Afr J Biochem Res 2:225–231
Verma N, Behera BC, Om Sharma B (2012) Glucosidase inhibitory and radical scavenging properties of lichen metabolites salazinic acid, sekikaic acid and usnic acid. Hacettepe J Biol Chem 40:7–21
Villhauer EB, Brinkman JA, Naderi GB et al (2003) 1-[(3-hydroxy-1 adamantyl) amino acetyl]-2-cyano-(S)-pyrrolidine: a potent, selective, and orally bioavailable dipeptidyl peptidase IV inhibitor with antihyperglycemic properties. J Med Chem 46:2774–2789
Vinayaka KS, Karthik S, Nandini KC et al (2013) Amylase inhibitory activity of some macrolichens of western ghats, Karnataka, India. Indian J Novel Drug Deliv 5:225–228
Yamamoto N, Ueda-Wakagi M, Sato T et al (2015) Measurement of glucose uptake in cultured cells. Curr Protoc Pharmacol 71:12.14.1–12.14.26
Zlatanović I, Stanković M, Stankov-Jovanović V et al (2017) Biological activities of *Umbilicaria crustulosa* (Ach.) Frey acetone extract. J Serb Chem Soc 82:141–150
Zrnzević I, Stanković M, Stankov Jovanovic V et al (2017) *Ramalina capitata* (Ach.) Nyl. acetone extract: HPLC analysis, genotoxicity, cholinesterase, antioxidant and antibacterial activity. EXCLI J 16:679–687

Future Directions in the Study of Pharmaceutical Potential of Lichens

9

Neeraj Verma and Bhaskar C. Behera

Abstract

Lichens are a stable self-supporting symbiotic organism, composed of a fungal and an algal partner. In this symbiotic form, lichens produce a number of unique secondary metabolites through various biosynthetic pathways, namely, acetyl polymalonyl, shikimic acid and mevalonic acid pathways. Most of the lichen substances are phenolic compounds and are reported to have wide variety of biological actions: antioxidant, antimicrobial, antiviral, anti-inflammatory, analgesic, antipyretic, antiproliferative and cytotoxic effects. Acetyl polymalonyl-derived polyketide compounds, depsides, depsidones, dibenzofuranes, xanthones and naphthaquinones, are of great interest. Compounds from other pathways are esters, terpenes, steroids, terphenylquinones and pulvinic acid. Although manifold biological properties of lichen secondary metabolites have been recognized, their pharmaceutical potential has not been fully explored due to their slow growing nature and difficulties in their artificial cultivation. Many researchers are still working hard to discover and identify the novel lead compounds from lichens. In this chapter, attention has been given to bring in notice some pharmaceutically important lichens and their secondary metabolites and to provide a direction for the study of lichen prospect.

9.1 Introduction

Lichens are unusual organisms composed of a fungus (mycobiont) with at least one symbiotic photosynthetic partner (photobiont). Lichens are found in almost all terrestrial habitats from the tropics to the polar region. The worldwide lichen

N. Verma (✉) · B. C. Behera
Mycology and Plant Pathology Group, Plant Science Division, Agharkar Research Institute, Pune, India

B. Ranković (ed.), *Lichen Secondary Metabolites*,
https://doi.org/10.1007/978-3-030-16814-8_9

biodiversity is estimated about 17,000 species (Nash 2008; Petrzik et al. 2014). Lichens are proven as the earliest colonizers of terrestrial habitats on earth with a worldwide distribution (Taylor et al. 1995; Mitrovic et al. 2011a). The adaptability of these organisms often adds a colourful aspect to their habitat growing either in extreme ecological environments or polar latitudes or at extreme altitudes (up to 7,400 m) (Boustie and Grube 2005). The distinct colours of these lichens are due to the massive accumulation of diverse secondary compounds, the "lichen substances". These substances are stress compounds, which help the organism for self-defence and to grow in very adverse and harsh environmental conditions (vide Huneck 1999). These lichen secondary metabolites are mainly phenolic compounds, depsides, depsidones, dibenzofuranes and other accessory pigments (Nash 1996), which are not known to occur in any other organism in nature (Huneck and Yoshimura 1996).

Throughout the ages, lichens and their secondary compounds are used for different purposes for humans and animals. Irrespective of the advances in medical sciences, the tribal peoples still utilize these organisms for different purposes depending on their nutritive and medicinal, decorative, brewing, distilling, dying, cosmetic and perfumery properties (Richardson 1988; Upreti and Chatterjee 2007; Karagoz et al. 2009). During the last five decades, exhaustive ethnobotanical work on lichens has been carried out by several workers in different parts of the world.

In Europe decoction of *Cetraria islandica* (Iceland moss) was recommended as a tonic for convalescents (Schneider 1904), in France the decoction of this species is used as pectoral and emollient (Novaretti and Lemordant 1990), while in Spain the same species is used for the treatment of catarrh and asthma and to treat mitigate inflammation (vide Upreti and Chatterjee 2007), and in Sweden, it is used for the treatment of asthma, diabetes, nephritis, lung diseases, whooping cough and colds (Airaksinen et al. 1986). In Alaska boiled juice of *Cladonia* species is drunk as medicine for diarrhoea and taken as tea for internal chest pains (Smith 1973; Kari et al. 1987). In the west coast of Vancouver Island in Canada, *Lobaria pulmonaria* is used as medicine against coughing up blood. *Lobaria retigera* and *Parmelia saxatilis* are used in China as Chinese medicine (Hu et al. 1980). Many species of *Usnea* like *Dolichousnea longissima* (*Usnea longissima*) in Chinese medicine are used as an expectorant and in the treatment of ulcers (Chopra et al. 1958), in East Africa *U. africana* is used to treat stomach ache (Kokwaro 1976), powder of *U. filipendula* is used in Russia to treat wounds and protection from bacterial infection (Moskalenko 1986), and *U. articulata* is used for wound and skin bruises in Auckland (Brooker et al. 1987).

In Asian countries lichens are still used as a folk and alternative medicine. Some workers, Brij et al. (1985), Brij (1988), Saklani and Upreti (1992) and Brij and Upreti (1995), have reported the information on ethnobotanical uses of some species of lichens used by different tribal communities of India. Lichen species *Lobaria pulmonaria* is used in Sikkim region for curing eczema, lung troubles, haemorrhages and asthma (Biswas 1956). In the same region another lichen species *Heterodermia diademata* is used to protect the wounds and cuts from infection (Saklani and Upreti 1992). In many other parts of India, different lichen species belonging to the family

Parmeliaceae viz. *Parmelia kamschadalis* are used to treat diarrhoea, dyspepsia, spermatorrhoea, amenorrhoea and dysentery, and the mixture of *Parmotrema sancti-angelii* ash and mustard/linseed oil is used for the treatment of ringworm-like skin disease (Brij and Upreti 1995). *Dolichousnea longissima* (*Usnea longissima*) is used for treating bone fractures in Garhwal and Himalayan regions of India (Brij 1988).

The genetic and phenotypic diversity of the organisms involved in lichen symbiosis represents a valuable source for commercially interesting compounds. With modern technology, the potential of discovering and utilizing useful lichen metabolites has increased. Till now 1,050 lichen compounds are reported (including those found in cultures), out of which few of them have shown various biological actions: antibiotic, antimycobacterial, antiviral, antitumour, analgesic, antipyretic and enzyme inhibitory (Matsubara et al. 1997; Huneck 1999; Muller 2001; Behera et al. 2006a; Oksanen 2006; Lopes et al. 2008; Molnar and Farkas 2010).

9.2 Class and Pathways of Lichen Compounds

Lichens are highly specialized fungi containing algal or cyanobacterial colonies in their thalli as a source of carbohydrates (Schwendener 1868). The majority of organic compounds found in lichens are secondary metabolites of the fungal component, which are exported outside the fungal cells to be found on cell surfaces as crystals in different parts of the thallus. They often accumulate in the upper cortex or in specialized structures such as fruiting bodies (Fahselt 1994; Elix 1996; Oksanen 2006). These compounds are usually insoluble in water and can only be extracted with organic solvents. Most of the lichen substances are phenolic compounds which act as antibiotics, pigments, enzyme inhibitors, immune-modulating agents, toxins, pesticides and antitumour agents. They have a major effect on the health, nutrition and economics of our society. A number of secondary metabolites are produced by lichens through biosynthetic pathways, mainly acetyl polymalonyl, shikimic acid and mevalonic acid pathways. Polymalonyl-derived polyketide compounds, depsides, depsidones, dibenzofuranes, xanthones and naphthaquinones, are of great interest. Compounds from other pathways are esters, terpenes, steroids, terphenylquinones and pulvinic acid (Fahselt 1994; Cohen and Towers 1995; Elix 1996; Muller 2001; Brunauer and Stocker-Worgotter 2005; Stocker-Worgotter 2005; Oksanen 2006).

9.3 Potential Biological Activities of Lichens of Pharmaceutical Interest

9.3.1 Antibacterial Activity

The first study on antibiotic properties of lichen was reported by Burkholder in 1944 (Burkholder et al. 1944). Later on Vartia (1973) reported the antimicrobial activity of several lichen species. After wide screening of antimicrobial activities of lichen extracts by various workers, it reflects that more than 50 % of lichen species that have been studied for various biological activities possess antimicrobial activity. Studies on lichen *Ramalina farinacea* and other 69 species of lichens from New Zealand showed inhibitory effect against *Bacillus*, *Pseudomonas*, *Escherichia coli*, *Streptococcus*, *Staphylococcus*, *Enterococcus* and *Mycobacterium* (Esimone and Adikwn 1999; Perry et al. 1999). Behera et al. (2005a) reported that acetone, methanol and light petroleum extracts of lichen *Usnea ghattensis* were effective against *Bacillus licheniformis*, *B. megaterium*, *B. subtilis* and *Staphylococcus aureus*. Rankovic et al. (2007a, b) tested aqueous, acetone and methanol extracts of *Cladonia furcata*, *Parmelia caperata*, *Parmelia pertusa*, *Hypogymnia physodes*, *Umbilicaria polyphylla*, *Lasallia pustulata*, *Parmelia sulcata*, *Umbilicaria crustulosa* and *Umbilicaria cylindrica* from Serbia on six species of bacteria and ten species of fungi. Many depsides, depsidones and dibenzofuranes have been isolated and characterized for antimicrobial activity. Mitrovic et al. (2011b) studied antibacterial activity of methanol extracts of five lichen species, *Parmelia sulcata*, *Flavoparmelia caperata*, *Evernia prunastri*, *Hypogymnia physodes* and *Cladonia foliacea*, against 15 bacterial strains. The results revealed the strongest inhibitory effect, especially on Gram-positive bacteria. In recent reports, Karagoz et al. (2009) evaluated aqueous and ethanol extracts of 11 different species from Turkey and determined potent antibacterial activity of aqueous extract of *Peltigera polydactyla* and ethanol extract of *Ramalina farinacea*. Chemical identification revealed evernic acid, vulpinic acid and hirtusneanoside as main components (Lawrey 1986; Rezanka and Sigler 2007). Verma et al. (2011) reported the bactericidal activity of lichens *Cladonia ochrochlora*, *Parmotrema nilgherrense*, *Parmotrema sancti-angelii* and their secondary compounds, alectoronic acid, atranorin, α-collatolic acid, fumarprotocetraric acid, hypoprotocetraric acid, lecanoric acid and protocetraric acid, with low MIC value against *Bacillus cereus*, *Bacillus subtilis*, *Escherichia coli*, *Klebsiella pneumoniae*, *Micrococcus luteus*, *Proteus vulgaris*, *Staphylococcus aureus*, *Streptococcus faecalis*, *Sarcina lutea* and yeast strains *Candida albicans* and *Cryptococcus* var. *diffluens* (Table 9.1).

9.3.2 Antioxidant Activity

The second most important biological activity of lichens reported so far is antioxidant activity. Lichens are the good source of natural antioxidants. They are highly resistant to potential damage caused by reactive oxygen species (ROS). Kranner et al. (2005) demonstrated the role of antioxidants in photoprotection and desiccation tolerance in lichen *Cladonia valcani* and its isolated symbionts. In the last few years, antioxidant activities of some depsides and depsidones isolated from several lichen species have been reported by many workers (Hidalgo et al. 1994; Gulcin et al. 2002; Odabasoglu et al. 2004; Choudhary et al. 2005; Devehat et al. 2007; Lopes et al. 2008; Paudel et al. 2008; Karunaratne et al. 2009). Several other biological activities of lichens have also been studied; Jayaprakash and Jaganmohan (2000) examined antioxidant capacities of methyl orsellinate, atranorin, orsellinic acid and lecanoric

Table 9.1 Lichen species and biological activities of their secondary metabolites

Lichen species	Metabolite	Class of compound	Production pathways[a]	Category of use	References
Alectoria ochroleuca	Usnic acid Vulpinic acid	Dibenzofuranes Pulvinic acid derivative	APP SAP	Antifungal activity	Lauterwein et al. (1995)
Arthothelium awasthii	Barbatic acid	Depsides	APP	Antioxidant activity Antityrosinase activity	Verma et al. (2008b, c)
Asashinea chrysantha	Chrysophanol Emodin	Anthraquinones Anthraquinones	APP APP	Anticancer activity	Mishchenko et al. (1980) Koyama et al. (1989)
Bulbothrix setschwanensis	Salazinic acid	Depsidones	APP	Anti-tyrosinase activity Xanthine oxidase inhibition	Behera and Makhija (2002)
Caloplaca aurantia	Emodin	Anthraquinones	APP	Antifungal activity	Manojlovic et al. (1998)
Caloplaca species	Anthraquinone	Anthraquinones	APP	Antifungal, antibacterial activity	Manojlovic et al. (2005)
Canoparmelia eruptens	Atranorin Lecanoric acid	Depsides Depsides	APP APP	Probiotic activity	Gaikwad et al. (2014)
Cetraria ericetorum *Cetraria islandica*	Fumarprotocetraric acid Naphthazarin Protocetraric acid Protolichesterinic acid	Depsidones Anthraquinones Depsidones Aliphatic and cycloaliphatic	APP APP APP APP	Inhibition of HIV-1 reverse transcriptase Inhibition of 5-lipoxygenase Antiproliferative activity Cytotoxic activity Antimicrobial activity Antimycobacterial activity Inhibition of DNA polymerase activity Inhibitor of human keratinocyte cell line	Pengsuparp et al. (1995) Borkowski et al. (1964) Ingolfsdottir et al. (1994, 1997a, 1998, 2000) Ogmundsdottir et al. (1998) Turk et al. (2003) Haraldsdottir et al. (2004)

(continued)

Table 9.1 (continued)

Lichen species	Metabolite	Class of compound	Production pathways[a]	Category of use	References
					Stepanenko et al. (1997) Paull et al. (1976) Muller et al. (1997)
Cladonia arbuscula	Usnic acid	Dibenzofuranes	APP	Antibiotic potential Antimycobacterial activity	Vartia (1973) Ingolfsdottir et al. (1998)
Cladonia confusa f. confusa (*Cladonia leptoclada*)	Usnic acid	Dibenzofuranes	APP	Antitumour activity	Kupchan and Kopperman (1975)
Cladonia ochrochlora	Fumarprotocetraric acid Hypoprotocetraric acid Protocetraric acid	Depsidones Depsidones Depsidones	APP APP APP	Bactericidal activity	Verma et al. (2011)
Dolichousnea diffracta (*Usnea diffracta*)	Decarboxy stenosporic acid Diffractaic acid Usnic acid	Depsides Depsides Dibenzofuranes	APP APP APP	Anti-inflammatory activity Analgesic activity Antipyretic activity Antimicrobial activity	Okuyama et al. (1995) Otsuka et al. (1972) Yamamoto et al. (1998)
Dolichousnea longissima (*Usnea longissima*)	Evernic acid Lichesterinic Protolichesterinic acid Longissiminone A Usnic acid	Depsides Aliphatic and cycloaliphatic Aliphatic and cycloaliphatic Dibenzofuranes	APP APP APP APP APP	Plant growth inhibition Inhibition Epstein–Barr virus activation Photosystem II inhibition Anti-inflammatory activity Inhibition of leukotriene B_4 biosynthesis	Nishitoba et al. (1987) Yamamoto et al. (1995) Endo et al. (1998) Kumar and Muller (1999) Choudhary et al. (2005)
Everniastrum cirrhatum	Atranorin Salazinic acid	Depsides Depsidones	APP APP	Probiotic activity	Gaikwad et al. (2014)

Flavocetraria nivalis *Flavocetraria cucullata*	Usnic acid	Dibenzofuranes	APP	Inhibition of tyrosine phosphatase Cytotoxic activity Apoptosis Antibiotic activity Antibacterial activity Antitrypanosomal activity Antiviral activity	Francolini et al. (2004) De Carvahlo et al. (2005) Mayer et al. (2005) Elo et al. (2007) Fazio et al. (2007) Bazin et al. (2008) Burlando et al. (2009)
Flavoparmelia euplecta	Euplectin	Naphthopyrones	APP	Cytotoxic agent	Ernst-Russell et al. (2000)
Graphis guimarana *Graphis nakanishiana*	Constictic acid Norstictic acid Stictic acid	Depsidones Depsidones Depsidones	APP APP APP	Scavenging of super oxide radical Inhibition of tyrosinase activity Inhibition of xanthine oxidase activity	Behera et al. (2006a, b)
Heterodermia obscurata	7,7′-Dichlorohypericin 5,7-Dichloroemodin	Anthraquinones Anthraquinones	APP APP	Inhibitory activity against herpes simplex virus type 1	Cohen and Towers (1995) Cohen et al. (1996)
Heterodermia podocarpa	Atranorin Zeorin	Depsides Terpenoids	APP MAP	Antioxidant activity Antityrosinase activity	Verma et al. (2008b, c)
Heterodea muelleri	Barbatic acid Diffractaic acid Usnic acid	Depsides Depsides Dibenzofuranes	APP APP APP	UV-C stress and cold temperature stress	Hager et al. (2008)
Hypotrachyna revoluta	8′-Methylmenegazziaic acid	Depsidones	APP	Radical scavenging activity	Papadopoulou et al. (2007)
Lecanora hybocarpa	Naphthazarin	Anthraquinones	APP	Cytotoxic activity	Ernst-Russell et al. (1999a)
Lobaria linita	Tenuiorin	Depsides	APP	Inhibition of 5-lipoxygenase activity	Ingolfsdottir and Gudmundsdottir (2002)

(continued)

Table 9.1 (continued)

Lichen species	Metabolite	Class of compound	Production pathways[a]	Category of use	References
Lobaria pulmonaria	Gyrophoric acid Melanin	Depsides Anthraquinones	APP APP	Light screening pigments	McEvoy et al. (2007)
Nephroma laevigatum	Emodin	Anthraquinones	APP	Antifungal activity	Manojlovic et al. (1998)
Parmelia nepalensis	Protolichesterinic acid	Aliphatic and cycloaliphatic	APP	Inhibition of 5-lipoxygenase activity	Kumar and Muller (1999)
Parmotrema austrosinense	Atranorin Lecanoric acid	Depsides Depsides	APP APP	Probiotic activity	Gaikwad et al. (2014)
Parmotrema cetratum (*Rimelia cetrata*)	Atranorin Consalazinic acid Salazinic acid	Depsides Depsidones Depsidones	APP APP APP	Probiotic activity	Gaikwad et al. (2014)
Parmotrema nilgherrense	Alectoronic acid Atranorin α-Collatolic acid	Depsidones Depsides Depsidones	APP APP APP	Bactericidal activity	Verma et al. (2011)
Parmotrema sancti-angelii	Atranorin α-Collatolic acid Lecanoric acid	Depsides Depsidones Depsides	APP APP APP	Bactericidal activity	Verma et al. (2011)
Parmotrema tinctorum	Atranorin Lecanoric acid	Depsides Depsides	APP APP	Antioxidant activity Antityrosinase activity	Verma et al. (2008b, c)
Peltigera aphthosa *Peltigera leucophlebia*	Tenuiorin	Depsides	APP	Inhibition of 5-lipoxygenase activity	Ingolfsdottir and Gudmundsdottir (2002)
Protousnea poeppigii	Divaricatinic acid Isodivaricatic acid Usnic acid	Depsides Cleavage product of depsides and depsidones Dibenzofuranes	APP APP APP	Antiprotozoal activity Antifungal activity Cytotoxic activity	Schmeda-Hirschmann et al. (2008) Bezivin et al. (2004)

Psoroma pallidum *Psoroma pulchrum* *Psoroma pholidotoides* (*Psoroma reticulatum*)	Pannarin	Depsidones	APP	Cytotoxic activity	Russo et al. (2006, 2008)
Pseudocyphellaria crocata	Tenuiorin	Depsides	APP	Inhibition of 5-lipoxygenase activity	Ingolfsdottir and Gudmundsdottir (2002)
Ramalina almquistii	Protolichesterinic acid Nephrosterinic acid	Aliphatic and cycloaliphatic	APP APP	Antitumour activity	Hirayama et al. (1980)
Ramalina celastri	Usnic acid	Dibenzofuranes	APP	Radical scavenging activity Glucosidase inhibitory activity	Verma et al. (2012a)
Ramalina farinacea	Evernic acid Norstictic acid Protocetraric acid Obtusatic acid Usnic acid	Depsides Depsidones Depsidones Depsides Dibenzofuranes	APP APP APP APP APP	Antimicrobial activity	Rastogi and Mehrotra (1993) Tay et al. (2004)
Ramalina nervulosa	Sekikaic acid Usnic acid	Depsides Dibenzofuranes	APP APP	Radical scavenging activity Glucosidase inhibitory activity	Verma et al. (2012a, b)
Ramalina pacifica	Salazinic acid Usnic acid	Depsidones Dibenzofuranes	APP APP	Radical scavenging activity Glucosidase inhibitory activity	Verma et al. (2012a, b)
Sphaerophorus globosus	Sphaerophorin	Depsides	APP	Cytotoxic activity Induction of apoptosis	Russo et al. (2006, 2008)
Stereocaulon alpinum *Stereocaulon azoreum*	Atranorin Lobaric acid	Depsides Depsidones	APP APP	Antiproliferative activity Inhibition of	Ingolfsdottir et al. (1996)

(continued)

Table 9.1 (continued)

Lichen species	Metabolite	Class of compound	Production pathways[a]	Category of use	References
Stereocaulon ramulosum *Stereocaulon sasakii* *Stereocaulon tomentosum*	Methyl haematommate Stereocalpin A	Depsides Didepsipeptide	APP APP	5-lipoxygenase activity Inhibition of platelet-type 12(S)-lipoxygenase Inhibition of arachidonate 5-lipoxygenase Inhibition of cyclooxygenase Antifungal activity Antimycobacterial activity Inhibition of tyrosine phosphatase Inhibition of cysteinyl leukotrienes Cytotoxicity	Haraldsdottir et al. (2004) Bucar et al. (2004) Hickey et al. (1990) Ingolfsdottir et al. (1998) Seo et al. (2009) Gissurarson et al. (1997) Seo et al. (2008)
Thamnolia vermicularis	Baeomycesic acid	Depsides	APP	Inhibition of 5-lipoxygenase activity	Ingolfsdottir et al. (1997b)
Usnea complanata	Psoromic acid Usnic acid	Depsidones Dibenzofuranes	APP APP	Antioxidant activity Antimicrobial activity Cardiovascular protective activity	Mahadik et al. (2011) Behera et al. (2012)
Usnea ghattensis	Norstictic acid Usnic acid	Depsidones Dibenzofuranes	APP APP	Antioxidant activity Hepatoprotective activity Antibacterial activity Antityrosinase activity	Verma et al. (2008a) Behera et al. (2004b, 2005a, b, 2006d, 2009a)
Umbilicaria sp.	Gyrophoric acid	Depsides	APP	Cytotoxicity activity Antitumour activity	Burlando et al. (2009)
Umbilicaria esculenta	Orcinol	Aromatic	APP	Anti-inflammatory activity	Kim et al. (1996)

Xanthoparmelia scabrosa	Scabrosin esters	Epipolythiodioxopiperazine	APP	Cytotoxic activity	Ernst-Russell et al. (1999b)
Xanthoria parietina	Emodin Fallacinal Parietin	Anthraquinones Anthraquinones Anthraquinones	APP APP APP	Antibacterial activity Antifungal activity Cytotoxic activity Virucidal activity	Manojlovic et al. (1998) Fazio et al. (2007) Ivanova et al. (2000) Manojlovic et al. (1998) Ivanova et al. (2000)

APP acetyl polymalonyl pathway, *SAP* shikimic acid pathway, *MAP* mevalonic acid pathway
[a]Production pathways of lichen secondary metabolites

acid. Behera and Makhija (2002) demonstrated the potential of biological activities like inhibition of tyrosinase and xanthine oxidase in *Bulbothrix setschwanensis*. In a further study, extracts of dried herbarium specimens of as many as 77 species belonging to the lichen family *Graphidaceae* have been screened for the bioactivities like inhibition of tyrosinase, xanthine oxidase and nitro blue tetrazolium (NBT) and scavenging of superoxide, and many of them have shown potential for these activities (Behera et al. 2003, 2004a, 2006a). Further they have also studied on lichen species *Usnea ghattensis*, *Arthothelium awasthii*, *Heterodermia podocarpa* and *Parmotrema tinctorum* as well as their secondary metabolites, atranorin, barbatic acid, lecanoric acid, norstictic acid, salazinic acid, usnic acid and zeorin, for the antityrosinase and antioxidant activities (Behera et al. 2004b; Verma 2011). Several clinical studies have shown that ethanol ingestion alters the prooxidant–antioxidant balance in the liver by the production of free radicals during its metabolism (Nordmann et al. 1992; Ishii et al. 1997; Aleynik et al. 1998; Mutlu-Turkoglu et al. 2000; Bailey et al. 2001; Balkan et al. 2002). Therefore, hepatoprotective activity in cultured lichen *Usnea ghattensis* has been screened and reported for the first time (Verma et al. 2008a). Bhattarai et al. (2008) reported stronger antioxidant activities in lichens from Antarctica. Mitrovic et al. (2011b) compared the chemical content of lichen extracts (*Flavoparmelia caperata*, *Evernia prunastri*, *Hypogymnia physodes*, *Cladonia foliacea*) and their free radical scavenging ability. Mahadik et al. (2011) reported the antioxidant activity of psoromic acid and usnic acid isolated from a lichen *Usnea complanata*. Later on, radical scavenging and glucosidase activity of lichen metabolites usnic acid, sekikaic acid and salazinic acid isolated from *Ramalina celastri*, *R. nervulosa* and *R. pacifica* have been reported (Verma et al. 2012a).

9.3.3 Antifungal Activity

Lichen species have also been reported for the antifungal activity. Recently, Manojlovic et al. (2005) reported antifungal activity of the anthraquinone parietin isolated from *Caloplaca cerina*. After that, Schmeda-Hirschmann et al. (2008) demonstrated the antifungal properties of Andean lichens *Protousnea poeppigii* and *Usnea rigida*, which contain divaricatinic acid, isodivaricatinic acid, usnic acid and 5-resorcinol. Later on Mitrovic et al. (2011b) reported strong antifungal effect of *Evernia prunastri* and *Hypogymnia physodes*. In the same year Mahadik et al. (2011) reported the antifungal activity of lichen species *Usnea complanata* against 14 fungal strains, *Alternaria alternata*, *Aspergillus flavus*, *A. fumigatus*, *A. niger*, *A. oryzae*, *A. terreus*, *Candida albicans*, *Cryptococcus albidus*, *Curvularia* sp., *Fusarium moniliformae*, *F. oxysporum*, *Rhizopus stolonifer*, *Trichoderma harzianum* and *T. viride*. Usnic acid and psoromic acid were found to be the main component for the activity.

9.3.4 Antiviral Activity

As far as antiviral properties of lichen secondary metabolites are concerned, Yamamoto et al. (1995) reported the inhibition of Epstein–Barr virus activation by lichesterinic acid and usnic acid from the lichen *Dolichousnea longissima* (*Usnea longissima*). Later on Perry et al. (1999) showed antiviral activity of usnic acid against herpes simplex type 1 and polio type 1 viruses. Lichenan, a widely distributed lichen compound, is reported for the inhibition of tobacco mosaic virus (Stubler and Buchenauer 1996). Anthraquinones are also important constituents of lichens and are of interest as antiviral agents against HIV (Schinazi et al. 1990). In particular, hypericin is of pharmaceutical relevance because of its dramatic antiretroviral activity (Lavie et al. 1989). The derivatives 7,7′-dichlorohypericin and 5,7-dichloroemodin isolated from the lichen *Heterodermia obscurata* (Cohen and Towers 1995) were shown to exhibit strong inhibitory activity against herpes simplex virus type 1 (Cohen et al. 1996). Parietin extracted from *Teloschistes chrysophthalmus* showed activity against Junin and Tacaribe arenaviruses (Fazio et al. 2007).

9.3.5 Antitumour Activity

Antitumour activities of lichen metabolites are of great interest for pharmaceutical industries. The most extensively studied lichen secondary metabolite is a dibenzofurane compound usnic acid, abundantly found in many lichen species and reported for the antitumour activity by many researchers (Cardarelli et al. 1997; Ingolfsdottir 2002; Kristmundsdottir et al. 2002) from the time of its first isolation in 1844 (Mitrovic et al. 2011a). Other compounds like pannarin isolated from *Psoroma pholidotoides* (*Psoroma reticulatum*), *P. pulchrum* and *P. pallidum* are reported to inhibit the cell growth and induce apoptosis in human prostate carcinoma and human melanoma cells (Russo et al. 2006, 2008). Euplectin from *Flavoparmelia euplecta* and naphthazarin, an anthraquinone from lichen *Lecanora hybocarpa*, have been reported for the cytotoxicity activity (Ernst-Russell et al. 1999a, 2000). Protolichesterinic acid of *Cetraria islandica* showed inhibition against the growth of breast cancer cell lines (Ogmundsdottir et al. 1998). A depside gyrophoric acid isolated from *Umbilicaria* sp. demonstrated the cytotoxic and antitumour activity (Burlando et al. 2009).

9.3.6 Antipyretic, Analgesic and Anti-inflammatory Activity

In recent studies, lichen compounds are proved to be a good source of antipyretic, analgesic and anti-inflammatory compounds. Diffractaic acid, usnic acid and decarboxy stenosporic acid isolated from the lichen *Dolichousnea diffracta* (*Usnea diffracta*) are reported for antipyretic, analgesic and anti-inflammatory activities (Otsuka et al. 1972; Okuyama et al. 1995). Kim et al. (1996) reported the anti-inflammatory activity of orcinol isolated from *Umbilicaria esculenta*. Longissiminone A isolated from *Dolichousnea longissima* (*Usnea longissima*) has shown anti-inflammatory activity (Choudhary et al. 2005).

9.3.7 Herbicidal Activity

Lichens are also a potential source of herbicidal compounds. The herbicidal effects of some lichen secondary metabolites have been reviewed by many workers. Barbatic acid, gyrophoric acid, haemathamnolic acid and lecanoric acid are reported as photosystem II-inhibiting herbicides (Lawrey 1995; Dayan and Romagni 2001, 2002). Evernic acid and protolichesterinic acid isolated from *Dolichousnea longissima* (*Usnea longissima*) are reported as plant growth and photosystem II inhibitor (Nishitoba et al. 1987; Endo et al. 1998). Several analogues of lichen-derived anthraquinones have strong herbicidal activity. Emodin and its analogues are highly active on grasses, causing malformation and bleaching in early seedlings. Usnic acid proved to be a strong herbicidal which has the potential to inhibit carotenoid biosynthesis through the enzyme 4-hydroxyphenylpyruvate dioxygenase (Dayan and Romagni 2001).

9.3.8 Insecticidal Activity

The unique secondary compounds present in lichens can also be used as novel pest control agents. Dayan and Romagni (2001) reviewed that the lichen feeding slug, *Pallifera varia*, when offered to eat some potential lichens, avoided those having wide variety of compounds and preferred those with the lowest number of lichen compounds. The slug-repelling activity of vulpinic acid is being investigated as a seed treatment for protecting higher plants. Likewise, the insect *Spodoptera littoralis* was strongly deterred by usnic acid and vulpinic acid (Dayan and Romagni 2001).

9.4 Harnessing of Pharmaceutically Important Metabolites from Lichen

Lichen species have shown a wide variety of biological activities, but they have been long neglected by mycologists and overlooked by pharmaceutical industries because of their slow growing nature and difficulties in their artificial cultivation (Crittenden and Porter 1991; Yamamoto et al. 1998; Behera et al. 2004a, b). Hale (1973) made the following generalization about growth rates of lichen species: foliose (0.5–4 mm year^{-1}), fruticose (1.5–5 mm year^{-1}) and crustose (0.5–2 mm year^{-1}). Therefore, industrial-scale harvesting of lichens is neither ecologically sensible nor sustainable and for many species is not feasible (Miao et al. 2001). To prevent the decline of lichen population in protected areas and harnessing of novel biologically active compounds, the only conceivable alternative is to in vitro culture the lichens in bulk quantities (Yamamoto et al. 1995; Brunauer and Stocker-Worgotter 2005).

Ahmadjian (1966) was the first lichenologist who succeeded in artificial reestablishment of the lichen *Cladonia cristatella*. Ahmadjian and Heikkila (1970) reported the successful culture of *Endocarpon pusillum* and *Staurothele clopima*. In the year 1985, Yamamoto et al. (1985) successfully cultured *Usnea rubescens* and *Ramalina yasudae* in vitro by thallus fragments and demonstrated the production of usnic acid in the cultures, and subsequently they patented the method. Yoshimura et al. (1987) reported the culture of *Cladonia vulcani*. Stocker-Worgotter and Turk (1988) cultured the cyanobacterial lichen species, *Peltigera didactyla*, from soredia. Yoshimura et al. (1989) again reported the successful culture of some *Umbilicaria* species of lichenized fungi and the production of secondary metabolites by cultured tissue of *Usnea flexilis*. In the year 1990, Yoshimura et al. (1990a, b) reported the tissue culture of some Antarctic lichens and also cultured *Usnea rubescens* and *Peltigera praetextata* to the vegetative thallus stage.

After resolving various difficulties in establishing lichen tissue culture technique, many researchers have started to screen biological activities of cultured lichens and their metabolites produced in vitro. In this direction, Higuchi et al. (1993) screened as many as 46 cultured lichen species for tyrosinase inhibitory activity and some of them (*Hypogymnia physodes*, *Letharia vulpina* and *Cetraria juniperina*) showed strong activity. Antibacterial, antifungal, superoxide scavenging and tyrosinase inhibitory activity of cultured lichens of families *Cladoniaceae*, *Graphidaceae*, *Parmeliaceae*, *Umbilicariaceae* and *Usneaceae* has been reported by Yamamoto et al. (1993). Behera and Makhija (2002) reported the tyrosinase and xanthine oxidase inhibition activity of cultured thallus of foliose lichen *Bulbothrix setschwanensis*. Further, inhibition of tyrosinase and xanthine oxidase and scavenging of superoxide anions of three cultured species of crustose lichen genus *Graphis*, *G. guimarana*, *G. nakanishiana* and *G. schizograpta*, producing secondary metabolites norstictic acid, constictic acid and stictic acid, have been reported (Behera et al. 2006b). In continuation to this the same workers cultured lichen species *Usnea ghattensis*, *Arthothelium awasthii*, *Heterodermia podocarpa* and *Parmotrema tinctorum*. Secondary metabolites usnic acid, norstictic acid, atranorin, zeorin, lecanoric acid and barbatic acid produced by the derived culture were studied for the antioxidant, antimicrobial, antityrosinase and hepatoprotective properties (Behera et al. 2005a, b, 2006c, d, 2009a; Verma et al. 2008a–c).

In laboratory conditions, lichen tissue grows much faster than natural thalli but more slowly than many other microorganisms (Yamamoto et al. 1993). To improvise the growth rate of lichen culture tissue for the production of lichen substances in high quantity, very few studies have been carried out so far. In a very recent study, Chooi et al. (2008) cloned and characterized the sequence of nonreducing polyketide synthase (PKS) gene from lichen *Xanthoparmelia semiviridis* and explored the suitability of *Aspergillus nidulans* for heterologous expression of the *X. semiviridis* PKS gene. However, further work is needed to understand the expression of PKS gene in surrogate host organism for desired lichen substances. Behera et al. (2009b) demonstrated the fusion of protoplast of isolated mycobiont of lichen *Usnea ghattensis* with the protoplast of fast-growing fungi *Aspergillus nidulans*. The result showed successful regeneration of fusant with the production of lichen secondary metabolite usnic acid. Further, three lichen species *Ramalina nervulosa*, *R. pacifica* and *Usnea complanata* were successfully cultured in bioreactor for the first time for the production of their secondary metabolites (psoromic acid, sekikaic acid, salazinic acid and usnic acid) in higher quantity (Verma et al. 2012b; Behera et al. 2012).

9.5 Future Prospects

Although the pace of work on lichens in recent days has been increased, it is still far behind with respect to their bioprospection. As of today with the established culturing methods, many lichens have been cultured. However, many more are not responded to the nutrient media and physiological parameters (temperature, pH); hence, they could not be brought under culture. In order to prospect the lichens or their secondary metabolites, we have to improve the conventional lichen tissue culture technique established with formulation of more appropriate nutrient media along with physiological parameter, which can support the symbiont growth with the production of secondary metabolites in larger quantity within a short span of time. Very recently it has been reported that the growth rate and the metabolite production at flask level are not sufficient to solve the problem of commercial exploitation of lichen secondary metabolites. For industrial-scale production of compounds salazinic acid, sekikaic acid and usnic acid found in lichen species *Ramalina nervulosa* and *Ramalina pacifica*, the optimized culture conditions (nutrient media, temperature, pH, stirrer speed and pO_2 level) are different even though they are from the same genus. In bioreactor, culture biomass of *R. nervulosa* and *R. pacifica* obtained 10.3–17.7 g along with the production of sekikaic acid 122.8 mg, salazinic acid 200 mg and usnic acid from 75.4 to 136.8 mg after 100 h batch running for each species (Verma et al. 2012b).

Bioreactor refers to an engineered device that supports a biologically active environment to carry out chemical process by which organisms enable to synthesize biochemically active substances at higher quantity in short duration of time (Mc Naught and Wilkinson 1997). Lichen is a dual organism consisting of photosynthetic partner (algae) and a mycobiont (fungi). The optimal physiological conditions for both the bionts in symbiosis is still unknown. Therefore, the culturing of lichens in bioreactor is very less. However, large-scale production of lichen culture biomass using bioreactors is a promising approach for industrial-scale utilization of lichen secondary compounds.

In the future, metabolic engineering and biotechnological approaches can be used as an alternative production system to overcome the limited availability of biologically active, commercially valuable and medicinally important secondary metabolite compounds. Advancement in culture techniques may provide new means for the culturing of non-cultured/exotic lichens and production of their secondary compounds as they produce in nature. This may create an ultimate advantage to provide a continuous, reliable source of natural products from this complex organism.

The increasing use of genetic tools in regulation of pathways in secondary metabolism will provide the basis for the production of commercially viable lichen substances. The knowledge of using precursor to activate biosynthetic pathways of desired lichen phytochemical in cultures is still in its infancy; hence, strategies are needed to develop an information on the requirement of precursor for lichen-derived compounds. The introduction of newer techniques of molecular biology, so as to produce transgenic cultures and to effect the expression and regulation of biosynthetic pathways, is also likely to be a significant step toward making cell cultures more generally applicable for the commercial production of secondary metabolites (Hussain et al. 2012). This probably has been the main impetus for understanding and manipulating for biosynthesis of various chemical, physiological and biotechnological pathways.

Recently, many devised techniques are being used for culturing of lichens to enhance their biomass and their unique secondary medicinally important compounds. These techniques include genetic engineering to find out the genes responsible for the production of lichen secondary compounds along with the expression of the gene in fast-growing fungi to reduce its slow growth rate. Furthermore chemistry along with computer-aided drug designing together may be able to develop novel synthetic analogues of lichen substances which can enlarge the access for new drug discovery by the pharmaceutical industries.

Acknowledgements We gratefully acknowledge the various financial supports by the Department of Biotechnology (DBT), Govt. of India, New Delhi [grant nos. BT/PR3133/BCE/08/237/2002; BT/PR8551/NDB/52/15/2006]; Council of Scientific and Industrial Research (CSIR), Govt. of India, New Delhi [grant nos. 09/670(034)/2006-EMR-I; 09/670(0046)2010/EMR-I]; and Science and Engineering Research Board (SERB), Govt. of India, New Delhi [grant no. SR/FT/LS-170/2009]. We are also thankful to the director of Agharkar Research Institute, Pune, for research facilities provided.

References

Ahmadjian V (1966) Lichens. In: Henry SM (ed) Symbiosis, vol I. Academic, New York, pp 36–97

Ahmadjian V, Heikkila H (1970) The culture and synthesis of *Endocarpon pusillum* and *Staurothele clopima*. Lichenologist 4:256–267

Airaksinen MM, Peura P, Ala-Fossi-Salokangas L et al (1986) Toxicity of plant material used as emergency food during famines in Finland. J Ethnopharmacol 18:273–296

Aleynik SI, Leo MA, Aleynik MA et al (1998) Increased circulating products of lipid peroxidation in patients with alcoholic liver disease. Alcohol Clin Exp Res 22:192–196

Bailey SM, Patel VB, Young TA et al (2001) Chronic ethanol consumption alters the glutathione/glutathione-peroxidase-1 system and protein oxidation status in rat liver. Alcohol Clin Exp Res 25:726–733

Balkan J, Kanbagli O, Aykac-Toker G et al (2002) Taurine treatment reduces hepatic lipids and oxidative stress in chronically ethanol-treated rats. Biol Pharm Bull 25:1231–1233

Bazin MA, Le Lamer AC, Delcros JG et al (2008) Synthesis and cytotoxic activities of usnic acid derivatives. Bioorg Med Chem 16:8737–8744

Behera BC, Makhija U (2002) Inhibition of tyrosinase and xanthine oxidase by lichen species *Bulbothrix setschwanesis*. Curr Sci 82:61–66
Behera BC, Adawadkar B, Makhija U (2003) Inhibitory activity of xanthine oxidase and superoxide-scavenging activity in some taxa of the lichen family *Graphidaceae*. Phytomedicine 10:536–543
Behera BC, Adawadkar B, Makhija U (2004a) Capacity of some *Graphidaceous* lichens to scavenge superoxide and inhibition of tyrosinase and xanthine oxidase activities. Curr Sci 87:83–87
Behera BC, Verma N, Sonone A et al (2004b) Determination of antioxidative and anti-tyrosinase potential of lichen *Usnea ghattensis* in-vitro. In: Reddy SM, Khanna S (eds) Biotechnological approaches for sustainable development. Allied, New Delhi, pp 94–103
Behera BC, Verma N, Sonone A et al (2005a) Antioxidant and antibacterial activities of lichen *Usnea ghattensis* in-vitro. Biotechnol Lett 27:991–995
Behera BC, Verma N, Sonone A et al (2005b) Evaluation of antioxidant potential of the cultured mycobiont of a lichen *Usnea ghattensis*. Phytother Res 19:58–64
Behera BC, Adawadkar B, Makhija U (2006a) Tyrosinase—inhibitory activity in some species of the lichen family *Graphidaceae*. J Herb Pharmacother 6:55–69
Behera BC, Adawadkar B, Makhija U (2006b) Tissue-culture of selected species of the Graphis lichen and their biological activities. Fitoterapia 77:208–215
Behera BC, Verma N, Sonone A et al (2006c) Experimental studies on the growth and usnic acid production in lichen *Usnea ghattensis* in-vitro. Microbiol Res 161:232–237
Behera BC, Verma N, Sonone A et al (2006d) Determination of antioxidative potential of lichen *Usnea ghattensis* in-vitro. LWT Food Sci Technol 39:80–85
Behera BC, Verma N, Sonone A et al (2009a) Optimization of culture conditions for lichen *Usnea ghattensis* G. Awasthi to increase biomass and antioxidant metabolite production. Food Technol Biotechnol 47:7–12
Behera BC, Sonone A, Makhija U (2009b) Protoplast isolation from cultured lichen *Usnea ghattensis*, their fusion with protoplasts of *Aspergillus nidulans*, fusant regeneration and production of usnic acid. Folia Microbiol 54:415–418
Behera BC, Mahadik N, Morey M (2012) Antioxidative and cardiovascular-protective activities of metabolite usnic acid and psoromic acid produced by lichen species *Usnea complanata* under submerged fermentation. Pharm Biol 50:968–979
Bezivin C, Tomasi S, Rouaud I et al (2004) Cytotoxic activity of compounds from the lichen: *Cladonia convoluta*. Planta Med 70:874–877
Bhattarai HD, Paudel B, Hong SG et al (2008) Thin layer chromatography analysis of antioxidant constituents of lichens from Antarctica. J Nat Med 62:481–484
Biswas K (1956) Common medicinal plants of Darjeeling and the Sikkim Himalayas. M/S Bengal Govt. Press, Calcutta, p 90
Borkowski B, Wozniak W, Gertig H et al (1964) Bacteriostatic action of some compounds from lichen *Cetraria islandica* and of usnic acid. Dissertationes Pharmaceuticae 16:189–194
Boustie J, Grube M (2005) Lichens—a promising source of bioactive secondary metabolites. Plant Genet Resour Char Utiliz 3:273–283
Brij L (1988) Traditional remedies for bone fracture among the tribals of Madhya Pradesh, India. Aryavaidyan 1:190–195
Brij L, Upreti DK (1995) Ethnobotanical notes on three Indian lichens. Lichenologist 27:77–79
Brij L, Upreti DK, Kalakoti BS (1985) Ethnobotanical utilization of lichen by the tribals of Madhya Pradesh. J Econ Taxon Bot 7:203–204
Brooker SG, Cambie RC, Cooper RC (1987) New Zealand medicinal plants. Heinemann, Auckland, p 63
Brunauer G, Stocker-Worgotter E (2005) Culture of lichen fungi for future production of biologically active compounds. Symbiosis 38:187–201
Bucar F, Schneider I, Ogmundsdottir H et al (2004) Antiproliferative lichen compounds with inhibitory activity on 12(S)-HETE production in human platelets. Phytomedicine 11:602–606

Burkholder PR, Evans AW, McVeigh I et al (1944) Antibiotic activity of lichens. Proc Natl Acad Sci USA 30:250–255

Burlando B, Ranzato E, Volante A et al (2009) Antiproliferative effects on tumour cells and promotion of keratinocyte wound healing by different lichen compounds. Planta Med 75:607–613

Cardarelli MA, Serino G, Campanella L et al (1997) Antimitotic effects of usnic acid on different biological systems. Cell Mol Life Sci 53:667–672

Chooi YH, Stalker DM, Davis MA et al (2008) Cloning and sequence characterization of a non-reducing polyketide synthase gene from the lichen *Xanthoparmelia semiviridis*. Mycol Res 112:147–161

Chopra RN, Chopra IC, Handa KL et al (1958) Indigenous drugs of India. Academic, Calcutta, p 816

Choudhary MI, Saima Jalil A and Atta-ur-Rahman (2005) Bioactive phenolic compounds from a medicinal lichen, *Usnea longissima*. Phytochem 66:2346–2350

Cohen PA, Towers GHN (1995) The anthraquinones of *Heterodermia obscurata*. Phytochemistry 40:911–915

Cohen PA, Hudson JB, Towers GHN (1996) Antiviral activities of anthraquinones, bianthrones and hypericin derivatives from lichens. Experientia 52:180–183

Crittenden PD, Porter N (1991) Lichen forming fungi: potential source of novel metabolites. Trends Biotechnol 9:409–414

Dayan FE, Romagni JG (2001) Lichens as a potential source of pesticides. Pestic Outlook 12:229–232

Dayan FE, Romagni JG (2002) Structural diversity of lichen metabolites and their potential for use. In: Upadhyaya R (ed) Advances in microbial toxin research and its biotechnological exploitation. Kluwer Academic/Plenum, New York, pp 151–169

De Carvahlo EAB, Andrade PP, Silva NH et al (2005) Effect of usnic acid from the lichen *Cladonia substellata* on Trypanosoma cruzi in-vitro: an ultrastructural study. Micron 36:155–161

Devehat FLL, Tomasi S, Elix JA et al (2007) Stictic acid derivatives from the lichen *Usnea articulata* and their antioxidant activities. J Nat Prod 70:1218–1220

Elix JA (1996) Biochemistry and secondary metabolites. In: Nash TH III (ed) Lichen biology. Cambridge University Press, Cambridge, pp 154–180

Elo H, Matikainen J, Pelttari E (2007) Potent activity of the lichen antibiotic (+)-usnic acid against clinical isolates of vancomycin-resistant enterococci and methicillin-resistant *Staphylococcus aureus*. Naturwissenschaften 94:465–468

Endo T, Takahagi T, Kinoshita Y et al (1998) Inhibition of photosystem II of spinach by lichen-derived depsides. Biosci Biotechnol Biochem 62:2023–2027

Ernst-Russell MA, Elix JA, Chai CLL et al (1999a) Hybocarpone, a novel cytotoxic naphthazarin derivative from mycobiont cultures of the lichen *Lecanora hybocarpa*. Tetrahedron Lett 40:6321–6324

Ernst-Russell MA, Chai CLL, Hurne AM et al (1999b) Structure revision and cytotoxic activity of the scabrosin esters, epidithiopiperazinediones from the lichen *Xanthoparmelia scabrosa*. Aust J Chem 52:279–283

Ernst-Russell MA, Chai CLL, Wardlaw JH et al (2000) Euplectin and coneuplectin, new naphthopyrones from the lichen *Flavoparmelia euplecta*. J Nat Prod 63:129–131

Esimone CO, Adikwn MU (1999) Antimicrobial activity of the cytotoxicity of *Ramalina farinacea*. Fitoterapia 7:428–431

Fahselt D (1994) Secondary biochemistry of lichens. Symbiosis 16:117–165

Fazio AT, Adler MT, Bertoni MD et al (2007) Lichen secondary metabolites from the cultured lichen mycobionts of *Teloschistes chrysophthalmus* and *Ramalina celastri* and their antiviral activities. Z Naturforsch C 62:543–549

Francolini I, Norris P, Piozzi A et al (2004) Usnic acid, a natural antimicrobial agent able to inhibit bacterial biofilm and formation on polymer surfaces. Antimicrob Agents Chemother 48:4360–4365

Gaikwad S, Verma N, Behera BC et al (2014) Growth promoting effects of some lichen metabolites on probiotic bacteria. J Food Sci Technol. 51:2624–2631

Gissurarson S, Sigurdsson S, Wagner H et al (1997) Effect of lobaric acid on cysteinyl-leukotriene formation and contractile activity of Guinea Pig *Taenia coli*. J Pharmacol Exp Ther 280:770–773

Gulcin I, Oktay M, Kufrevioglu OI et al (2002) Determination of antioxidant activity of lichen *Cetraria islandica* (L) Ach. J Ethnopharmacol 79:325–329

Hager A, Brunauer G, Turk R et al (2008) Production and bioactivity of common lichen metabolites as exemplified by *Heterodea muelleri* (Hampe) Nyl. J Chem Ecol 34:113–120

Hale ME (1973) Growth. In: Ahmadjian V, Hale ME (eds) The lichens. Academic, London, pp 473–492

Haraldsdottir S, Guolaugsdottir E, Ingolfsdottir K et al (2004) Anti-proliferative effects of lichen-derived lipoxygenase inhibitors on twelve human cancer cell lines of different tissue origin in-vitro. Planta Med 70:1098–1100

Hickey BJ, Lumsden AJ, Cole ALJ et al (1990) Antibiotic compounds from New Zealand plants: methyl haematommate, an anti-fungal agent from *Stereocaulon ramulosum*. N Z Nat Sci 17:49–53

Hidalgo ME, Fernandez E, Quilhot W et al (1994) Antioxidant activity of depsides and depsidones. Phytochemistry 37:1585–1587

Higuchi M, Miura Y, Boohene J et al (1993) Inhibition of tyrosinase activity by cultured lichen tissues and bionts. Planta Med 59:253–255

Hirayama T, Fujikawa F, Kasahara T et al (1980) Anti-tumor activities of some lichen products and their degradation products. Yaku Zasshi 100:755–759

Hu S-y, Kong YC, But PPH (1980) An enumeration of the Chinese materia medica. The Chinese University, Hong Kong, p 59

Huneck S (1999) The significance of lichens and their metabolites. Naturwissenschaften 86:559–570

Huneck S, Yoshimura I (1996) Identification of lichen substances. Springer, Berlin, p 493

Hussain MS, Fareed S, Ansari S et al (2012) Current approaches toward production of secondary plant metabolites. J Pharm Bioallied Sci 4:10–20

Ingolfsdottir K, Gudmundsdottir GF (2002) Effects of tenuiorin and methyl orsellinate from the lichen *Peltigera leucophlebia* on 5-/15-lipoxygenases and proliferation of malignant cell lines in-vitro. Phytomedicine 9:654–658

Ingolfsdottir K, Breu W, Huneck S et al (1994) In-vitro inhibition of 5-lipoxygenase by protolichesterinic acid from *Cetraria islandica*. Phytomedicine 1:187–191

Ingolfsdottir K, Gissurarson SR, Muller-Jakic B et al (1996) Inhibitory effects of the lichen metabolite lobaric acid on arachidonate metabolism in-vitro. Phytomedicine 2:243–246

Ingolfsdottir K, Hjalmarsdottir MA, Sigurdsson A et al (1997a) In vitro susceptibility of *Helicobacter pylori* to protolichesterinic acid from the lichen *Cetraria islandica*. Antimicrob Agents Chem 41:215–217

Ingolfsdottir K, Wiedemann B, Birgisdottir M et al (1997b) Inhibitory effects of baeomycesic acid from the lichen *Thamnolia subuliformis* on 5-lipoxygenase in-vitro. Phytomedicine 4:125–128

Ingolfsdottir K, Chung GAC, Skulason VG et al (1998) Antimycobacterial activity of lichen metabolites in-vitro. Eur J Pharm Sci 6:141–144

Ingolfsdottir K, Lee SK, Bhat KPL et al (2000) Evaluation of selected lichens from Iceland for cancer chemopreventive and cytotoxic activity. Pharm Biol 38:313–317

Ingolfsdottir K (2002) Usnic acid. Phytochemistry 61:729–736

Ishii H, Kurose I, Kato S (1997) Pathogenesis of alcoholic liver disease with particular emphasis on oxidative stress. J Gastroenterol Hepatol 12(Suppl):S272–S282

Ivanova V, Schlegel R, Grafe U (2000) 2-Methoxy-4,5,7-trihydroxy-anthraquinone, a new lichen metabolite produced by *Xanthoria parietina*. Pharmazie 55:785–786

Jayaprakash GK, Jaganmohan RL (2000) Phenolic constituents from the lichen *Parmotrema stuppeum* hale and their antioxidant activity. Zeitsch Fur Naturf 55:1018–1022

Karagoz A, Dogruoz N, Zeybek Z et al (2009) Antibacterial activity of some lichen extracts. J Med Plants Res 3:1034–1039

Kari PR, Tanaina Plantlore, Denaina Ketuna (1987) An ethnobotany of the Denaina Indians of South-Central Alaska. USDI, National park service, Alaska Region, p 176

Karunaratne V, Choudhary MI, Ali S et al (2009) Natural novel antioxidants. US Patent 20090048332. Publication info: http://www.fags.orglpatents/app/20090048332

Kim JW, Song KS, Chang HW et al (1996) Two phenolic compounds isolated from *Umbilicaria esculenta* as phospholipase A2 inhibitors. Hanguk Kyunhakhoechi 24:237–242

Kokwaro JO (1976) Medicinal plants of East Africa. East African Literature Bureau, Lampala, Nairobi, Dar es Salaam. p 384

Koyama M, Takahashi K, Chou T-C et al (1989) Intercalating agents with covalent bond forming capability. A novel type of potential anticancer agents. 2. Derivatives of chrysophanol and emodin. J Med Chem 32:1594–1599

Kranner I, Cram WJ, Zorn M et al (2005) Antioxidant and photoprotection in a lichen as compared with its isolated symbiotic partners. Proc Natl Acad Sci USA 102:3141–3146

Kristmundsdottir T, Aradottir HA, Ingolfsdottir K et al (2002) Solubilization of the lichen metabolite (+)-usnic acid for testing in tissue culture. J Pharm Pharmacol 54:1447–1452

Kumar KCS, Muller K (1999) Lichen metabolites. 2. Antiproliferative and cytotoxic activity of gyrophoric, usnic, and diffractaic acid on human keratinocyte growth. J Nat Prod 62:821–823

Kupchan SM, Kopperman HL (1975) L-Usnic acid: tumor inhibitor isolated from lichens. Experientia 31:625

Lauterwein M, Oethinger M, Belsner K et al (1995) In vitro activities of the lichen secondary metabolites vulpinic acid, (+)-usnic acid, and (-)-usnic acid against aerobic and anaerobic microorganisms. Antimicrob Agents Chemother 39:2541–2543

Lavie G, Valentine F, Levin B et al (1989) Studies of the mechanisms of action of the antiretroviral agents hypericin and pseudohypericin. Proc Natl Acad Sci USA 86:5963–5967

Lawrey JD (1986) Biological role of lichen substances. Bryologist 89:11–122

Lawrey JD (1995) The chemical ecology of lichen mycoparasites. Can J Bot 73:603–608

Lopes TIB, Coelho RG, Yoshida NC et al (2008) Radical scavenging activity of Orsellinates. Chem Pharm Bull 56:1551–1554

Mahadik N, Morey MV, Behera BC et al (2011) Cardiovascular-protective, antioxidative and antimicrobial properties of natural thallus of lichen *Usnea complanata*. LAJP 30:220–228

Manojlovic NT, Solujic S, Sukdolak S et al (1998) Anthraquinones from the lichen *Xanthoria parietina*. J Serb Chem Soc 63:7–11

Manojlovic NT, Novakovic M, Stevovic V et al (2005) Antimicrobial metabolites from three Serbian *Caloplaca*. Pharm Biol 43:718–722

Matsubara H, Kinoshita K, Koyama K et al (1997) Antityrosinase activity of lichen metabolites and their synthetic analogues. J Hatt Bot Lab 83:179–185

Mayer M, O'Neill MA, Murray KE et al (2005) Usnic acid: a non-genotoxic compound with anti-cancer properties. Anticancer Drugs 16:805–809

Mc Naught A, Wilkinson A (1997) IUPAC compendium of chemical terminology, 2nd edn. Royal Society of Chemistry, Cambridge, p 148

McEvoy M, Gauslaa Y, Solhaug KA (2007) Changes in pools of depsidones and melanins, and their function, during growth and acclimation under contrasting natural light in the lichen *Lobaria pulmonaria*. New Phytol 175:271–282

Miao V, Coffet-LeGal MF, Brown D et al (2001) Genetic approaches to harvesting lichen products. Trends Biotechnol 19:349–355

Mishchenko NP, Stepanenko LS, Krivoshchekova OE et al (1980) Anthraquinones of the lichen *Asahinea chrysantha*. Chem Nat Compd (English Translation) 16:117–121

Mitrovic T, Stamenkovic S, Cvetkovic V et al (2011a) Lichens as source of versatile bioactive compounds. Biol Nyss 2:1–6

Mitrovic T, Stamenkovic S, Cvetkovic V et al (2011b) Antioxidant, antimicrobial and antiproliferative activities of five lichen species. Int J Mol Sci 12:5428–5448

Molnar K, Farkas E (2010) Current results on biological activities of lichen secondary metabolites: a review. Z Naturforsch 65:157–173

Moskalenko SA (1986) Preliminary screening of far-Eastern ethnomedicinal plants for antibacterial activity. J Ethnopharmacol 15:221–259

Muller K (2001) Pharmaceutically relevant metabolites from lichens. Appl Microbiol Biotechnol 56:9–16

Muller K, Prinz H, Gawlik I et al (1997) Simple analogues of anthralin: unusual specificity of structure and antiproliferative activity. J Med Chem 40:3773–3780

Mutlu-Turkoglu U, Dogru-Abbasoglu S, Aykac TG et al (2000) Increased lipid and protein oxidation and DNA damage in patients with chronic alcoholism. J Lab Clin Med 136:287–291

Nash TH III (ed) (1996) Lichen biology, Cambridge University Press, Cambridge

Nash TH III (ed) (2008) Lichen biology, 2nd edn. Cambridge University Press, Cambridge

Nishitoba Y, Nishimura H, Nishiyama T et al (1987) Lichen acids, plant growth inhibitors from *Usnea longissima*. Phytochemistry 26:3181

Nordmann R, Ribiere C, Rouach H (1992) Implication of free radical mechanisms in ethanol-induced cellular injury. Free Radic Biol Med 12:219–240

Novaretti R, Lemordant D (1990) Plants in the traditional medicine of the Ubaye Valley. J Ethnopharmacol 30:1–34

Odabasoglu F, Aslan A, Cakir A et al (2004) Comparison of antioxidant activity and phenolic content of three lichen species. Phytother Res 18:938–941

Ogmundsdottir HM, Zoega GM, Gissurarson SR et al (1998) Anti-proliferative effects of lichen-derived inhibitors of 5-lipoxygenase on malignant cell-lines and mitogen-stimulated lymphocytes. J Pharm Pharmacol 50:107–115

Oksanen I (2006) Ecological and biotechnological aspects of lichens. Appl Microbiol Biotechnol 73:723–734

Okuyama E, Umeyama K, Yamazaki M et al (1995) Usnic acid and diffractaic acid as analgesic and antipyretic components of *Usnea diffracta*. Planta Med 61:113–115

Otsuka H, Komiya T, Tsukumi M et al (1972) Studies on anti-inflammatory drugs. Anti-inflammatory activity of crude drugs and plants. (II). J Takeda Res Lab 31:247–254

Papadopoulou P, Tzakou O, Vagias C et al (2007) b-Orcinol metabolites from the lichen *Hypotrachyna revoluta*. Molecules 12:997–1005

Paudel B, Bhattarai HD, Lee JS et al (2008) Antioxidant activity of polar lichens from King George Island (Antarctica). Polar Biol 31:605–608

Paull KD, Zee Cheng RK, Cheng CC (1976) Some substituted naphthazarins as potential anticancer agents. J Med Chem 19:337–339

Pengsuparp T, Cai L, Constant H et al (1995) Mechanistic evaluation of new plant-derived compounds that inhibit HIV-1 reverse transcriptase. J Nat Prod 58:1024–1031

Perry NB, Benn MH, Brennan NJ et al (1999) Antimicrobial, antiviral and cytotoxic activity of New Zealand lichens. Lichenologist 31:627–636

Petrzik K, Vondrak J, Bartak M et al (2014) Lichens—a new source or yet unknown host of herbaceous plant viruses? Eur J Plant Pathol 138:549–559

Rankovic B, Misic M, Sukdolak S (2007a) Antimicrobial activity of the lichens *Cladonia furcata*, *Parmelia caperata*, *Parmelia pertusa*, *Hypogymnia physodes* and *Umbilicaria polyphylla*. Br J Biomed Sci 64:143–148

Rankovic B, Misic M, Sukdolak S (2007b) Evaluation of antimicrobial activity of the lichens *Lasallia pustulata*, *Parmelia sulcata*, *Umbilicaria crustulosa* and *Umbilicaria cylindrica*. Microbiology 76:723–727

Rastogi RP, Mehrotra BN (1993) Compendium of Indian medicinal plants, Vol II (1970–1979). Central Drug Research Institute, Lucknow and Publications and Information Directorate, CSIR, New Delhi, pp 169–170

Rezanka T, Sigler K (2007) Hirtusneanoside, an unsymmetrical dimeric tetrahydroxanthone from the lichen *Usnea hirta*. J Nat Prod 70:1487–1491

Richardson DHS (1988) Medicinal and other economic aspects of lichens. In: Galun M (ed) CRC handbook of lichenology, vol III. CRS, Boca Raton, FL, pp 93–108

Russo A, Piovano M, Lombardo L et al (2006) Pannarin inhibits cell growth and induces cell death in human prostate carcinoma DU145 cells. Anticancer Drugs 17:1163–1169
Russo A, Piovano M, Lombardo L et al (2008) Lichen metabolites prevent UV light and nitric-oxide mediated plasmid DNA damage and induce apoptosis in human melanoma cells. Life Sci 83:468–474
Saklani A, Upreti DK (1992) Folk uses of some lichens in Sikkim. J Ethnopharm 37:229–233
Schinazi RF, Chu CK, Babu JR et al (1990) Anthraquinones as a new class of antiviral agents against human immunodeficiency virus. Antiviral Res 13:265–272
Schmeda-Hirschmann G, Tapia A, Lima B et al (2008) A new antifungal and antiprotozoal depside from the Andean lichen *Protousnea poeppigii*. Phytother Res 22:349–355
Schneider A (1904) A guide to the study of lichens. Knight and Miller, Boston, MA, p 234
Schwendener S (1868) Uber die Beziehungen zwischen Algen und Flechtengonidien. Botanische Zeitung (Berlin)\Bot Zeitung 26:289–292
Seo C, Yim JH, Lee HK et al (2008) Stereocalpin A, a bioactive cyclic depsipeptide from the Antarctic lichen *Stereocaulon alpinum*. Tetrahedron Lett 49:29–31
Seo C, Sohn JH, Ahn JS et al (2009) Protein tyrosine phosphatase 1B inhibitory effects of depsidone and pseudodepsidone metabolites from the Antarctic lichen *Stereocaulon alpinum*. Bioorg Med Chem Lett 19:2801–2803
Smith GW (1973) Arctic pharmacognosia. Arctic 26:328–333
Stepanenko LS, Krivoshchekova OE, Dmitrenok PS et al (1997) Quinones of *Cetraria islandica*. Phytochemistry 46:565–568
Stocker-Worgotter E (2005) Approaches to a biotechnology of lichen forming fungi: induction of polyketide pathways and the formation of chemosyndromes in axenically cultured mycobionts. Recent Res Dev Phytochem 9:115–131
Stocker-Worgotter E, Turk R (1988) Culture of the cyanobacterial lichen *Peltigera didactyla* from soredia under laboratory conditions. Lichenologist 20:369–376
Stubler D, Buchenauer H (1996) Antiviral activity of the glucan lichenan (poly-$\beta\{1\rightarrow3,\ 1\rightarrow4\}$ D-anhydroglucose) 1. Biological activity in tobacco plants. J Phytopathol 144:37–43
Tay T, Turk AO, Yılmaz M et al (2004) Evaluation of the antimicrobial activity of the acetone extract of the lichen *Ramalina farinacea* and its (+)-usnic acid, norstictic acid and protocetraric acid constituents. Z Naturforsch 59c:384–388
Taylor TN, Hass H, Remy W et al (1995) The oldest fossil lichen. Nature 378:244–1
Turk AO, Yilmaz M, Kivanc M et al (2003) The antimicrobial activity of extracts of the lichen *Cetraria aculeata* and its protolichesterinic acid constituent. Z Naturforsch C 58:850–854
Upreti DK, Chatterjee S (2007) Significance of lichens and their secondary metabolites: a review. In: Ganguli BN, Deshmukh SK (eds) Fungi multifaceted microbes. Anamaya, New Delhi, pp 169–188
Vartia KO (1973) Antibiotics in lichens. In: Ahmadjiian V, Hale ME (eds) The lichens, 3rd edn. Academic, New York, pp 547–561
Verma N (2011) Studies on antioxidant activities of some lichen metabolites developed in-vitro, Shodh ganga, Indian ETD Repository, Issue 2 Sept 2011
Verma N, Behera BC, Makhija U (2008a) Antioxidant and hepatoprotective activity of a lichen *Usnea ghattensis* in-vitro. Appl Biochem Biotechnol 151:167–181
Verma N, Behera BC, Sonone A et al (2008b) Cell aggregates derived from natural lichen thallus fragments: antioxidant activities of lichen metabolites developed in-vitro. Nat Prod Commun 3:1911–1918
Verma N, Behera BC, Sonone A et al (2008c) Lipid peroxidation and tyrosinase inhibition by lichen symbionts grown in-vitro. Afr J Biochem Res 2:225–231
Verma N, Behera BC, Parizadeh H et al (2011) Bactericidal activity of some lichen secondary compounds of *Cladonia ochrochlora*, *Parmotrema nilgherrensis* and *Parmotrema sancti-angelii*. Int J Drug Dev Res 3:222–232

Verma N, Behera BC, Sharma BO (2012a) Glucosidase inhibitory and radical scavenging properties of lichen metabolites salazinic acid, sekikaic acid and usnic acid. Hacettepe J Biol Chem 40:7–21

Verma N, Behera BC, Joshi A (2012b) Studies on nutritional requirement for the culture of lichen *Ramalina nervulosa* and *Ramalina pacifica* to enhance the production of antioxidant metabolites. Folia Microbiol 57:107–114

Yamamoto Y, Mizuguchi R, Yamada Y (1985) Tissue culture of *Usnea rubescens* and *Ramalina yasudae* and production of usnic acid in their cultures. Agric Biol Chem 49:3347–3348

Yamamoto Y, Miura Y, Higuchi M et al (1993) Using lichen tissue culture in modern biology. Bryologist 96:384–393

Yamamoto Y, Miura Y, Kinoshita Y et al (1995) Screening of tissue cultures and thalli of lichens and some of their active constituents for inhibition of tumor promoter induced Epstein-Barr virus activation. Chem Pharm Bull 43:1388–1390

Yamamoto Y, Kinoshita Y, Matsubara H et al (1998) Screening of biological activities and isolation of biological active compounds from lichens. Recent Res Dev Phytochem 2:23–24

Yoshimura I, Kurokawa T, Nakno T et al (1987) A preliminary report of cultures of *Cladonia vulcani* and the effects of the hydrogen ion concentration on them. Bull Kochi Gakuen Coll 18:335–343

Yoshimura I, Kurokawa T, Nakno T et al (1989) Production of secondary metabolic substances by cultured tissue of *Usnea flexilis*. Bull Kochi Gakuen Coll 20:535–540

Yoshimura I, Kurokawa T, Kanda H (1990a) Tissue culture of some Antarctic lichens preserved in the refrigerator. Proc NIPR Symp Polar Biol 3:224–228

Yoshimura I, Kurokawa T, Yamamoto Y et al (1990b) Thallus formation of *Usnea rubescens* and *Peltigera praetextata* in-vitro. Bull Kochi Gakuen Coll 21:565–576

CPSIA information can be obtained
at www.ICGtesting.com
Printed in the USA
LVHW012042150819
627783LV00002B/9/P

9 783030 168131